Piezoelectric Materials:
Advances in Science, Technology and Applications

NATO Science Series

A Series presenting the results of activities sponsored by the NATO Science Committee. The Series is published by IOS Press and Kluwer Academic Publishers, in conjunction with the NATO Scientific Affairs Division.

A. Life Sciences	IOS Press
B. Physics	Kluwer Academic Publishers
C. Mathematical and Physical Sciences	Kluwer Academic Publishers
D. Behavioural and Social Sciences	Kluwer Academic Publishers
E. Applied Sciences	Kluwer Academic Publishers
F. Computer and Systems Sciences	IOS Press

1. Disarmament Technologies	Kluwer Academic Publishers
2. Environmental Security	Kluwer Academic Publishers
3. High Technology	Kluwer Academic Publishers
4. Science and Technology Policy	IOS Press
5. Computer Networking	IOS Press

NATO-PCO-DATA BASE

The NATO Science Series continues the series of books published formerly in the NATO ASI Series. An electronic index to the NATO ASI Series provides full bibliographical references (with keywords and/or abstracts) to more than 50000 contributions from international scientists published in all sections of the NATO ASI Series.
Access to the NATO-PCO-DATA BASE is possible via CD-ROM "NATO-PCO-DATA BASE" with user-friendly retrieval software in English, French and German (WTV GmbH and DATAWARE Technologies Inc. 1989).

The CD-ROM of the NATO ASI Series can be ordered from: PCO, Overijse, Belgium

Series 3. High Technology – Vol. 76

Piezoelectric Materials: Advances in Science, Technology and Applications

edited by

Carmen Galassi
National Research Council,
Research Institute for Ceramics Technology,
Faenza, Italy

Maria Dinescu
Institute of Atomic Physics,
Bucharest, Romania

Kenji Uchino
The Pennsylvania State University,
Pennsylvania, U.S.A.

and

Michael Sayer
Queen's University,
Department of Physics,
Kingston, Ontario, Canada

Kluwer Academic Publishers

Dordrecht / Boston / London

Published in cooperation with NATO Scientific Affairs Division

Proceedings of the NATO Advanced Research Workshop on
Piezoelectric Materials: Advances in Science, Technology and Applications
Predeal, Romania
24-27 May 1999

A C.I.P. Catalogue record for this book is available from the Library of Congress.

ISBN 0-7923-6212-8 (HB)
ISBN 0-7923-6213-6 (PB)

Published by Kluwer Academic Publishers,
P.O. Box 17, 3300 AA Dordrecht, The Netherlands.

Sold and distributed in North, Central and South America
by Kluwer Academic Publishers,
101 Philip Drive, Norwell, MA 02061, U.S.A.

In all other countries, sold and distributed
by Kluwer Academic Publishers,
P.O. Box 322, 3300 AH Dordrecht, The Netherlands.

Printed on acid-free paper

Printed in the Netherlands.

CONTENTS

III MICROSTRUCTURAL AND PIEZOELECTRIC CHARACTERISATION

PREFACE

In this couple of decades, applications of piezoelectrics to sensors and actuators have been dramatically accelerated, in addition to the discovery of new materials and devices. Some of the highlights include electrostrictive materials for positioners, relaxor-normal ferroelectric single crystals with very high electromechanical couplings for medical transducers, thin/thick PZT films for Micro Electro-Mechanical Systems starting from a sophisticated chemical technology, and multilayer type actuators fabricated by cofiring technique. All will provide a remarkable industrial impact in the 21st century.

This volume contains the Proceedings of the NATO Advanced Research Workshop PIEZOELECTRIC MATERIALS: ADVANCES IN SCIENCE, TECHNOLOGY AND APPLICATIONS held in Predeal, Romania, May 24-27, 1999. This meeting was supported by NATO, Italian National Research Council, Romanian National Agency for Science, Technology and Innovation and UNESCO. The Workshop had 45 participants in total, 37 oral and 8 poster presentations. The meeting was successfully organised by Drs. Carmen Galassi (IRTEC-CNR, Italy) and Maria Dinescu (Institute of Atomic Physics, NILPRP, Romania)

All the papers were refereed, mostly during the Workshop, by the Key Speakers and few other participants and this strong spirit of collaboration gave a further improvement to the quality of the Proceedings and is here especially acknowledged.

The Proceedings are categorised in accordance with:
- Fundamentals on ferroelectrics piezoelectrics and relaxors
- Powder preparation and processing
- Microstructural and piezoelectric characterisation
- Processing and characterisation of films
- Applications of bulk devices and thin films.

We believe that this volume will provide the better understanding on the recent science and technology in the piezoelectric area.

Last, but not least, the Editors would like to appreciate all the scientists and engineers who attended the workshop and made it fruitful and enjoyable

Carmen Galassi (IRTEC-CNR, Italy) *Principal Editor*

Maria Dinescu (NILPRP, Romania)
Michael Sayer (Queen's University, Kingston Canada)
Kenji Uchino (Penn State University USA) *Co-Editors*

LIST OF PARTICIPANTS

Co-Directors

GALASSI Dr. Carmen
CNR-IRTEC
Research Institute for Ceramics
Technology
Faenza (RA) ITALY

DINESCU Dr. Maria
Inst. for Atomic Physics, NILPRP.
Bucharest ROMANIA

Key Speakers

BAUER Dr. Sigfrid
Angewandte Physik
Johannes-Kepler-Universität
Linz AUSTRIA

BAUERLE Prof. Dieter
Angewandte Physik
Johannes-Kepler Universitaet Linz
Linz AUSTRIA

SAYER Prof. Michael
Dept. of Physics, Queen's University
Kingston, Ontario CANADA

BASTIEN Dr. Francis
LPMO-CNRS
Besancon Cedex FRANCE

GONNARD Dr. Paul
LGEF INSA
Lyon FRANCE

HAUDEN Prof. Daniel
L.P.M.O. - CNRS
Besançon FRANCE

LEJEUNE Dr. Martine
ENSCI, L.M.C.T.S.
Limoges FRANCE

JANOCHA Prof. Hartmut
Dienstleistungzentrum neue Aktoren
mit MikroSystem- und
Signalverarbeitungskonzepten
(D*ASS)
University of the Saarland Saarbrucken
GERMANY

STERNBERG Dr. Andris
Inst. for Solid State Physics
Riga LATVIA

LEMANOV Prof. Vladislav A. F.
IOFFE Physico-Technical Institute
St. Petersburg RUSSIA

KOSEC Prof. Maria
Jozef Stefan Institute, Univ. Ljubliana,
Ljubljana, 3000 SLOVENIA

DAMJANOVIC Dr. Dragan
Laboratory of Ceramics
EPFL-DMX-LC
Lausanne SWITZERLAND

DOGAN Prof. Aydin
Ceramics Eng. Dept.
Anadolu University
Eskisehir TURKEY

UCHINO Prof. Kenji
The Pennsylania State University
134 Mat. Res. Lab.
PA USA

Other Participants

HRISTOV Alexandre
CLMI-BAS
Acad. G.Bonchev
Sofia BULGARIA

ALEXE Dr. MARIN
Max Plank Institute of
Microstructure Physics
Halle(Saale) GERMANY

HAMMER Dr. Marianne
Robert Bosch GmbH
Stuttgart GERMANY

HARNAGEA Dr. Catalin
Max Plank Institute of Materials
Physics
Halle Saale GERMANY

SCHUH Dr. Carsten
SIEMENS AG
Munchen GERMANY

SPORN Dr. Dieter
Ceramics Department Fraunhofer-
Institut füer Silicatforschung
Wuerzburg GERMANY

BABINI Dr. Gian Nicola
CNR-IRTEC
Research Institute for Ceramics
Technology
Faenza(RA) ITALY

VERARDI Dr. Patrizio
Istituto di Acustica "O. M. Corbino"
CNR
Roma ITALY

WATTS Dr. Bernard E.
MASPEC CNR
Parma ITALY

CIPLYS Dr. Daumantas
Laboratory of Physical Acoustics
Faculty of Physics
Vilnius LITHUANIA

FERREIRA Dr. Jose M.F.
Departemento de Engeharia e do Vidro
Universidad de Aveiro
Aveiro PORTUGAL

AMARANDE Dr. Lumita
National Institute for Materials Physics
Magurele-Bucharest ROMANIA

BAUER-GOGONEA Dr. Simona
Institute of Atomic Physics
Bucharest ROMANIA

BOERASU Dr. Iulian
National Institute for Materials Physics
Bucharest-Magurle ROMANIA

BOROICA Dr. Lucica
National Institute of Glass S.A,
Bucharest ROMANIA

CHILIBON Dr. Irinela
National Institute of R&D for
Optoelectronics, INOE
Bucharest ROMANIA

DINU Dr. Raluca
Institute of Atomic Physics
Bucharest ROMANIA

LUPEI Dr. Aurelia
National Institute of Materials Physics
Magurele
Bucharest ROMANIA

MATEESCU Dr. Irina
National Institute of Materials Physics
Magurele Bucharest ROMANIA

MOISIN Dr.AnaMaria
Inst. for Electrical Engineering,
Bucharest ROMANIA

ROBU Dr. Maria
Institute of Optoelectrics
Magurele-Bucharest ROMANIA

ROBU O.
Institute for Marine Research,
Constanta, ROMANIA

ANISIMKIN Dr. Vladimir
Institute of Radioengineering and
Electronics (RAS)
Moscow RUSSIA

IVANOV Prof. S.N.
Institute of Radioengineering and
Electronics (RAS)
Moscow RUSSIA

YARMARKIN Dr. V. K
A.F. IOFFE – Physico-Technical
Institute
St. Petrsburg RUSSIA

VILLEGAS Dr. Marina Gracia
Instituto de Ceràmica y Vidrio (CSIC)
Arganda del Rey Madrid SPAIN

ERKALFA Dr. Hilkat
Materials and Chemical Technologies
Research Institute
Tubitak Gebze Kocaeli TURKEY

PODLIPENETS Prof. Alexander N.
Faculty of Air-Spacecraft Systems –
National University of Ukraine
Kyiv UKRAINE

HALL Prof. David A.
Materials Science Centre
University of Manchester (UMIST)
Manchester UNITED KINGDOM

LOWRIE Dr. Fiona
DERA
Farnborough
Hampshire UNITED KINGDOM

MALONEY Dr. Dom
DERA
Farnborough
Hampshire UNITED KINGDOM

FITZGERALD Prof. John J.
Depart. of Chem. and Biochemistry
South Dakota University
Brooklings USA

JOHNSON Dr. Michael
The Penn. State University
University Park, USA

PIEZO-, PYRO-, AND FERROELECTRICITY
IN BIOLOGICAL MATERIALS

V.V. LEMANOV
A.F.Ioffe Physical-Technical Institute
194021 St.Petersburg, Russia

You may say anything you like but
we all are made up of ferroelectrics

B.T. Matthias

1. Introduction

The physical properties of all the materials are well known to be determined by their symmetry. The lower the symmetry, the richer the palette of material physical properties. On the other hand, the lower is the symmetry of a system, the more ordered it is. Certainly, living organic materials are highly ordered, and one may expect properties characteristic for low-symmetry materials, such as, for example, piezoelectric and pyroelectric effects. People began to be interested in this problem very long ago. Pasteur was probably the first to suggest over 100 years ago that biological systems have chiralic dissymmetric properties and that these properties are important for the functioning of the biological systems. Much later, researchers began to study piezolectric, pyroelectric, and ferroelectric properties of biological materials.

In the first part of this short review, we discuss results of these studies in hard and soft animal and plant tissue. The second part is devoted to the protein amino acids out of which all living creatures build their proteins.

2. Hard and Soft Biological Tissue

In the 1960th it was suggested that piezoelectricity is a fundamental property of biological materials [1-3] and is observed in different soft and hard tissues: human femur, skin, Achilles tendon of ox and horse, and so on. The piezoelectric effect is supposed to be associated with the presence of oriented fibrous proteins such as collagen (a botanical counterpart of collagen is cellulose). Dry tendon is almost pure collagen. The piezoelectric signal appeared to be rather large: a segment of dried horse tendon, for example, produces electric signals on bending as large as half a volt.

For biological materials to possess piezoelectric properties, the materials should have crystalline structure or at least some texture with lack of inversion symmetry. The

1

C. Galassi et al. (eds.), Piezoelectric Materials: Advances in Science, Technology and Applications, 1–9.

symmetry of many biological materials is supposed to be D_∞ or $C_{\infty v}$ or a superposition of the both. For D_∞ group there is one independent component of the piezoelectric tensor, $d_{14} = -d_{25}$, for $C_{\infty v}$ group one has three independent components, $d_{15} = d_{24}$, $d_{31} = d_{32}$, d_{33} . For crystalline collagen the symmetry was supposed to be hexagonal, C_6 [1] with four independent components, $d_{14} = -d_{25}$, $d_{15} = d_{24}$, $d_{31} = d_{32}$, and d_{33} .

The piezoelectric coefficients of some biological materials are given in Table 1.

For comparison, the piezoelectric coefficients of 'good' inorganic piezoelectrics, ZnS (C_{6v}) , $Bi_{12}SiO_{20}$ (T) and SiO_2 (D_3) are also given in Table 1. One can see that the piezoelectric coefficients of biological materials are in general less than, but in some cases comparable to, those of strong inorganic piezoelectrics.

TABLE 1. Piezoelectric coefficients of some biological and inorganic materials (in 10^{-12} m/V)

Material	d_{14}	d_{15}	d_{31}	d_{33}
Bovine Achilles tendon	-2.7	1.4	0.09	0.07
Horse femur	-0.2	0.04	0.003	0.003
Silk	-1.1	0.25	0.02	0.023
ZnS	-2.8		-1.1	3.2
$Bi_{12}SiO_{20}$	40			
SiO_2	0.8	$d_{11} = 2.2$		

Among 20 piezoelectric point groups with no inversion symmetry there are 10 polar groups which allow the existence of a polar vector. If biological materials have a crystalline component with polar symmetry or a texture with $C_{\infty v}$ polar group then such materials should possess pyroelectric properties and, at least in principle, they may be ferroelectrics.

Lang [4] was the first to measure pyroelectric coefficients in animal bone and tendon. The pyroelectric coefficient in the region of room temperatures $p=4 \cdot 10^{-9}$ C/m^2 K was observed. Much larger pyroelectric coefficients were later found in plant leaves [5], collagen structures and nervous tissue [6], in different type receptors of living organisms [7], in wheat grains [8]. In the thorax of life insect the p coefficient as large as $3.5 \cdot 10^{-6}$ C/m^2 K was found [9]. A review paper on pyroelectricity in biological materials was published by Lang [9].

In Table 2 the pyroelectric coefficients of some biological materials are presented along with those of well-known inorganic pyroelectrics (non-ferroelectrics) CdS and ZnS. Analyzing the pyroelectric effect it is very important to distinguish between the primary and the secondary effects. The both can exist only in crystals of 10 polar point groups and in polar textures, the latter is determined by the thermal expansion coefficients α_{ik} and the piezoelectric coefficient d_{ikl}. Lang [10] studied this problem for the pyroelectric effect in animal bone.

TABLE 2. Pyroelectric coefficients of some biological and inorganic materials

Material	$p_i{}^X$ (10^{-6} C/m^2 K)
Hoof tendon	0.004
Insect thorax	3.5
Wheat	4.6
Plant leaves	0.15
Tourmaline	4
CdS	6
ZnS	0.4

The text-book expression for the pyroelectric coefficient says:

$$p_3{}^X = p_3{}^x + d_{3ik}c_{iklm}\,\alpha_{lm} \qquad (1)$$

where $p_3{}^X$ is the total pyroelectric coefficient (at constant stress); $p_3{}^x$, the primary pyroelectric coefficient (at constant strain) and the 3 axis is along the polar direction.

The second term in the right-hand part of Eq.1 describes the secondary pyroelectric effect.

For materials with $C_{\infty v}$ symmetry (the same holds for C_{6v} symmetry) one has

$$p_3{}^X - p_3{}^x = 2d_{31}(c_{11}+c_{12})\alpha_1 + 2d_{31}c_{13}\alpha_3 + 2d_{33}c_{13}\alpha_1 + d_{33}c_{33}\alpha_3 \qquad (2)$$

According to Lang [10], for animal bone $p_3{}^X = 25$; $p_3{}^x = -92$; $(p_3{}^X - p_3{}^x) = 117$ (in 10^{-10} C/m^2 K).

From this it follows that the main contribution into the measured pyroelectric effect comes from the secondary effect. Thus, the presence of the pyroelectric effect in general cannot be an evidence of the polar nature of materials.

As mentioned above, all the polar pyroelectric materials may be ferroelectrics which possess a spontaneous electric polarization that can be reversed by an external electric field. So the question arises whether the biological materials can be ferroelectrics.

There is a long story in this field. B.T. Matthias and A. von Hippel were probably the first to suggest in the late 1960th that the ferroelectricity may play an essential role in the functioning of living organisms. Later on many authors claimed that biological materials (bone, tendon, cell membranes) are ferroelectrics (see [3] and references therein). According to authors cited in Reference 3, dry bone is a weakly ferroelectric material consisting of domains which can be oriented by an applied electric field. With the electric field about $6 \cdot 10^5$ V/m, an induced polarization was measured to be about $4 \cdot 10^{-5}$ C/m^2. If one considers this as a ferroelectric spontaneous polarization it appears to be at least 3 order of magnitude less than in conventional ferroelectrics. However, the observed dielectric hysteresis loops do not look like real ferroelectric hysteresis loops.

There are also claims in the literature [11,12] concerning the ferroelectric properties of DNA and RNA. And again, quasi-ferroelectric hysteresis loops observed in the experiments cannot be considered as an evidence of the ferroelectric properties of the materials.

A ferroelectric model of ion channels in membranes of cells and neurones has been proposed and developed [13,14].

However, all these results on the ferroelectric behavior of biological materials need to be rigorously re-examined and re-measured. A lot of artefacts are known to play an important role in such experiments, in particular, hysteresis loops may be observed in any nonlinear dielectrics with losses.

So we can conclude that piezoelectric and pyroelectric effects are a fundamental property of biological materials but up to now we are not sure whether this property is essential for living beings. Piezoelectricity and pyroelectricity may be simply an accompany phenomena but not determinative. Nevertheless, Mother-Nature should be sufficiently economic to create something and not to use this. As Russian poet Mayakovsky wrote: 'If the stars are switching on, it means that somebody needs them.'

As for ferroelectric properties of biological materials, this is now quite an open question.

3. Protein Amino Acids

If primary pyroelectricity and possible ferroelectricity (if any) in biological materials are determined by proteins, as people believe, but one can neither prove these properties nor understand the underlying physics, it is important to find another approach to tackle the problem. The best approach seems to start from the very beginning, namely from protein amino acids (or primary amino acids). These amino acids (20 in numbers) are building blocks of proteins of all living creatures. Studies of physical properties at first of the protein amino acids, then of their derivatives, and at last of proteins seem to be the best rout to solve the problem of piezo-, pyro-, and ferroelectric properties of biological materials and their role in the functioning of living matter.

The general molecular structure of the protein amino acids can be described as follows:

$$
\begin{array}{c}
\text{H} \\
| \\
\text{H}_2\text{N} - \text{C} - \text{COOH} \\
| \\
\text{R}
\end{array}
$$

where R is the radical which varies from R = H in the simplest amino acid glycine to, for example, R = $CH_2C_6H_5$ in the aromatic phenylalanine or R = $CH_2C_6H_4OH$ in the aromatic tyrosine.

Single crystals practically of all the primary amino acids may be grown from aqueous (or other) solution. They are transparent in the visible spectral range with refractive index between 1.5 and 1.7.

Amino acid single crystals are of low symmetry without an inversion center, and many of them belong to the polar symmetry groups. Thise means that these protein amino acids are pyroelectrics and possibly ferroelectrics.

The structure of the protein amino acids is known from X-ray diffraction measurements [15-28]. Below we give only some general summarized data. When it was impossible to distinguish between centrosymmetric and non-centrosymmetric groups from X-ray diffraction measurements we used the date on optical second harmonic generation [29] and on piezoelectric measurements both from the literature [30] and from our experiments.

All the protein amino acids with the exception of α-glycine (point group C_{2h}) belong to enantomorphic groups, i.e. they are optically active and can exist in the levorotated (L) and dextrorotated (D) molecular configurations. Proteins of all the living creatures are known to be made up of only L-amino acids (this L-asymmetry is still a mystery).

Being enantiomorphic all the L-amino acids (as well as their D-enantiomers) are non-centrosymmetric, that is, they possess properties described by polar tensors of the odd rank such as piezoelectricity or nonlinear optics. Some of the amino acids crystallize in both monoclinic and orthorhombic forms. At least 10 amino acids are polar with the space group $P2_1 - C_2^2$. These are valine, leucine, isoleucine, aspartic acid, lysine, arginine, cysteine, methionine, tyrosine and histidine. L-cystine crystallizes in hexagonal and tetragonal forms. The tetragonal form belong to the polar group $P4_1 - C_4^2$. All the other amino acids are of $P2_12_12_1 - D_2^4$ orthorhombic space group except of hexagonal L-cystine which has $P6_122 - D_6^2$ space group.

Enantomorphic crystals can exist in a racemic, or DL-form. As a rule, the racemic crystals built up of L- and D-isomers should be centrosymmetric. But this is not always the case for DL-amino acids. Among 19 DL-amino acids , 10 amino acids are centrosymmetric (mainly C_i or C_{2h} groups), DL-alanine and DL-tyrosine have polar symmetry group C_{2v} (D_{2h} group was proposed for DL-tyrosine [29]), for 5 amino acids X-ray, piezoelectric and SHG measurements are contradictory. For DL-cysteine and DL-cystine the X-ray data are not avaible (as far as we can know).

The piezoelectric properties of the protein amino acids were studied only qualitatively [30]. According to our measurements, γ-glycine (C_3 point group) and DL-alanine (C_{2v}) are the most strong amino acid piezoelectrics (comparable to, or even stronger than, quartz crystals). L-alanine, L-valine, L-glutamic acid and DL-tyrosine have much smaller piezoelectric effect.

Optical second harmonic generation (which has symmetry properties similar to the piezoelectric effect) was observed in many amino acids along with dipeptides, proteins and viruses [29]. In Reference 29, 71 α-amino acids (both the protein amino acids and other) as well as 22 dipeptides, 6 tripeptides, 16 proteins, 5 viruses were studied. All the amino acids except α-glycine, all the peptides containing at least one amino acid enantiomer, all the proteins, and viruses appeared to be non-centrosymmetric with optical second harmonic generation. On the other hand, most of DL-, racemic amino acids were shown to be centrosymmetric.

To the best of our knowledge the data on pyroelectric properties of amino acids (except γ-glycine) are not available in the literature. Now, such measurements as well as the search of ferroelectric properties are under way in our laboratory at Ioffe Institute.

Derivatives of the protein amino acids, in particular compounds with inorganic components (including metals of life) are of great interest both for biology and for physics of solid state (or physics of soft state, to be more precise) and physics of ferroelectrics [31,32]. Suffice it to say that triglycine sulfate (TGS), a compound of the simplest protein amino acid, is one of the best pyroelectric material and a model object in physics of ferroelectricity. Amino acid derivatives are rather strong piezoelectrics. For example, L-arginine phosphat (point group C_2) has piezoelectric coefficient d_{25} $=8.6 \ 10^{-12}$ C/N, 4 times larger than d_{11} in quartz crystals [33].

Data on pyroelectric and ferroelectric properties of some amino acid derivatives are presented in Table 3.

The derivatives of two non-protein amino acids, sarcosine and betaine, are also included in Table 3. This may be interesting because sarcosine is an isomer of alanine, and betaine is an isomer of valine. This fact can be of importance since sarcosine and betaine compounds are ferroelectrics with interesting properties [40,41].

It should be noted that in general studies of physical properties of the protein amino acid crystals are now rather scarce. Most of them are devoted to L-alanine crystals.

TABLE 3. Pyroelectric and ferroelectric properties of aminoacid-based compounds

Material	$p \ (10^{-6} C/m^2 K)$	T_c (K)	$P_s \ (10^{-2} C/m^2)$	Reference
γ-glycine	13			34
$GlyMnCl_2 \ 2H_2O$	38	≥ 350	1.07	35
$GlyCoCl_2 \ 2H_2O$	21	≥ 350	0.48	35
$Gly_2 \ HNO_3$		206	0.6	36
$GlyH_3PO_3$		224		37
$GlyD_3PO_3$	≈ 100	322	0.5	37
Gly-L-Ala HI H_2O	10.7			38
$Gly_3 \ H_2SO_4$ (TGS)	350	320	2.2	36
$Sarc_3 \ CaCl_2$ (TSCC)	≈ 50	127	0.27	39
Betaine H_3AsO_4 (BA)		119		40
Betaine H_3PO_3 (BPI)		216		40

Raman spectra were studied in single crystals of L-alanine, and experimental findings were interpreted as a result of a mode instability with possible dynamic localization of vibrational energy [42].

Ultrasonic velocities and thermal conductivity were measured in L-alanine between helium temperatures and a room temperature. The Debye temperature was determined to be 205 K. From the thermal conductivity temperature dependence it was concluded that the lattice modes are strongly anharmonic [43].

Several phase transitions were observed in asparagine monohydrate (orthorhombic $P2_12_12_1$ - $D_2{}^4$) under action of hydrostatic pressure up to 2 GPa [44].

Luminescence, EPR, and resonant Raman scattering was studied in electron irradiated crystals of L-alanine to obtain information on electronic structure of colour centers [45].

As mentioned above, studies of dielectric, piezoelectric, and pyroelectric properties of the protein amino acids and their derivatives as well as the search for possible ferroelectric phase transitions are now in progress in our laboratory.

Single crystals of many amino acids have been grown: glycine, glycine phosphate, L-alanine, DL-valine, DL-serine, DL-serine sulfat.

Most of amino acids possess piezoelectric properties and are soft and labile objects apt to phase transitions. In crystalline powder of α-glycine some unusual piezoelectric effects have been observed [46]. In the experiment fine-grain crystalline powder of commercial α-glycine was used. The sample was placed in a capacitor, and a strong radio-frequency electric field was applied to the capacitor as a pulse with duration of $(1-6)10^{-6}$ s with a carrier frequency of about 10 MHz. Crystalline α-glycine is centrosymmetric (C_{2h} point group) and has no piezoelectric effect. A piezoelectric response of the sample observed in the experiment may be due to small contamination of γ-glycine (C_3 point group). Usually, for piezoelectric powders, the piezoelectric response and its temporal evolution after the application of an electric pulse has a chaotic random shape. In the case of glycine powder the piezoelectric response has a quite regular character ('coherent' signal) : one observes periodic oscillations of the signal with the period of the order of 10^{-5} s with monotonous exponential decrease of the signal determined by the ultrasonic attenuation. Specific angular dependence of the signal shows that some periodic spatial structures are formed in the powder under action of high radio-frequency electric field. The periodic character of the piezoelectric response may be ascribed to the presence of weakly coupled elastic oscillators.

According to preliminary dielectric and piezoelectric measurements, there is a phase transition at about 250 K in α-glycine and some of its derivatives.

A phonon echo in L-alanine single-crystalline powder (D_2 group) was observed [47]. Echo signal appears at the moment t = 2τ where τ is the delay time between two electric pulses applied to the sample. The pulse duration was 6 10^{-6} s with a carrier frequency of 10 MHz. The pulse separation time τ varied between 55 10^{-6} s and 165 10^{-6} s. The relaxation time T_2 of the phonon echo and its temperature dependence was measured, and an abrupt change of T_2 near 170 K was observed. Dielectric measurements in L-alanine single crystals showed that a step-like anomaly in the dielectric constant about 15 % occurred at the same temperature. These results indicate a phase transition at 170 K in L-alanine crystals. To understand the nature of this phase transition more research should be done.

8

4. Conclusions

In conclusion, we believe that studies of physical properties of the protein amino acids and their derivatives with future expansion of the studies to proteins open new possibilities both for biology and for physics of polar crystals.

As for a possible role of piezo-, pyro-, and ferroelectricity in the functioning of living creatures, this may most likely occur only in the form of ferroelectric liquid crystals. Many of the protein primary amino acids or their derivatives seem to be a base for constituting such biological ferroelectric liquid crystals. Much more research should be done to understand all these phenomena.

This work was partly supported by the Russian Foundation for Basic Research (grant 99-02-18307).

5. References

1. Fukuda, E. and Yasuda, I. (1964) Piezoelectric effect in collagen, *Jap. J. Appl. Phys.* **3**, 117-121.
2. Shamos, M.H. and Lavine, L.S. (1967) Piezoelectricity as a fundamental property of biological tissues, *Nature* **213**, 267-269.
3. Williams, W.S. (1982) Piezoelectric effects in biological materials, *Ferroelectrics* **41**, 225-246.
4. Lang, S. (1966) Pyroelectric effect in bone and tendon, *Nature* **212**, 704-705.
5. Lang, S.B. and Athenstaedt, H. (1978) Anomalous pyroelectric behavior in theleaves of the palm-like plant, *Ferroelectrics* **17**, 511-519.
6. Athenstaed, H. (1970) Permanent longitudinal electric polarization and pyroelectric behavior of collagenous structures and nervous tissue in man and other vertebrates, *Nature* **228**, 830-834.
7. Anhenstaed, H. (1976) Pyroelectric sensors of organisms, *Ferroelectrics* **11**, 365-369.
8. Anthenstaed, H. (1976) Pyroelectric properties of wheat, *Ferroelectrics* **14**, 753 -759.
9. Lang, S. (1981) Pyroelectricity: occurrence in biological material and possible physiological applications, *Ferroelectrics* **34**, 3-9.
10. Lang, S.B. (1969) Thermal expansion coefficients and the primary and secondary pyroelectric coefficients of animal bone, *Nature* **224**, 798-799.
11. Polonsky, J., Douzou, P., and Sadron, C. (1960) Mise en évidence de propriétés ferroélectriques dans DNA, *C.R. Acad. Sci.* **250**, 3414-3416.
12. Stanford, A.L. and Lorey, R.A. (1968) Evidence of ferroelectricity in RNA, *Nature* **219**, 1250-1251.
13. Leuchtag, H.R. (1988) A proposed physical explanation of the activation of sodium channels, *Ferroelectrics* **86**, 105-113.
14. Tokimoto, T. and Shirane, K. (1993) Ferroelectric diffused electrical bilayer model for membrane excitation, *Ferroelectrics* **146**, 73-80.
15. Gurskaya, G.V. (1968) *The molecular structure of amino acids: determination by X-ray diffraction analysis*, Consultant Bureau, New York.
16. Simpson,H.J. and Marsh, R.E. !1966) The crystal structure of L-alanine, *Acta Cryst.* **20**, 550-555.
17. Derissen, J.L., Endeman, H.J., and Peerdeman, A.F. (1968) The crystal and molecular structure of L-aspartic acid, *Acta Cryst.* **B24**, 1349-1354.
18. Khavas, B. (1970) The unit cell and space group of L-methionine, L-β-phenylalanine, and DL-tyrosine, *Acta Cryst.* **B26**, 1919-1922.
19. Khavas, B. (1985) X-ray study of L-phenylalanine dimorph and D-tryptophane, *Ind.J.Phys.* **59A**, 219-226.
20. Chaney, M.O. and Steinranf, L.K. (1974) The crystal and molecular structure of tetragonal L-cystine, *Acta Cryst.* **B30**, 711-716.
21. Maddin, J.J., McGandy, E.L., and Seeman, N.C. (1972) The crystal structure of the orthorhombic form of L-(+)Histidine, *Acta Cryst.* **B28**, 2377-2382.

22. Maddin, J.J., McGandy, E.L., and Seeman, N.C. (1972) The crystal structure of the monoclinic form of L-histidine, *Acta Cryst.* **B28**, 2382-2389.
23. Harding, M.M. and Long, H.A. (1968) The crystal structure of L-cysteine, *Acta Cryst.* **B24**,1096-1102.
24. Kerr, K.A., Ashmore, J.P., and Koetzie, T.F. (1975) A neutron diffraction study of L-cysteine, *Acta Cryst.* **B31**, 2022-2026.
25. Torii, K. and Iitaka, Y. (1971) The crystal structure of L-isoleucine, *Acta Cryst.* **B27**, 2237-2246.
26. Khawas, B. (1971) X-ray study of L-arginine HCl, L-cysteine, DL-lysine, and DL-phenylalanine, *Acta Cryst.* **B27**, 1517-1520.
27. Benedetti, E., Pedone, C., and Sirigu, A. (1973) The crystal structure of DL-isoleucine and structural relation between racemic and optically active pairs in some aminoacids, *Acta Cryst.* **B29**, 730-733.
28. Harding, M.M. and Howieson, R.M.(1976) L-leucine, *Acta Cryst.* **B32**, 633-634.
29. Delfino, M. (1978) A comprehencive optical secod harmonic generation study of the non-centrosymmetric character of biological structures, *Mol.Cryst.Liq. Cryst.* **52**, 271-284.
30. Vasilescu, D., Cornillon, R., and Mallet, G. (1970) Piezoelectric resonances in amino-acids, *Nature* **225**, 635.
31. Fousek, J. (1991) Ferroelectricity: remarks on hystorical aspects and present trends, *Ferroelectrics* **113**, 3-20.
32. Sworakowski, J. (1992) Ferroelectricity and related properties of molecular solids, *Ferroelectrics* **128**, 295-306.
33. Silvestrova, I.M., Nabakhtiani, G.N., Kozin, V.B., Kuznetsov, V.A., and Pisarevsky, Y.V. (1992) Elastic, piezoelectric, and dielectric properties of LAP crystals, *Kristallographia* **37**, 1535-1541.
34. Gladky, V.V. and Zholudev, I.S. (1965) Pyroelectric properties of some single crystals, *Kristallographia* **10**, 63-67.
35. Barlew, C., Spasov, V., and Teravitcharova, S. (1994) Pyro- and ferroelectric properties of nGly MeCl$_2$ 2H$_2$O , *Ferroelectrics* **158**, 157-162.
36. Pepinsky, R., Vedam, K., Hoshino, S., and Okaya, Y. (1958) Ferroelectricity in di-glycine nitrate, *Phys. Rev.* **111**, 430-431.
37. Baran, J., Sledez, M., Jakubas, R., and Bator, G. (1997) Ferroelectric phase transition in deuterated glycinium phosphate, *Phys. Rev.* **B51**, 169-172.
38. Kehrer, A. and Weiss, A. (1990) The pyroelectric coefficient of Gly-L-Ala HBr H$_2$O and Gly-L-Ala HI H$_2$O , *Ferroelectrics* **106**, 405-410.
39. Makita, Y. (1965) Ferroelectricity in TSCC, *J. Phys. Soc. Jap.* **20**, 2073-2080.
40. Schaak, G. (1990) Betaine compounds, *Ferroelectrics* **104**, 147-158.
41. Balashova, E.V., Lemanov, V.V.,Albers, J., and Kloepperpieper, A. (1998) Ultrasonic study of betaine compounds, *Ferroelectrics* **208-209**, 63-81.
42. Miglory, A., Maxton, P.M., Clogston, A.M., Zirngiebl, E., and Lowe, M. (1998) Anomalous Temperature dependence in the Raman spectra of L-alanine: evidence for dynamic localization, *Phys. Rev.* **B38**, 13464-13467 .
43. Kwok, R.S., Maxton, P., and Miglory, A. (1990) Thermal conductivity of single crystal L-alanine, *Sol. St. Commun.* **74**, 1193-1195.
44. Moreno, J.D., et al. (1997) Pressure induced phase transitions in monohydrate L-asparagine amino acid crystals, *Sol. St. Commun.* **103**, 655-657.
45. Winkler, E., Etchegon, P., Feinstein, A., and Fainstein, C. (1998) Luminescence and resonant Raman scattering of colour centers in irradiated crystalline L- alanine, *Phys. Rev.* **B57**, 13477-13483.
46. Lemanov, V.V. and Popov, S.N. (1998) Unusual electromechanical effects in glycine, *Fiz. Tverd. Tela* **40**, 1086-1089. (*Phys. Sol. State* **40**, N 6.)
47. Lemanov, V.V. and Popov, S.N. (1998) Phonon echo in L-alanine, *Fiz. Tverd. Tela* **40**, 2119-2120. (*Phys.Sol.State* **40**, 1921-1922.)

PIEZO-, PYRO- AND FERROELECTRIC POLYMERS

S. BAUER, S. BAUER-GOGONEA, M. LINDNER, AND
K. SCHRATTBAUER

Angewandte Physik, Johannes-Kepler-Universität Linz,

Altenbergertsr. 69, A-4040 Linz, Austria

Abstract

Piezo, pyro, and ferroelectric polymers are known since 30 years. With piezo- and pyroelectric coefficients being less than that of crystalline or ceramic piezoelectrics, polymers have found niche applications in diverse fields, ranging from sensor systems and nondestructive testing to fundamental research applications, such as photopyroelectric spectroscopy and microcalorimetry. This survey discusses the technologically important polymers polyvinylidene fluoride {PVDF} and its copolymers with trifluoroethylene {P(VDF-TrFE)}. Recent developments include the preparation and characterization of ultra-thin ferroelectric polymer films and relaxor-type ferroelectric polymers with large electrostrictive responses. Special emphasis is given to techniques for measuring the piezo- and pyroelectric activity, and to their use for the nondestructive probing of nonuniform space-charge and polarization distributions in polymer films. A few recent applications are selected which display the large potential of exploiting ferroelectric polymers.

1. Introduction

1999 marks the 30th anniversary of the discovery of strong piezoelectricity in the polymer polyvinylidene fluoride -$(CH_2CF_2)_n$- {PVDF} [1], now known to be a ferroelectric material. Since then, ferroelectricity has been identified in copolymers of vinylidene fluoride and trifluoroethylene -$(CH_2CF_2)_x$-$(CHFCF_2)_{1-x}$- {P(VDF-TrFE)}, in disubstituted diacetylenes, odd numbered nylons, polyureas, polyurethane and several liquid crystalline polymers [2]. So far, only PVDF and its copolymers P(VDF-TrFE) are technologically important.

PVDF is semicrystalline with at least four crystal phases and a crystallinity of around 50% [3]. In the crystalline form, polymer chains have regular conformations, the most favorable torsional bond arrangements have substituents at 180° to each other (*trans* or *t*)

C. Galassi et al. (eds.), Piezoelectric Materials: Advances in Science, Technology and Applications, 11–19.
© 2000 *Kluwer Academic Publishers. Printed in the Netherlands.*

or at ±60° (*gauche*$^±$ or *g*$^+$). The most stable crystalline phase of PVDF is the nonpolar α-phase with a *tg*$^+$*tg*$^-$ chain conformation. The α-phase can be transformed into the ferroelectric β-phase with an all-*trans* chain conformation by mechanical stretching. The copolymers of PVDF with TrFE crystallize directly into the polar β-phase. The ferroelectric crystallites are embedded in an amorphous, polar matrix. For PVDF and its copolymers the glass transition of the amorphous phase is around -40°C, so that at room temperature the polar, amorphous matrix is in the rubber state. In poled ferroelectric polymers, the ferroelectric crystallites are no longer split in domains and trapped charges at the interface between crystallites and amorphous matrix are essential for compensating the depolarizing field. A schematic sketch of the various morphological elements in a ferroelectric polymer is depicted in Fig. 1 [4]. Dielectric, ferro-, piezo- and pyroelectric properties of ferroelectric polymers strongly depend on the crystallinity and morphology of samples, so that even after 30 years of research there is still a debate on the relative magnitudes of the various physical mechanisms that contribute to the material parameters [2,5].

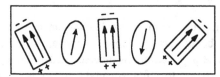

Figure 1: Polarized state of a semicrystalline, ferroelectric polymer at room temperature. The ferroelectric polarization is compensated and stabilized by charges at the interfaces between crystallites and amorphous matrix. The dipoles in the amorphous matrix, which is in the rubber state above the glass-transition are statistically oriented (after [5]).

Ferroelectric polymers are now available in a broad range of thickness, from sub-micron thick films [6] up to 1mm thick plates [7]. They are easily cast in unconventional geometries, for example as a piezoelectric dielectric in a functional coaxial cable [8]. Ferroelectric polymers have found various applications, e.g. as hydrophones, shock gauges, IR-sensors, impact and vibration sensors and as traffic monitors [9,10].

In the present survey, some of the more recent developments in the fields of fundamentals and applications of ferroelectric polymers are briefly reviewed, special emphasis is given to similarities and differences to ferroelectric ceramics.

2. Ferroelectric properties

The phase diagram of VDF-TrFE copolymers, including the melting temperature T_m, the crystallization temperature T_{cr} and the Curie temperature T_c (determined upon heating and cooling) is depicted in Fig.2 versus the VDF content [11]. From 50 to approximately 80 mol% VDF, the copolymers show a ferroelectric phase transition from a high-temperature para- to a low temperature ferroelectric phase. Above 80 mol% VDF content, the phase transition is not observed, as melting occurs prior to the Curie transition. The extrapolated Curie temperature for PVDF is 205°C, 25°C above the melting temperature. VDF-rich copolymers (between 60 and 80 mol%) show first-order phase transitions, with large thermal hysteresis in the Curie temperature upon heating and cooling. Between 50 and 60% the thermal hysteresis disappears, and the phase transition changes from first- to second-order (see below). Copolymers with a VDF content below 50% show a phase transition from the high-temperature para- to a low-temperature antiferroelectric or antiferroelectric-like state [11].

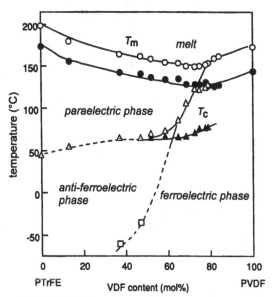

Figure 2: Phase diagram for P(VDF-TrFE) copolymers. (o: melting temperature; •: crystallization temperature; Δ: Curie temperature on heating, ▲: Curie temperature on cooling, □: ferroelectric-to-antiferroelectric transition temperature) (redrawn after [11]).

The phase transitions of P(VDF-TrFE) copolymers are broad and characterized by a distribution of Curie temperatures, as expected for a semicrystalline material containing crystallites with large distributions in size. The thermal hysteresis in a 70/30 copolymer is obvious in the temperature-dependent dielectric function (Fig. 3 left) and extends over 40°C. The absence of a thermal hysteresis in a 56/44 copolymer is evident from Fig. 3 (right) [12, 13]. Thermal hysteresis is a direct proof for a first-order phase transition, whereas the absence of thermal hysteresis may be associated to a second-order phase transition. There is still a controversial discussion on the order of the phase-transition for copolymers with a VDF content between 50 and 60 mol%. DSC measurements indicate a weak first-order phase transition [14], whereas small-signal nonlinear dielectric measurements show evidence for a second-order phase transition [12, 13].

Small-signal dielectric nonlinearities were found to be versatile for determining the coefficients in a Landau-Devonshire free-energy expansion [12, 15]. Additionally, it was shown that the third-order nonlinear dielectric constant allows for a discrimination between first- and second-order phase transitions [12,15]. The second-order nonlinear dielectric function appears only in the noncentrosymmetric state. It has been found to be a sensitive means for detecting non-switchable polarization in ferroelectric polymers [16], a technique which also might become interesting for ferroelectric ceramics.

14

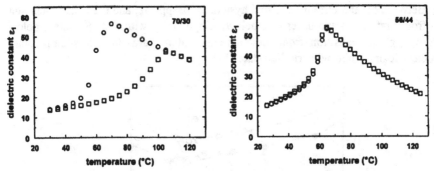

Figure 3: Dielectric constant ε' measured at 6kHz for a 70/30 and 56/44 mol% P(VDF-TrFE) copolymer as a function of temperature ($\square$: heating cycle, o: cooling cycle) (redrawn after [12])

PVDF ferroelectric polymers show polarization hysteresis loops with a remanent polarization $P_r=7\mu C/cm^2$ and coercive fields $E_c=100V/\mu m$. The remanent polarization is comparable with that of $SrBi_2Ta_2O_9$ (SBT) ceramic films, one of the most promising candidates for ferroelectric memory applications. However, the application of ferroelectric polymers for memory devices is still hindered due to the very large coercive fields, as switching voltages on the order of 10V are necessary for 100nm thick films [17]. Methods for reducing the coercive field of ferroelectric polymers are thus highly demanded.

A narrowing of the hysteresis loop and a significant reduction of the coercive field is possible after irradiation of P(VDF-TrFE) films with 3MeV electron beams [18]. The reduction in size of the ferroelectric crystallites results in relaxor-like behavior of such copolymer films. The findings were explained by a size-reduction of the ferroelectric crystallites, which results in relaxor-like behavior of the copolymers. Switching can be performed very fast, but the switched polarization is not stable. Relaxation of the polarization in the switched state is observed, similar to ferroelectric ceramic films [19]. The relaxation was ascribed to the delayed charge compensation of polarization gradients, which leads to large internal depolarizing fields in the crystallites.

3. Piezo- and pyroelectricity

3.1 PIEZO- AND PYROELECTRIC RELAXATION

Piezo- and pyroelectricity is the electrical response of a material to a change in pressure or temperature, respectively. In ferroelectric polymers, the pyroelectric coefficient p_{exp} is defined as the temperature derivative of the charge Q induced on the sample electrodes of area A upon heating or cooling: $p_{exp}=(1/A)dQ/dT$ (piezoelectric coefficients are defined in a similar way). Pyroelectricity in semicrystalline polymers arises from electrostriction, dipole libration, dimensional changes in both amorphous and crystalline changes, as well as from reversible changes of crystallinity and motion of charges necessary to compensate the polarization of the crystallites.

The importance of space charges in ferroelectric polymers became evident only recently [5]. An important consequence of the interplay between polarization and compensation charges is piezo- and pyroelectric relaxation [20]. The pyroelectric response of PVDF films irreversibly decreases well below the Curie temperature during cyclic increase and decrease of the temperature [21,22] as a result of de-trapping of compensation charges. The interplay between polar crystallites and compensation charges has another important consequence. Piezo- and pyroelectric coefficients exhibit relaxation and must be represented by complex quantities [20]. The low-frequency (quasi-static) pyroelectric coefficient is larger than the high-frequency (dynamic) pyroelectric coefficient [23], as a result of the additional movement of compensation charges at low modulation frequencies (Fig. 4).

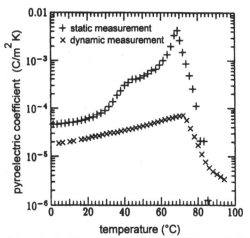

Figure 4 Quasi-statically and dynamically determined pyroelectric coefficients of a 70/30 P(VDF-TrFE) copolymer, demonstrating pyroelectric relaxation (after [23]).

An observation of pyroelectric relaxation in PVDF at room temperature has been reported in [24]. It must thus be noted that a comparison of material data from different sources must be made carefully, especially if the measurement conditions are not specified precisely. Similar effects may be observed in polycrystalline ceramics, where charges are necessary for the compensation of polarization gradients between different grains.

3.2 POLING AND POLARIZATION DISTRIBUTIONS

In semicrystalline polymers, space charges may pre-exist in the polymer, may be injected or may be internally generated during the application of a poling field. In the presence of a spatially and temporally varying space-charge distribution $\rho(x,t)$, the poling field $E(x,t)$ and thus the resulting polarization $P(x,t)$ in the polymer are spatially and temporally nonuniform. From a practical point of view, knowledge of the dependence of $P(x)$ on the various poling variables is sufficient. Techniques for the probing of electric field or polarization distributions across the film thickness of dielectrics are well established [2]. Polarization distributions in „well" poled ferroelectric polymer films of different thickness are shown in Fig. 5 [25]. Near surface depolarized layers are typically observed in ferroelectric polymers [26]. The polarization distribution is an important factor in high frequency pyroelectric sensor applications, since near surface depolarized layers drastically decrease the response speed. Polarization distributions are also likely to appear in ferroelectric ceramics, since charges are essential for compensating polarization gradients in adjacent grains.

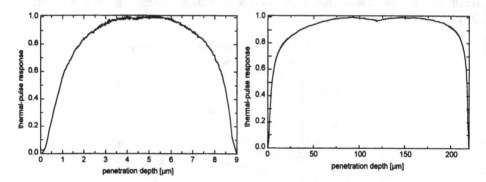

Figure 5: Polarization distributions of „well" poled 9μm thick PVDF polymer (left) and 220 μm thick P(VDF-TrFE) copolymer films (right). Note the near surface depolarized layers.

3.3 MATERIALS CHARACTERIZATION

In order to compare materials for piezo- and pyroelectric applications, material characteristics such as thermal, elastic, dielectric, piezo- and pyroelectric coefficients must be known. Techniques for material characterization, especially developed or adapted for thin films are therefore highly demanded. Electro-acoustic techniques have been employed for the determination of the velocity of sound, piezoelectric coefficients and polarization distributions [3]. Pulsed electro-thermal techniques have proved to be useful for the determination of pyroelectric, piezoelectric, thermal (specific heat and thermal diffusivity) and elastic (coupling factors and velocity of sound) material parameters, as well as polarization distributions [27-30]. In electro-thermal techniques, the electrical response of the polymer capacitor is measured after transient thermal excitation with a short heat pulse. The technique is reliable, simple-to-use and can be easily adapted for the investigation of unconventional sample geometries, e.g. cables and hollow spheres. Fig. 6 shows the combined oscillatory piezo- and pyroelectric response of a 500μm thick free-standing piezoelectric polymer after heat pulse excitation [30]. The oscillatory component of the signal in Fig. 6 can be used to extract piezoelectric material parameters [29,30], coupling factors [29] and the velocity of sound (Fig. 7) [30]. It must be noted that electro-thermal techniques are not necessarily restricted to polymers, they may also be implemented for the characterization of other materials, such as ferroelectric ceramics.

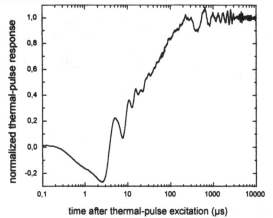

Figure 6: Combined piezo- and pyroelectric response after excitation of a free-standing piezoelectric polymer film with a short heat pulse (after [30]).

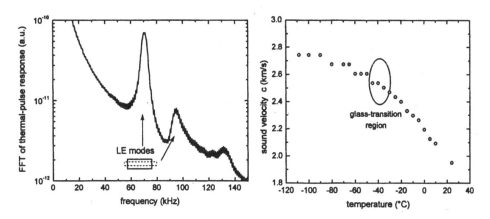

Figure 7 (left): Resonance frequencies of the two LE-modes of a 1mm thick, rectangulary shaped 75/25 P(VDF-TrFE) copolymer sample and (right) sound velocity versus temperature for the same sample.

A comparison of material data for ferroelectric PVDF and $PbZr_xTi_{1-x}O_3$ (PZT) ceramics is provided in Table 1. The low acoustical impedance, the fairly high hydrostatic coefficients, and the easy processing into different shapes make ferroelectric polymers interesting for applications.

4. Applications

Piezoelectric polymers are now well established materials for shock-gauges, hydrophones, piezocables etc. [3]. Highlights for the use of ferroelectric polymers in fundamental research applications are extremely sensitive microcalorimeters for

18

measuring the specific heat of samples with a mass of a few µg [31] or for the measurement of the heat of adsorption with a sensitivity of 100nJ [32].

Table I: Material properties of the ferroelectric polymer PVDF and the ceramic PbZr$_x$Ti$_{1-x}$O$_3$ (PZT) near the morphotropic phase boundary. Material data are compiled from [2,33].

	Unit	PVDF	PZT
Density ρ	gcm^{-3}	1.78	7.75
specific heat c	10^6Jm^{-3}K^{-1}	2.3	2.5
thermal conductivity κ	Wm^{-1}K^{-1}	0.14	0.18
velocity of sound v_s	ms^{-1}	2150	4350
acoustic impedance	kgm^{-2}s^{-1}	3.8	33.8
Curie temperature T_c	°C	205>T$_m$*	365
relative dielectric constant ε_{33}	–	9	200-600
dielectric loss tan δ	–	0.03	0.003
remanent polarization P_r	µCcm^{-2}	7	30-50
pyroelectric coefficient p_{exp}	µCcm^{-2}K^{-1}	25	380
piezoelectric strain constants: d_{33}	pCN^{-1}	-31	374
d_{31}	pCN^{-1}	21	-171
Hydrostatic coefficients: d_h	pC/N	-9	32
g_h		-130	2

*: extrapolated Curie temperature, above the melting temperature T$_m$.

5. Acknowledgments

Work supported by the Fonds zur Förderung der wissenschaftlichen Forschung (FWF) in Austria.

6. References

1. Kawai, H. (1969) The piezoelectricity of poly(vinylidene fluoride), Jpn. J. Appl. Phys. 8, 975-976.
2. Gerhard-Multhaupt, R. (ed.) (1999), Electrets, 3rd ed., Vol. II, Laplacian Press, Morgan Hill, chapters 11 and 12.
3. Lovinger, A. J. (1983) Ferroelectric polymers, Science 220, 1115-1121.
4. Rollik, D., Bauer, S., and Gerhard-Multhaupt, R. (1999) Separate contributions to the pyroelectricity in polyvinilidene fluoride from the amorphous and crystalline phases as well as from their interface, J. Appl. Phys. 85, 3282-3288.
5. Sessler, G. M., Das-Gupta, D. K., DeReggi, A. S., Eisenmenger, W., Furukawa, T., Giacometti, J. A., and Gerhard-Multhaupt, R., (1992) Piezo- and pyroelectricity in electrets: Caused by charges, dipoles or both? IEEE Trans. Electr. Insul. 27, 872-897.
6. Bune, A.V., Fridkin, V.M., Ducharme, S., Blinov, L. M., Palto, S. P., Sorokin, A. V., Yudin, S. G., Zlatkin, A. (1998) Two-dimensional ferroelectric films, Nature 391, 874-877.
7. Isner-Brom, P., Brissand, M., Heintz, R., Eyrand, L., and Bauer, F. (1995) Intrinsic piezoelectric characterization of PVDF copolymers: Determination of elastic constants, Ferroelectrics 171, 271-279.
8. Fox, D. (1991) A high performance piezoelectric cable, Ferroelectrics 115, 215-224.
9. Nalwa, H. S. (ed.) (1995) Ferroelectric polymers: Chemistry, physics and applications, Marcel Dekker, New York, Basel, Hong-Kong.
10. Fraden, J. (1996) Handbook of modern sensors: Physics, designs and applications 2nd ed., American Institute of Physics, Woodbury, New York.
11. Furukawa, T. (1997) Structure and functional properties of ferroelectric polymers, Adv. Colloid Interface Sci. 71-72, 183-208.

12. Heiler B. and Ploss B. (1994) Dielectric nonlinearities of P(VDF-TrFE), *Proceedings, 8th International Symposium on Electrets ISE8*, 662-667 (Paris 1994).
13. Bauer S. (1996) Poled polymers for sensors and photonic applications, *J. Appl. Phys.* **80**, 5531-5558.
14. Balta Calleja, F. J., Gonzales-Arche, A., Ezquerra, T. A., Santa-Cruz, C., Batallan, F., Frick, B., and Lopez-Cabarcos, E. (1993) Structure and properties of ferroelectric copolymers of poly(vinylidene fluoride), *Adv. Polym. Sci.* **108**, Springer, Berlin, Heidelberg.
15. Ikeda, I., Kominami, K., Koyama, K., and Wada, Y. (1987) Nonlinear dielectric constant and ferroelectric-to-paraelectric phase transition in copolymers of vinylidene fluoride and trifluoroethylene, *J. Appl. Phys.* **62**, 3339-3942.
16. Ploss, B., and Ploss, B. (1998) Influence of poling and annealing on the nonlinear dielectric permittivity of P(VDF-TrFE) copolymers, *IEEE Trans. Diel. Electr. Insul.* **5**, 91-95.
17. Ducharme, S., Bune, A. V., Blinov, L. M., Fridkin, V. M., Palto, S. P., Sorokin, A. V., and Yudin, S., G. (1998) Critical point in ferroelectric Langmuir-Blodgett films, *Phys. Rev. B* **57**, 25-28.
18. Zhang, Q. M., Bharti, V., and Zhao, X. (1998) Giant electrostriction and relaxor ferroelectric behavior in electron-irradiated poly(vinylidene fluoride-trifluoroethylene) copolymer, *Science* **280**, 2101-2104.
19. Eberle, G., Schmidt, H. and Eisenmenger (1996) W., Piezoelectric polymer electrets, *IEEE Trans. Diel. Electr. Insul.* **3**, 624-646.
20. Bauer S. (1995) The pyroelectric response of polymers and its applications, *Trends Polym. Sci.* **3**, 288-296.
21. De Rossi, D., DeReggi, A. S., Broadhurst, M. G., Roth, S. C., and Davis, G. T. (1982) Method of evaluating the thermal stability of the pyroelectric properties of polyvinylidene-fluoride: Effects of poling temperature and field, *J. Appl. Phys.* **53**, 6520-6525.
22. Fedosov, S. N. and Sergeeva, A. E. (1989) Nature of pyroelectricity in polyvinylidene fluoride, *Sov. Phys. Solid State* **31**, 503-505.
23. Ruf. R., Bauer, S., and Ploss, B. (1992) The ferroelectric phase transition of P(VDF-TrFE) polymers, *Ferroelectrics* **127**, 209-214.
24. Ploss, B. and Domig, A. (1994) Static and dynamic pyroelectric properties of PVDF, *Ferroelectrics* **159**, 263-268.
25. M. Lindner, K. Schrattbauer, unpublished results.
26. Ploss, B., Emmerich, R., and Bauer, S. (1992) Thermal wave probing of pyroelectric distributions in the surface region of ferroelectric materials: A new method for the analysis, *J. Appl. Phys.* **72**, 5363-5370.
27. Bauer, S. and DeReggi, A. S. (1996) Pulsed electro-thermal technique for measuring the thermal diffusivity of dielectric films on conducting substrates, *J. Appl. Phys* **80**, 6124-6128.
28. Bauer, S. and Bauer-Gogonea, S. (1998) Pyroelectric investigations: Tool for the study of thermal, elastic, and electric properties, *Dielectrics Newsletter*, september issue 1-3.
29. Takahashi, Y., Hiraoka, K., and Furukawa, T. (1998) Time evolution of laser induced pyroelectric responses in a VDF/TrFE copolymer, *IEEE Trans. Diel. Electr. Insul.* **5**, 957-960.
30. Bauer-Gogonea, S., Bauer, S., and Wirges, W. (1999) Pulsed electrothermal technique for the characterization of dielectric films, Proc. SPIE, in press.
31. Bauer, S., and Ploss, B. (1992) Design and properties of a pyroelectric microcalorimeter, *IEEE Trans. Electr. Insul.* **27**, 861-866.
32. Stuckless, J. T., Frei, N. A., and Campbell, C. T. (1998) A novel single-crystal adsorption calorimeter and additions for determining metal adsorption and adhesion energies, *Rev. Sci. Instrum.* **69**, 2427-2438.
33. Wang, T. T., Herbert, J. M., and Glass, A. M. (eds.) (1988) *The applications of ferroelectric polymers*, Blackie and Son, London.

THE KINETIC AND SCATTERING OF NONEQUILIBRIUM PHONONS BY DOMAIN WALLS AND GRAIN BOUNDARIES IN MONOCRYSTAL AND CERAMICS OF BATIO₃

S.N. lvanov, A.V. Taranov, and E.N. Khazanov
Institute of Radio-Engineering and Electronics. Russian Academy of Sciences, 103907 Moscow, Russia
E.P. Smirnova
A. F. Ioffe Physical-Technical Institute, Russian Academy of Sciences. 194021 St. Petersburg, Russia

Phonon transport in ferroelectric ceramics and single crystals BaTiO₃ has been experimentally investigated.. The scattering by domain walls in BaTiO3 single crystals and by boundary grains in ceramic samples have been identified, and our results are in quantitative agreement with calculations.

1. Introduction

In our experiments we have studied propagation of weakly nonequilibrium phonons at $\Delta T = T_h - T_o << T_o$, where T_h is the heater temperature and T_o is the ambient temperature, in one of the classical ferroelectric materials in single-crystal and ceramic modifications BaTiO₃. When the thermal pulse technique [1] is used, i.e., the size of the samples in the phonon flux direction is larger than their diffusion length, the propagation of nonequilibrium phonons is described by the conventional heat equation. The peak amplitude of the bolometer signal as a function of temperature yields information about the mechanism of phonon scattering in the material, and in ceramics about the structure of grain boundaries [2,3].

The interest in studies of ferroelectric materials is stimulated by at least two objectives.

1. Identification of the contribution of phonon scattering on grain boundaries. At present numerous theoretical calculations of coherent ultrasound reflection and refraction on grain boundaries are available [4,5]. On the other hand, no experimental data that are at all trustworthy have been obtained to date, primarily because the reflection of coherent ultrasound waves from domain boundaries is very low [6,7]. Suslov and Kagan [8,9] attempted to determine the contribution of domain boundaries to scattering of phonons in a thermal pulse in virtual ferroelectrics SrTaO₃ and KTaO₃ in an applied electric field, but the interpretation of the experimental data proved to be very complicated and unreliable.

C. Galassi et al. (eds.), Piezoelectric Materials: Advances in Science, Technology and Applications, 21–26.

2. Determination of the main phonon scattering mechanism in a thermal pulse in ceramic and single crystal of BaTiO₃ and comparison with the expected results for ferroelectric materials such as PLZT which have a diffusive phase transition and are characterized by anomalies in their low-temperature specific heat and thermal conductivity that can be interpreted in terms of the glass model [10]. With a view to interpreting measurements by the thermal pulse method, the transit time of a nonequilibrium phonon pulse as a function of temperature in such materials was calculated by Kozub et al.[3,11].

Thus, the aim of this work was the experimental investigation of phonon scattering by domain walls in single crystals and by grain boundaries in ceramics of the conventional BaTiO₃ ferroelectric with a narrow phase transition.

2. Experimental techniques

All BaTiO₃ ferroelectric ceramic and single-crystals samples used in our experiments were fabricated by conventional technologies. The ceramics were densely packed with a density of 97-98% of the theoretical single-crystal density at an average grain dimension of about 10^{-3} cm.

The structure of a sample cleaved surface was tested by a Joel JSM-840 scanning electron microscope. The micrographs characterized on the qualitative level a single-crystal domain structure or a pattern of grains and grain boundaries in a ceramic sample. The phonon kinetics at helium temperature was studied by the thermal pulse technique described in detail in [1]. Thin (~1000A) gold film is deposited on one side of a plate of tested material, and this film is heated by a short ($\approx 10^{-7}$ s) current pulse, so it acts as an injector of nonequilibrium phonons into the sample. On the opposite sample surface, a tin bolometer of a meandered shape with area 0.3x0.25 mm is fabricated. If the bolometer characteristic is biased by a weak magnetic field of $\approx 2 \cdot 10^2$Oe, one can measure nonequilibrium phonon scattering versus temperature over a range of 1.7-3.8 K. The power dissipated in the heater is set at a sufficiently low level that the injected phonons could be described in terms of the ambient temperature in analyzing experimental data.

The basic parameter measured in the experiments is the transit time t_{max} of the nonequilibrium phonon peak detected by the bolometer. Its dependence on the temperature and features of the tested sample structure has been analyzed.

3. Results and discussion

Typical micrographs of sample surfaces are shown in Fig.1. The BaTiO₃ single crystal is characterized by a fairly regular pattern of a-type domains (Fig.1a). The boundary between the domains is a region where the order parameter is inhomogeneous, [12] and its width r_c can be treated as a domain wall boundary thickness. There is good reason to assume that r_c equals several lattice constants, i.e., in BaTiO3 $r_c \approx 20$-30A, so that the phonon wavelength in our experiments is $\lambda_{ph} \gg r_c$. In real crystals a domain wall is pinned by defects, crystal inhomogeneities on crystal boundaries and in the wall region, i.e., elastic stress is present in the layer of thickness ~r_c. The domain structure determines most of the ferroelectric properties of practical significance, and one aim of the reported work was to estimate its effect on the phonon

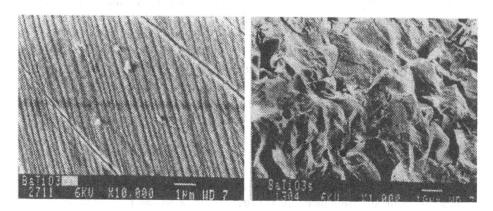

Figure1. Electron micrographs of (a) strip domains in the *xy* plane of a BaTiO₃ single crystal and (b) cleaved surface of a ceramic sample fabricated by the hot-pressing technique.

Ceramic samples under study were characterized by dense packing of most of their grains, of which the majority looked like crystallites (Fig. 1b). The statistical analysis of large sets of grains yielded the mean grain size *R* for a specific sample.

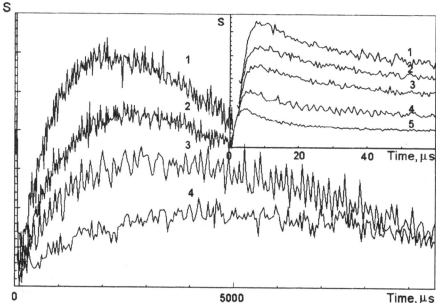

Figure 2. Signals due to nonequilibrium phonons in a BaTiO₃ ceramic sample with *L=0.1* cm. 1-T=3.81K, 2-3.5K, 3-2.97K, 4-2.56K. The inset shows nonequilibrium phonon signals in a BaTiO₃ single-crystal sample with L= 0.043 cm. 1-T=3.8K, 2-3.77K, 3-3.49K, 4-2.99K, 5-2.56K.

24

Before proceeding to the analysis of nonequilibrium phonon propagation, let us discuss features common to all the samples. Thus, the curves of bolometer signal versus time are bell-shaped with clearly defined peaks, which is typical of the diffusion propagation mode (Fig.2). The transit time t_{max} of the nonequilibrium phonon peak to the bolometer is proportional to L^2 with good accuracy, where L is the plate size in the phonon flux direction.

Given t_{max}, one can estimate the effective diffusion coefficient of the dominant phonon group [2], $D_{eff} \approx L^2/ t_{max}$. On the other hand, the thermal conductivity κ and specific heat C_υ at helium temperature are known. This allowed us to calculate the phonon diffusion coefficient by the standard formula $D=\kappa/C_\upsilon$ and compare with our measurements. It turned out that the discrepancy of D in BaTiO$_3$ single crystals and in BaTiO$_3$ ceramic was within 10% [10,13].

In our opinion, observation of the proportionality $t_{max} \propto L^2$ and absolute measurements of the diffusion coefficient for the dominant group of phonons in fairly good agreement with independent measurements provide solid evidence in favor of the applicability of the diffusion model [2,3] to our results, so we can proceed to the analysis of the t_{max} temperature dependence with a view to determining the phonon scattering mechanism in the ferroelectric BaTiO$_3$ studied. Measurements of the typical BaTiO$_3$ samples are shown in Fig.2.

The inset shows series of curves of the nonequilibrium phonon signal for a BaTiO$_3$ single crystal, for which we have $\partial t_{max}/\partial T > 0$ and $t_{max} = f(T)$. The dependence is approximated by curve $t_{max} = A + BT^4$ (see Fig.3a).

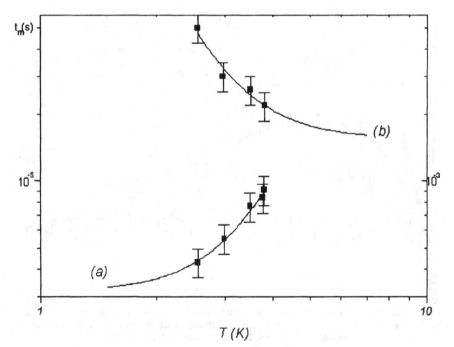

Fig.3. Temperature dependences of t_{max} for single crystal sample BaTiO$_3$ (curve a) and ceramic sample (curve b). Solid lines - approximation (see text).

This result indicates that in the BaTiO3 single crystal the phonon scattering is controlled by domain wall boundaries. Let us make some quantitative estimates. In the model with diffusion of nonequilibrium phonons [2,3] t_{max} is related to the effective phonon free path l_{eff} by the simple formula

$$t_{max} = \frac{3}{2} \frac{L^2}{l_{eff} \upsilon_s}$$ (1)

where υ_s is the mean phonon velocity in the sample. As we note above the dependence of t_{max} on T has two regions and we can assume:

$$l_{eff}^{-1} = l_1^{-1} + l_2^{-1}$$ (2)

It can be supposed that the plateau region of t_{max} is controlled by the acoustic mismatch laws (l_1) and the $t_{max} \sim T^4$ region by the Rayleigh scattering (l_2). Using the data plotted in Fig.3a, we obtain $l_1 \cong 6{,}5 \cdot 10^{-3}$ cm, $l_2^{-1} \cdot T^{-4} \cong 5{,}8$ cm$^{-1} \cdot$K^{-4}.

Two neighboring domains in a ferroelectric are regions of the same crystal with different orientations related to one another by one or more symmetry operations. A phonon flux can undergo reflection on their boundary. This problem was solved [9] for the case of heat transfer in a system of N plane-parallel layers of thickness d. The effective free path of phonons is expressed by the formula [9]

$$l_{ph} = \frac{L \cdot (1-s)}{1 + sL/d}$$ (3)

where $L=Nd$ is the sample thickness, s is the phonon reflectivity on the boundary between two layers. At room temperature our samples have $d \approx 3 \cdot 10^{-5}$ cm (see Fig.1a), and we assume that this parameter is slightly changed as a result of the phase transition taking place as the temperature drops from room to helium. Assuming the experimental result $l_1 = l_{ph}$, we estimate the reflectivity $s \cong 0{,}5 \cdot 10^{-2}$. This measurement seems to be fairly reasonable in view of the results of [6,7] and an estimate of the phonon reflectivity on boundaries between grains in the ceramic corundum, $s = 2.6 \cdot 10^{-2}$ [14]. The difference of the value s in BaTiO$_3$ and corundum ceramics caused by difference in physical constants of these materials.

These estimates have been made under the assumption that phonon scattering due to changes in the density and stiffness in domain walls can be neglected. Inclusion of reflections due to fluctuations in density $\Delta\rho/\rho$ and elasticity $\Delta c/c$ in the domain wall region [5,15] accounts for the temperature dependence of t_{max} observed in our samples, whereas quantitative estimates are difficult owing to the lack of data on $\Delta\rho/\rho$ and $\Delta c/c$ in the domain wall region.

The main set of the curves Fig.2 was obtained in a ceramic sample of BaTiO$_3$ and we see the change the sign the dependence t_{max} on temperature: $\partial t_{max}/\partial T < 0$. The quantity of t_{max} for a ceramic sample is larger by two orders than in monocrystal sample. Such result is typical when the grains are not packed closely together and the boundaries are loose. If the linear size of the contact area r is less than the characteristic phonon wavelengh λ the diffraction processes control the pass of phonons from one grain to another. In this situation phonons pass through grain boundaries more readily the higher the phonon frequences. In our specific case, the sample temperature [2] and the dependence $t_{max}=f(T)$ is approximated by the curve $t_{max}=C+DT^{-4}$ (see Fig.3b).

In the model with diffusion of phonons in ceramic material [2] we have:

$$t_{max} = \frac{L^2}{\upsilon_s \cdot R \cdot f_\omega (\Sigma/S)} \tag{4}$$

where S is the average grain surface area, Σ is the total area of contacts (junctions) with neighboring grains, f_ω is the transition factor for a phonon with a frequency ω across a contact surface, R is the mean grain size. Assuming $\Sigma/S \cong 0,5$ and $R \cong 3\cdot10^{-4}$ (see Fig.1b) we obtain from the temperature dependence part of curve Fig.3b $f_\omega \cong 10^{-2}$ ($T=3,8$K) and the size contacts between grains about $0.05\div0.1\mu$m. Such assesment seems to be fairly reasonable. Some decay of the temperature dependence with T increasing is explained by including other scattering mechanisms. But the question about these mechanisms is open.

In summary, a technique based on the propagation of nonequilibrium phonons in ferroelectrics allows us to obtain quantitative estimates of phonon scattering by domain boundaries in single crystals BaTiO$_3$. A boundary model of ceramics grains qualitatively explains the phonon kinetics in BaTiO$_3$ ceramics. It is also important that the $t_{max}(T)$ dependences in BaTiO$_3$ and PZLT ceramics have the opposite sign [3,11,16].

The work was supported by the Russian Fund for Fundamental Research (Grant 97-02-16810).
E-mail: ivanov@mail.cplire.ru

References

1. S. N. Ivanov, A. V. Taranov, and E. N. Khazanov, (1991) Zh. Eksp. Teor. Fiz. **99**,1311 [Sov. Phys. JETP **72**, 731 (1991)].
2. S. N. Ivanov, A. G. Kozorezov, E. N. Khazanov, and A.V. Taranov, (1992) Solid State Commun. **83**, 365.
3. V.I. Kozub, A. M. Rudin, and H. Schober, (1994) Phys. Rev. B **50**,6032.
4. Kessenikh, D. G. Sannikov, and L. A. Shuvalov, (1971) Kristallografiya **16**,350; **17**, 345 (1972) [Sov. Phys. Crystallogr. **16**, 287 (1971); **17**, 291 (1972)].
5. G. Kessenikh and L. A. Shuvalov, (1984) Izv. Akad. Nauk SSSR, Ser. Fiz. **48**, 1168.
6. S. Kh. Esayan, V. V. Lemanov, and G. A. Smolenskii, (1974) Dokl. Akad. Nauk SSSR **217**, 83 [Sov. Phys. Dokl. **19**, 393 (1975)].
7. V. V. Belov, 0. Yu. Serdobol'skaya, and M. A. Suchkova, (1984) Fiz. Tverd. Tela **26**, 556 [Sov. Phys. Solid State **26**, 334 (1984)].
8. A. V. Suslov, (1992) Fiz. Tverd. Tela **34**, 319 [Sov. Phys. Solid State **34**,170 (1992)].
9. V. D. Kagan and A. V. Suslov, (1994) Fiz. Tverd. Tela **36**, 2672 [Phys. Solid State **36**, 1457 (1994)].
10. J. J. De Yoreo and R. 0. Pohl, (1985) Phys. Rev. B **32**, 5780.
11. V. 1. Kozub and A. M. Rudin, (1996) Fiz. Tverd. Tela **38**, 337 [Phys.Solid State **38**, 189 (1996)].
12. B. A. Strukov and A. P. Levanyuk, (1998) *Ferroelectric Phenomena in Crystals: Physical Foundations*, Springer, New York [Russian original, Nauka, Moscow (1983)].
13. W. N. Lawless, (1976) Phys. Rev. B **14**, 134.
14. A. A. Kaplyanskii, M. B. Mel'nikov, and E. S. Feofilov, (1996) Fiz. Tverd. Tela **38**, 1434 [Phys. Solid State **38**, 792 (1996)].
15. A. G. Kozorezov, J. K. Wigmore, C. Erd *el al.*, (1998) Phys. Rev. B **57**, 7411.
16. S.N. Ivanov, E.N. Khazanov , E.P. Smirnova, A.V. Taranov (1999) JEPT, , **88**, N2, p. 342-346.

FERROELECTRICS WITH PERIODICALLY LAYERED DOMAIN STRUCTURE: COMPUTER SIMULATION OF ACOUSTICAL PROPERTIES

A.N. PODLIPENETS
Institute for Physics of Metals, UNAS
36 Vernadsky Bul., Kyiv 252142, Ukraine

AND

L.P. ZINCHUK
Institute of Mechanics, UNAS
3 Nesterov Str., Kyiv 252057, Ukraine

1. Introduction

Researches exploring the dynamical properties of such smart materials as multilayer structures (MLS's) attract the increasing attention. Particularly, the MLS's consisting of coherent array of antiphase domains have been reported [1-3]. Their acoustical properties and possible applications have been studied both theoretically and experimentally [3-8]. This paper is organized to present a detailed computer simulation study of shear wave propagation in periodic MLS's consisting of infinite, semi-infinite or finite number of antiphase domains. These MLS's in ferroelectric crystals of hexagonal symmetry are assumed to be formed by the alternation of two domain layers with different thicknesses and opposite directions of the spontaneous polarization. Shear waves, which are polarized parallel to the domain walls and propagate along the perpendicular to the spontaneous polarization direction, are considered. The corresponding dispersion equations are particular cases of the general equations derived previously for composite piezoelectric structures by using the mathematical approach named by the periodic Hamiltonian system formalism [7, 9-14]. The computer experiments have been conducted for the domain MLS's in $BaTiO_3$ over a wide range of frequencies and wavelengths. The dispersion relations are discussed as functions of the ratio between thicknesses of two adjacent domains. Results demonstrate the great diversity of spatial localization of

27

C. Galassi et al. (eds.), Piezoelectric Materials: Advances in Science, Technology and Applications, 27–38.
© 2000 *Kluwer Academic Publishers. Printed in the Netherlands.*

diferent modes and their "sensitivity" to the ratio of thicknesses. The correlation of dispersion spectra for infinite, semi-infinite and finite MLS's is demonstrated to predict the field profiles for the surface and plate modes in relation to their dispersion curves location. The existence of backward plate modes in the finite MLS is described for the first time. The first part of paper describes studies that have been performed for perfect periodic systems. But it is well known that even a small concentration of artificial or technological disruptions of a crystal order can drastically alter the acoustical properties. Effects of this sort occur when the layers with different thickness or different constituents (impurity cells, growth defects, coating, etc) appear within or on the perfect periodic MLS's (see [13,14] and references there). The another aim of this paper is to present a computer simulation study of the dispersion equations for surface modes, propagating in systems consisting of a periodic properties of MLS and an additional layer. The additional surface layer that modifies the acoustical properties of the MLS may be considered either as a technological planar defect (domain of another thickness) or as an artificial film (domain or piezoelectric/metal layer). The relations for surface modes are discussed as functions of the nature and the relative parameters of additional layer.

2. Dispersion equations for perfect MLS

We consider a ferroelectric crystal of hexagonal symmetry, belonging to the 6mm class with its polar axis along the x_3 axis of a reference orthonormal coordinate system. The crystal MLS is modelled either by an unbounded medium (for bulk waves) or a semibounded medium $x_2 \geq 0$ (for surface waves) or a bounded medium $0 \leq x_2 \leq H$, $Nh \leq H \leq Nh + h$, $N = 0, 1, 2, \ldots$ (for plate waves), formed respectively by the infinite, semi-infinite or finite repetition of a unit cell of thickness h. We suppose that a unit cell occupies the region $0 \leq x_2 \leq h, h = h_1 + h_2$, and consists of two domains with thicknesses h_1, h_2 and opposite directions of the spontaneous polarization. The 180° variations of the crystal polarization direction are equivalent to changes in sign of the effective piezoelectric moduli e_{15} in adjacent domains. The propagation wavevector $\vec{k}$, parallel to the domain walls, is along x_1. In this particular geometry, the transverse vibration (parallel to x_3), in which we are interested here, is accompanied by an electric potential and decouples from pure elastic sagittal vibrations (polarized in the x_1x_2 plane).

Within the framework of Hamiltonian system formalism [9–12], it has been shown that the simultaneous solution of piezoacoustics equations for transverse waves within MLS is conveniently sought in the form

$$\{D_2(x_1, x_2, t), \sigma_{23}(x_1, x_2, t), \varphi(x_1, x_2, t), u_3(x_1, x_2, t)\} = \\ = \{q_1(x_2), q_2(x_2), p_1(x_2), p_2(x_2)\} \exp(ikx_1 - i\omega t), \tag{1}$$

where $D_2, \sigma_{23}, \varphi$, and u_3 are respectively the normal component of electric displacement, the stress component, the electric potential and the displacement vector component, k is the wavenumber, ω is the circular frequency. Then for determination of column vectors $\vec{q} = \mathrm{col}\,(q_1, q_2)$ and $\vec{p} = \mathrm{col}\,(p_1, p_2)$ we obtain h-periodic Hamiltonian system of linear differential equations

$$\frac{d}{dx_2}\begin{bmatrix} \vec{q} \\ \vec{p} \end{bmatrix} = A(x_2)\begin{bmatrix} \vec{q} \\ \vec{p} \end{bmatrix}, \quad A(x_2) = \begin{bmatrix} 0 & Q(x_2) \\ -P(x_2) & 0 \end{bmatrix} \tag{2}$$

with Hamilton function

$$[H(x_2, \vec{q}, \vec{p}) = [\vec{q}^{\mathsf{T}} P(x_2)\vec{q} + \vec{p}^{\mathsf{T}} Q(x_2)\vec{p}]/2. \tag{3}$$

The symmetrical matrices-functions $Q(x_2)$ and $P(x_2)$ have the following structure

$$Q(x_2) = \begin{bmatrix} -\varepsilon_{11}(x_2)k^2 & e_{15}(x_2)k^2 \\ e_{15}(x_2)k^2 & c_{44}(x_2)k^2 - \rho(x_2)\omega^2 \end{bmatrix},$$

$$P(x_2) = \begin{bmatrix} c_{44}(x_2)d(x_2) & -e_{15}(x_2)d(x_2) \\ -e_{15}(x_2)d(x_2) & -1/\overline{c}_{44}(x_2) \end{bmatrix}, \tag{4}$$

$$\overline{c}_{44}(x_2) = c_{44}(x_2) + [e_{15}(x_2)]^2/\varepsilon_{11}(x_2),$$
$$d(x_2) = 1/[\varepsilon_{11}(x_2)\overline{c}_{44}(x_2)],$$

where $\rho(x_2), c_{ij}(x_2), e_{ij}(x_2)$ and $\varepsilon_{ij}(x_2)$ are respectively the mass density, the elastic, the piezoelectric and the dielectric characteristics of the MLS's. In general, the characteristics are assumed to be arbitrary, absolutely integrable, periodic functions

$$\rho(x_2) = \rho(x_2 + h), c_{ij}(x_2) = c_{ij}(x_2 + h), \\ e_{ij}(x_2) = e_{ij}(x_2 + h), \varepsilon_{ij}(x_2) = \varepsilon_{ij}(x_2 + h). \tag{5}$$

However, in the present context $\rho(x_2), c_{ij}(x_2)$ and $\varepsilon_{ij}(x_2)$ are constant and only $e_{ij}(x_2)$ is piecewise-constant, periodic function.

The vectors $\vec{q}$ and $\vec{p}$ for the n-th unit cell $(n-1)h \le x_2 \le nh$ can be represented as

$$\begin{bmatrix} \vec{q}(x_2 + nh - h) \\ \vec{p}(x_2 + nh - h) \end{bmatrix} = \sum_{\ell=1}^{4} K_\ell \text{æ}_\ell^n U(x_2)\vec{Y}_\ell, 0 \le x_2 \le h, n = 0, \pm1, \pm2, \ldots, \tag{6}$$

where K_ℓ are the unknown coefficients, $U(x_2)$ is a matrixant, æ_ℓ and $\vec{Y}_\ell = \text{col}\,(Y_\ell^{(1)}, \dots, Y_\ell^{(4)})$ are eigenvalues and their associated eigenvectors of the monodromy matrix $U(h)$ [9,10]. The characteristic equation of $U(h)$ admits reciprocal roots and upon substitution of $\text{æ} + \text{æ}^{-1} = 2b$ is reduced to equations

$$\text{æ}^2 - 2b_j\text{æ} + 1 = 0, 2b_j = -a_1/2 - (-1)^j a_1^2/4 - 2a_2 + 2)^{1/2}, \qquad (7)$$
$$j = 1, 2,$$

where a_j are expressed through the elements of $U(h)$.

The general solution (6) is a superposition of the linearly independent partial solutions of the Hamiltonian system of ordinary differential equations [10], and for every partial solution there is a related eigenvalue æ_ℓ. When $n \to \pm\infty$, only some of these solutions are bounded at fixed values of parameters k and ω, that is only some of them describe physically realizable propagating modes. For this reason the summands in (6), which do not satisfy the condition of boundness at infinity, are to be rejected when motions in unbounded or semibounded media are considered. Since equations (7) has two pairs of reciprocal roots, for existence of bulk modes, at least two eigenvalues must be complex conjugates and their moduli must be equal to unity. This occurs if at least one of the values b_j meets the requirements

$$\text{Im}\,b_j(k,\omega) = 0, \qquad |\,\text{Re}\,b_j(k,\omega)\,| \leq 1. \qquad (8)$$

Conditions (8), relating the frequency ω to the wave number k implicitly, determine the bulk wave existence bands, that is to say, are the dispersion relations for bulk waves.

Dispersion equations for surface waves in the half-space $x_2 > 0$ are derived on the basis of the general representation (6) and the corresponding mechanical and electrical boundary conditions on the surface $x_2 = 0$, taking into account the requirement that the amplitude decays away from the surface $x_2 = 0$. For example, in the case of a metallized surface $x_2 = 0$, the propagation of surface waves is described by the equation

$$Y_1^{(2)}Y_2^{(3)} - Y_1^{(3)}Y_2^{(2)} = 0, \, |\,\text{æ}_j\,| < 1, \, j = 1, 2, \qquad (9)$$

and in the case of a non-metallized surface $x_2 = 0$ we have

$$Y_1^{(2)}Y_2^{(1)} - Y_1^{(1)}Y_2^{(2)} + k\varepsilon_0(Y_1^{(2)}Y_2^{(3)} - Y_1^{(3)}Y_2^{(2)}) = 0, \, |\,\text{æ}_j\,| < 1, \, j = 1, 2, (10)$$

where ε_0 is the dielectric permittivity of vacuum.

The dispersion equation for plate waves in a finite MLS with metallized free surfaces $x_2 = 0$ and $x_2 = H, Nh \leq H \leq Nh + h$, has the form

$$\det\{\, \text{col}\,[Y_\ell^{(2)}; \, Y_\ell^{(3)}; \, \text{æ}_\ell^N \sum_{j=1}^4 [U(H - Nh)]_{2j} Y_\ell^{(j)};$$
$$\text{æ}_\ell^N \sum_{j=1}^4 [U(H - Nh)]_{3j} Y_\ell^{(j)}]\}_{\ell=\overline{1,4}} = 0. \qquad (11)$$

The dispersion equation for MLS with non-metallized faces is derived similarly [9].

3. Dispersion equations for MLS with additional layer

The additional layer of thickness h_0 is bonded to MLS along the plane $x_2 = 0$ and is contiguous with vacuum along the plane $x_2 = -h_0$. Here this layer is presumed to be either a homogeneous domain with the same $6mm$ symmetry as the MLS layers or a metal film. Its parameters are indicated by index 0 for the domain and by m for metal layer. The usual interface conditions expressing the continuity of the mechanical displacement, the normal stress, the normal electric displacement and the potential are assumed to be fulfilled at $x_2 = 0$ and all interfaces of perfect MLS. We distinguish two types of boundary conditions at the interface $x_2 = 0$ depending on whether the additional layer is of piezoelectric or metallic material. When the additional layer is of piezoelectric homogeneous material with the same $6mm$ symmetry and orientation as the MLS layers, we suppose that the free surface $x_2 = -h_0$ is either metallized or non-metallized. For the first case we have obtained the dispersion equation

$$
\begin{vmatrix}
\alpha_1 \alpha_2^{-1} C_1 & \alpha_3 S_1 & Y_1^{(1)} & Y_2^{(1)} \\
\alpha_1 (C_2 - C_1) & \beta_2 & Y_1^{(2)} & Y_2^{(2)} \\
\beta_1 & \beta_3 & Y_1^{(3)} & Y_2^{(3)} \\
S_2 & C_2 & Y_1^{(4)} & Y_2^{(4)}
\end{vmatrix} = 0, \ | \ae_1 | < 1, \ | \ae_2 | < 1, \quad (12)
$$

where

$$C_1 = \cosh(kh_0), \ C_2 = \mathrm{Cs}(\Omega_0 h_0), \ S_1 = \sinh(kh_0), \ S_2 = \mathrm{Ss}(\Omega_0 h_0),$$
$$\alpha_1 = \bar{c}_{44,0}\Omega_0, \ \alpha_2 = e_{15,0}\varepsilon_{11,0}^{-1}, \ \alpha_3 = ke_{15,0}, \ \alpha_4 = -k\varepsilon_{11,0},$$
$$\bar{c}_{44,0} = c_{44,0} + e_{15,0}^2\varepsilon_{11,0}^{-1}, \ \beta_1 = \alpha_2 S_2 - \alpha_1\alpha_3^{-1}S_1,$$
$$\beta_2 = \gamma_0\alpha_1 S_2 - \alpha_2\alpha_3 S_1, \ \beta_3 = \alpha_2(C_2 - C_1), \ [\gamma_0; \mathrm{Ss}(\cdot); \mathrm{Cs}(\cdot); \Omega_0^2] =$$

$$
= \begin{cases}
[\ -1; \ \sin(\cdot); \ \cos(\cdot); \ \rho_0\omega^2/\bar{c}_{44,0} - k^2, \ ], & \rho_0\omega^2/\bar{c}_{44,0} > k^2; \\
[\ +1; \ \sinh(\cdot); \ \cosh(\cdot); \ k^2 - \rho_0\omega^2/\bar{c}_{44,0}, \ ], & \rho_0\omega^2/\bar{c}_{44,0} < k^2,
\end{cases}
$$

$\ae_\ell$ and $Y_\ell^{(j)}$ are the eigenvalues and the associated components of eigenvectors of the monodromy matrix [10]. For the non-metallized free surface we have derived

$$
\begin{vmatrix}
-\varepsilon_{11,0} & 0 & 0 & 0 & \varepsilon_0 \\
\alpha_2^{-1} C_1 & \alpha_2^{-1}\alpha_3 S_1 & Y_1^{(1)} & Y_2^{(1)} & 0 \\
\beta_4 & \beta_5 & Y_1^{(2)} & Y_2^{(2)} & \beta_6 \\
\beta_7 & C_2 - C_1 & Y_1^{(3)} & Y_2^{(3)} & -C_2 \\
\alpha_1^{-1}\alpha_3 S_2 & \alpha_2^{-1} C_2 & Y_1^{(4)} & Y_2^{(4)} & -\alpha_2^{-1} C_2
\end{vmatrix} = 0, \quad (13)
$$

$$| \ae_1 | < 1, | \ae_2 | < 1,$$

where $\beta_4 = \alpha_3(C_2 - C_1)$, $\beta_5 = \gamma_0\alpha_1\alpha_2^{-1}S_2 - \alpha_3 S_1$, $\beta_6 = -\gamma_0\alpha_1\alpha_2^{-1}S_2$, $\beta_7 = \alpha_1^{-1}\alpha_2\alpha_3 S_2 - S_1$, ε_0 is the dielectric permittivity of vacuum.

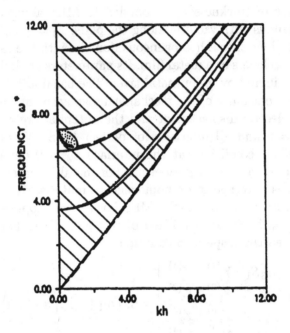

Figure 1. Bulk bands (areas shaded by oblique lines) and surface wave dispersion curves (dashed) for perfect MLS with $h_1/h_2 = 2/3$. The area shaded by dots corresponds to region where values b_j [7] are complex.

When the additional layer is of isotropic metallic material, we can write [11,15] the dispersion equation for the surface modes as

$$\begin{vmatrix} \gamma_m d_m \mathrm{Ss}(\Omega_m h_0) & Y_1^{(2)} & Y_2^{(2)} \\ 0 & Y_1^{(3)} & Y_2^{(3)} \\ \mathrm{Cs}(\Omega_m h_0) & Y_1^{(4)} & Y_2^{(4)} \end{vmatrix} = 0, \; | \ae_1 | < 1, | \ae_2 | < 1, \qquad (14)$$

where $v_m^2 = c_{44,m}/\rho_m$, $\alpha_m = c_{44,m}\Omega_m$, $[\gamma_m; \mathrm{Ss}(\cdot); \mathrm{Cs}(\cdot); \Omega_m^2] =$

$$= \begin{cases} [-1; & \sin(\cdot); & \cos(\cdot); & \omega^2 v_m^{-2} - k^2,], & \omega^2 v_m^{-2} > k^2; \\ [+1; & \sinh(\cdot); & \cosh(\cdot); & k^2 - \omega^2 v_m^{-2},], & \omega^2 v_m^{-2} < k^2. \end{cases}$$

The dispersion equations (12) and (14) represent the implicit relations between the frequency ω and wave number k and are valid for any relative

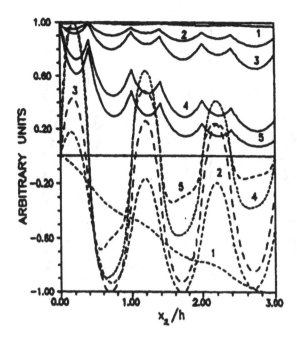

Figure 2. The profiles of the mechanical displacement (solid) and the electrical potential (dased) for fundamental surface mode at: 1 – $\omega^* = 0.4$, 2 – $\omega^* = 2.5$, 3 – $\omega^* = 4.4$, 4 – $\omega^* = 8.5$, 5 – $\omega^* = 12$.

value of the wavelengths compared to the period h of the MLS (assuming that we remain in the range of the elasticity theory where the atomic character of the vibrations does not play a role [16]).

4. Computer results

Fig.1 represents the bulk bands for the infinite MLS and the surface modes for the semi–infinite MLS of the poled $BaTiO_3$ crystal with $h_1/h = 0.4$. Here we assumed that the free surface is metallized and thus used the equation (9). The dimensionless quantities $k^* = kh$ and $\omega^* = \omega h(\overline{\rho}/\overline{c})^{1/2}$ are used on both axes, $\overline{\rho} = 10^3 kg/m^3$, $\overline{c} = 10^{10} N/m^2$.

Fig. 2 shows the typical profiles of the mechanical displacement (solid) and the electric potential (dashed) for fundamental surface mode at different ω^*. The displacement corresponding $\omega^* = 0.4$ and 2.5 has maximum at $x_2/h = 0.4$, i.e. on the interface of two upper domains, as in the Maerfeld – Tournois wave [6,17].

Fig. 3 show the existence of surface modes and the width of bulk bands in relation to h_1/h at the different constant ω^*. One can see that the number

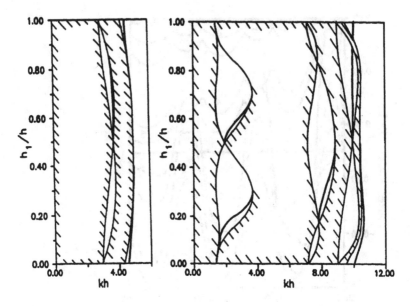

Figure 3. The existence of surface modes(solid curves) and the width of bulk bands (shaded areas) in relation to h_1/h at the constant ω^*: $\omega^* = 5$ (left part) and $\omega^* = 11$ (right part).

of existing surface modes for a given frequency is very sensitive to the ratio of domain thicknesses.

Fig. 4 illustrate the case of finite MLS considered also in [4]. The dispersion curves (dashed) for plate waves are shown together with the bulk bands of infinite MLS. For purposes of graphical clarity, we took $H = 2h$ here. The 1st mode and part of the 5th mode are very close to the surface modes of fig.1, the field profiles coincide practically with surface mode ones, i.e. these modes are localised near the surface $x_2 = 0$. The 3rd and 7th curves correspond to the modes localised near the surface $x_2 = H$. It is interesting to note for the first time, that the 4th, 6th and 7th modes of the multilayer waveguide have the minimum not at $k^* = 0$. This means that backward waves exist for these modes at small values of k^*.

In this section we also present a few illustrations of the dispersion curves for the surface waves in the $BaTiO_3$ MLS's with different additional layers and discuss their behaviour as functions of the nature and thickness of these layers. Fig.5 represents the bulk bands for the infinite MLS ($h_1/h = 0.4$) and the surface modes for the semi-infinite MLS with an additional domain layer. Here we assumed that the free surface is metallised and thus used the equation (12). Figs. 5–6 show that, with the variation of additional domain

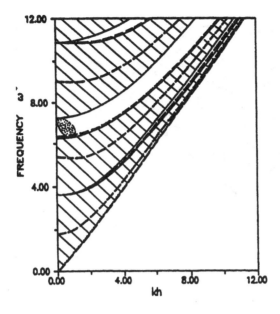

Figure 4. The dispersion curves (dashed) for plate waves in the finite MLS: $H = 2h$

thickness, surface modes move within their frequency gaps and even may disappear in one gap and appear in another one. For example, the surface mode does not exist below the bulk bands at $h_0 = 0$, but appears when $0 < h_0 < h_1$. As the thickness h_0 is increased within this interval, the starting point of the curve moves from larger values of k^* and ω^* on the band edge to smaller ones: $(k^*, \omega^*) \approx (7.2, 7.4)$ for $h_0 = 0.1h$, $(5.5, 4.2)$ and $(3.5, 3.4)$ for $0.2h$ and $0.3h$ relatively. At $h_0 = h_1$ the curve starts at the point $(k^* = 0, \omega^* = 0)$. The localisation of surface modes is also changed with the variation of additional layer. When a dispersion curve is near a bulk band boundary, the corresponding surface wave is quasibulk one and deeply penetrated into the MLS. The surface modes become more localised with increasing distance from the bulk band boundary. The wave amplitude in the n-th unit cell of MLS is proportional to $| æ_j |^n$, $j = 1, 2$. In fig.7, the dispersion curves are shown for larger values of h_0. One can see that the surface mode does not exist below the bulk bands for these values of h_0. It is interesting to note, that at $h_0 = 0.9h$ the dispersion curve in the second frequency gap has a minimum not at $k^* = 0$. This means that the backward surface wave exist for this mode at small values of k^*. Our computer experiments, conducted for the MLS's with metal layer and

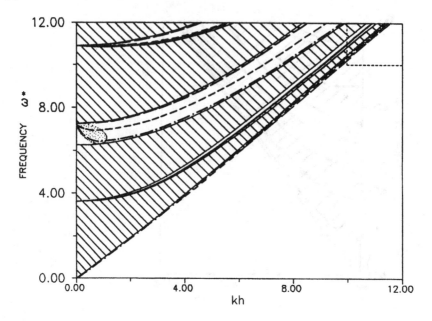

Figure 5. Bulk bands (areas shaded by oblique lines) and surface wave dispersion curves for MLS with $h_1/h_2 = 2/3$ and additional domain layer of thickness h_0: $h_0 = 0$ (solid curves), $h_0 = 0.1h$ (dotted), $0.2h$ (--- dashed), $0.3h$ (- - - dashed), $0.4h$ (dot-dashed). The area shaded by dots corresponds to region where values b_j [7,10] are complex.

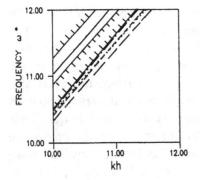

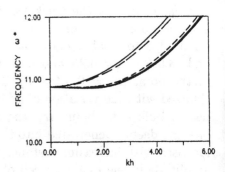

Figure 6. Two parts of fig. 5 on an enlarged scale. The left part corresponds to the dashed frame in the right top corner of fig. 5.

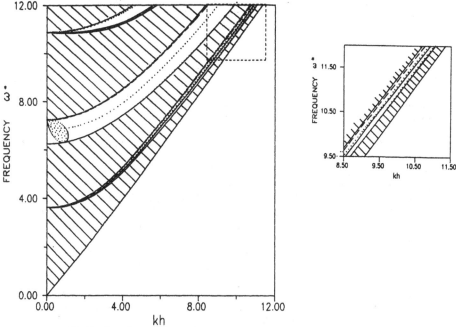

Figure 7. Bulk bands and surface wave dispersion curves for MLS with $h_1/h_2 = 2/3$ and additional domain layer of thickness h_0: $h_0 = 0.6h$ (solid curves), $h_0 = 0.7h$ (dashed), $0.9h$ (dotted). The right part corresponds to the dashed frame in the right top corner of the left part.

the equation (14), show that, in contrast to the homogeneous crystal, the considered system can support shear surface waves even if the metal layer velocity is greater than the velocity of $BaTiO_3$. Because of space limitation, a more detailed discussion of the results for this as well as the additional results for MLS of finite number of domains will be presented in forthcoming papers.

In conclusion, we have shown that all dispersive characteristics of the domain structures are very sensitive to the inclusion of the additional surface layer and its nature and relative parameters (particulary, thickness). The ferroelectric crystals with MSL may exhibit a great variability in dispersion laws, mode localisations and wave velocities suggesting that these easy changing heterostructures may have applications in acoustic wave devices.

Acknowledgements

The research described in this publication was made possible in part by Grants Nos. UB5000 and UB5200 from the International Science Founda-

38

tion. The authors are grateful to Dr. V.G. Mozhaev (Moscow University) for useful discussions.

References

1. Antipov, V.V., Blistanov, A.A., Sorokin, N.F., and Chizhikov, S.I. (1985) Preparation of regular domain structure in ferroelectrics $LiNbO_3$ and $LiTaO_3$, *Kristallografiya*, **30**, 734–738.
2. Meeks, S.W. and Auld, B.A. (1985) Periodic domain walls and ferroelectric bubles in neodymium pentaphosphate, *Appl. Phys. Lett.* **47**, 102–104.
3. Zhu, Y.Y., Ming, N.B., Jiang, W.H. and Shui, Y.A. (1988) Acoustic superlattice of $LiNbO_3$ crystals and its applications to bulk-wave transducers for ultrasonic generation and detection up to 800MHz, *Appl. Phys. Lett.* **53**, 1381–1383.
4. Li Xingjiao, Li Yibing, Li Shaoping, Cross, L.E. and Newnham, R.E. (1990) Excitations in laminar domain systems in ferroelectrics, *J. Phys.: Condens. Matter* **2**, 9577–9588.
5. Mozhaev, V.G., Morozova, G.P. and Serdobolskaya, O.Yu. (1990) Propagation of acoustic waves along the walls of ferroelectric domain, *Sov. Phys. Solid State* **32**, 1872-1873.
6. Mozhaev, V.G. and Weihnacht, M. (1995) On the possibility of existence of a new type of interface acoustic waves at 180 degrees ferroelectric domain boundary, *Proc. 1995 IEEE Ultrasonics Symp.* **1**, 649–652.
7. Zinchuk, L.P. and Podlipenets, A.N. (1996) Bulk, surface and plate shear waves in periodically layered domain structure, *Proc. 14th Int. Conf. Utilization of Ultrasonic Methods in Condensed Matter, Aug. 30 – Sept. 2, 1995, Zilina, Slovak Republic, Part 1*, 190–193.
8. Shuvalov, A.L. and Gorkunova, A.S. (1999) Transverse acoustic waves in piezoelectric and ferroelectric antiphase superlattices, *Phys. Rev. B* **59**, to be published.
9. Zinchuk, L.P., Podlipenets, A.N. and Shulga, N.A. (1988) Electroelastic shear waves in a stratified periodic medium, *Sov. Appl. Mech.* **24**, 245–250.
10. Zinchuk, L.P., Podlipenets, A.N. and Shulga, N.A. (1990) Dispersion relations for electroelastic shear waves in a periodically layered medium, *Sov. Appl. Mech.* **26**, 1092-1099.
11. Podlipenets, A.N. and Zinchuk, L.P. (1994) Surface shear wave propagation in the system of piezoelectric superlattice and additional metal layer, *Proc. 1994 IEEE Ultrasonics Symp.* **1**, 347–350.
12. Podlipenets, A.N. (1995) Wave propagation in periodically layered elastic and electroelastic media, in D.F.Parker and A.H.England (eds.), *Anisotropy, Inhomogeneity and Nonlinearity in Solid Mechanics*, Kluwer Academic Publishers, Dordrecht, pp. 469–474.
13. Podlipenets, A.N. and Zinchuk, L.P. (1995) Surface shear modes in piezoelectric superlattice with additional layer, *Functional Materials* **2**, 179–186.
14. Podlipenets, A.N. and Zinchuk, L.P. (1996) Acoustic shear wave propagation in superlattices with planar surface defects, *Ultrasonics World Congress 1995 Proc., Part 1*, 85–88.
15. Zinchuk, L.P. and Podlipenets, A.N. (1989), Shear surface waves in laminated "metal – piezoceramic" composites, *Sov. Appl. Mech.* **25**, 1107–1111.
16. Nougaoui, A. and Djafari Rouhani, B. (1987) Dynamics of infinite and semi–infinite piezoelectric superlattices: shear horizontal waves and effective medium approximation, *Surf. Sci.* **185**, 154–174.
17. Maerfeld, C. and Tournois, P. (1971) Pure shear elastic surface wave guided by the interface of two semi-infinite media, *Appl. Phys. Lett.* **19**, 117–118.

STRUCTURE AND PROPERTIES OF NOVEL Pb(B'$_{1/2}$Nb$_{1/2}$)O$_3$-PbTiO$_3$ BINARY SYSTEMS WITH HIGH PIEZOELECTRIC COUPLING

A. STERNBERG, L. SHEBANOVS, M. ANTONOVA, M. LIVINSH, I. SHORUBALKO
Institute of Solid State Physics University of Latvia,
Kengaraga str.8, Riga LV-1063, Latvia
J.Y.YAMASHITA
Toshiba Corporation, Kawasaki, Japan

Abstract

The (1-x)Pb(Lu$_{1/2}$Nb$_{1/2}$)O$_3$–xPbTiO$_3$ (PLuNT) and (1-x)Pb(Er$_{1/2}$Nb$_{1/2}$)O$_3$–xPbTiO$_3$ (PErNT) binary systems have been originally synthesized. Pure lutecium niobate (PLuN) (x = 0) has a pronounced long-range order in the B-sublattice and an antiferroelectric to paraelectric phase transition at ~258°C. The phase structure of the PLuNT system, at room temperature, changes from a pseudomonoclinic (psd-M, space group Bmm2) to tetragonal (T, space group P4mm). The pseudomonoclinic phase extends over the 0 ≤ x ≤ 0.38 interval within which the monoclinic angle β proceeds a minimum near to 90° at x≅0.2. The morphotropic region covers over the interval x = 0.38-0.49, the concentration ratio psd-M:T≅1 (the morphotropic phase boundary - MPB) corresponds to x = 0.41. Dielectric dispersion and broadening of the phase transition - features typical to relaxors are observed within the concentration interval of 0.1 ≤ x ≤ 0.3. The highest electromechanical coupling coefficients: k$_p$ = 0.66, k$_t$ = 0.48, k$_{31}$ = 0.34 of (1-x)PLuN - xPT ceramics are attained in compositions near the MPB at x≈0.41. Non-isovalent doping of PLuNT with La^{3+} in Pb sublattice shifts the MPB to lower values of x.

The unit cell of erbium niobate (PErN) is described as pseudomonoclinic of orthorhombic Bmm2 symmetry: a=c=4.2161 Å; b=4.0869 Å; β=90.55° and composition is characterized with antiferroelectric phase transition at 305°C. The PErNT system has the morphotropic phase region extending over the x=0.4-0.6 interval.

1. Introduction

Recently much attention has been paid to electromechanical (piezoelectric, electrostrictive) materials because of their promising application in electronics, ultrasound transduction, acoustic sensing, microelectromechanical systems (MEMS). Substitutions in A and B sites of the perovskite (ABO$_3$) structure in the most known

39

C. Galassi et al. (eds.), Piezoelectric Materials: Advances in Science, Technology and Applications, 39–48.

Pb(Zr,Ti)O_3 (PZT) [1] piezoelectric ceramics (compositions near morphotropic phase boundary (MPB), which divides the rhombohedral (R) and tetragonal (T) phases) have been one of ways to optimize and to improve the electromechanical (EM) properties of the materials. An alternative approach is creating complex perovskites - binary, ternary and many-component systems, including Pb(B',B'')O_3 compounds, mainly relaxors, and lead titanate PbTiO_3 (PT). Important systems reported are: [Pb(Mg$_{1/3}$ Nb$_{2/3}$)O_3 (PMN) - PT] (PMNT) [2,3], [Pb(Sc$_{1/2}$ Nb$_{1/2}$)O_3 (PSN) - PT] (PSNT) [4-6] , [Pb(Sc$_{1/2}$ Ta$_{1/2}$)O_3 - PT] [7], [PSN - PMN - PT] (PSMNT) [8], PMN - PSN and [Pb(Zn$_{1/3}$ Nb$_{2/3}$)O_3 (PZN) - PSN] [9], PZN - PMN - PT [10], Pb(Yb$_{1/2}$ Nb$_{1/2}$)O_3 – PT [11], Pb(In$_{1/2}$Nb$_{1/2}$) O_3 - PT [12] and [Pb(Ni$_{1/3}$Nb$_{2/3}$)O_3 (PNN) – PT - PZ] [13]. Strong EM coupling (k_p=0.76, k_t=0.56, k_{31}=0.46, k_{33}=0.79) has been reported by T.Yamamoto [14] in hot-pressed PSNT ceramic samples with grain size ~1 μm and by Y.Yamashita [8] in PSMNT ternary ceramic material - k_p=0.72, k_{31}=0.45, k_{33}=0.77, the piezoelectric coefficient d_{33} reaching 640 pC/N. The value of $d_{33} \approx 815$ pC/N is reported by Tai-Bor Wu [15] in 5 mol.% La modified PMNT ceramics, the maximum value of k_{33} in ceramics up to 0.8 and d_{33}=900 pC/N were attained in the 0.5 PNN – 0.345 PT – 0.155 PZ composition [13]. A large effective piezoelectric coefficient $d_{33} \approx 1800$ pC/N, unusual in ceramics, has been achieved by J.Zhao [16] in the electric field induced state of PMNT (~0.2 PT) compositions (bias field 2.5 kV/cm, T=80°C); the strain at 10 kV/cm - 0.15% (in composition near 0.3 PT).

Furthermore, compared to conventional PZT compositions complex perovskite single crystals have the advantage of being easy to grow [17-19]. Large values of k_{33} and d_{33} have been reported in single crystals - $k_{33} \approx 0.92$ in PZNT [17], d_{33}=1500 pC/N in PMNT [20]. S.E.Park has observed ultrahigh strain levels up to 1.7% (under electrical bias 120 kV/cm) and k_{33} >0.9, $d_{33} \geq 2000$ pC/N in PZNT (x=0.08 PT) crystals grown by flux technique [19].

The most pronounced piezoelectric properties of ferroelectric two(multi)-component perovskite solid solutions are characteristic to the compositions close to the MPB. However, there are few studies of the morphotropic phase transition between Bmm2 (orthorhombic symmetry with a pseudomonoclinic unit cell) and P4mm structures. Such a phase transition might be of certain interest because of the 12 equivalent polar directions of the orthorhombic phase. Possible coexistence of the structures is suggested by the P4mm→Bmm2 transition in BaTiO_3 where both the phases are observed in a certain interval of temperature [21].

The above consideration taken into account the solid solution systems (1-x)Pb(Lu$_{1/2}$ Nb$_{1/2}$)O_3 –xPbTiO_3 and (1-x) Pb (Er$_{1/2}$ Nb$_{1/2}$)O_3 –xPbTiO_3 (further abbreviated as PLuNT and PErNT) were examined with the aim to investigate structural, dielectric and piezoelectric properties of ceramics, especially in the MPB region.

2. Experimental

In order to obtain the solid solution series both conventional mixed oxides route and the wolframite precursor method were used. In case of the PLuNT system the latter provided better results, but in case of the PErNT system the mixed oxide techniques appeared to be more successful: in the 30-80 mol.% concentration interval of $PbTiO_3$ it provided substitutional solid solutions. The technological parameters and distinctions of PLuNT and PErNT synthesis are reported elsewhere [22].

Crystallographic studies were made by X-ray diffraction maxima 200, 220, 222 analysis using a DRON-UM1 diffractometer with Co K_α radiation, and Fe β filter. Dielectric permittivity and loss were measured using a HP4284 LCR instrument, the dielectric hysteresis loops were obtained by the Sawyer-Tower circuit in quasistatic regime.

The piezoelectric measurements were made on samples poled for 15 min. at 25 kV/cm and 125^0C and cooled under field to room temperature. A HP4194 impedance analyzer was used to determine the piezoelectric characteristics by the IEEE resonance method [23].

3. Results and Discussion

3.1. CERAMIC PROPERTIES

The hot-pressing temperature was different for different PLuNT compositions: 950^0C to 1100^0C (x<0.2 and x>0.7) and $930-1050^0C$ within the interval 0.2<x<0.7. To compensate the loss a 1 wt% of excess PbO was used. However, this turned out not to be the optimum amount, being specific to each of the compositions, since excess PbO was found in particular samples after hot-pressing as well. To decrease the amount of over-stoichiometric lead oxide the hot-pressed ceramics were fired in air at 1200^0C for 2 hours.

Remaining PbO, on the other side, is responsible for poor polarization (leaky hysteresis loops) and lower electromechanical coupling. Polarization and piezoelectric properties of the PLuNT samples measured during the last step of the working program display this unwanted effect. Specific resistance of the PLuNT samples at room temperature is $3 \cdot 10^8 - 10^9$ $\Omega \cdot m$. At poling temperatures of 120 -130^0C it decreases to 10^7 $\Omega \cdot m$.

Ceramic samples of PLuNT obtained from compositions with 1 wt.% of excess PbO have a dense structure and comparatively large grains of the average size of 5-6 μm while many small pores and average size of grains 1-2 μm and lower density are characteristic to samples from stoichiometric compositions.

The density of ceramics ρ gradually decreases if PT concentration increases as expected from the density values calculated from X-ray data: for "end" compositions PLuN and PT $\rho_{calculated}$ is 9.1 g/cm^3 and 7.9 g/cm^3, correspondingly.

In case of PErNT system the amount of excess PbO added was even higher - up to 3 wt.% because of a higher calcining temperature (850^0C) and, therefore, higher volatility of PbO was expected in case of PErNT compared to PLuNT.

3.2. PHASE DIAGRAM

The following features of phase diagram of the $(1-x)$ $Pb(Lu_{1/2}Nb_{1/2})O_3 - x$ $PbTiO_3$ solid solution system can be revealed (Fig.1a,b).

Pure lutecium niobate (PLuN) has a pronounced long-range order in the B sublattice (the X-ray diffraction pattern suggests a superstructure) and a pseudomonoclinic (psd-M) cell with linear parameters a=c=4.150 Å, b=4.119 Å and an angle β=90.43°. The space group of PLuN and compositions at small x is Bmm2 and the true unit cell is orthorhombic. A pseudomonoclinic primitive cell was used to analyze the X-ray diffraction data and to calculate parameter values. The orthorhombic (a_{orth}; b_{orth}; c_{orth}) and pseudomonoclinic (a,b,β) sets of parameters are associated by relations: a_{orth}=2a sin β/2; b_{orth}=2a cos β/2; c_{orth}=b.

With the increase of x (the concentration of $PbTiO_3$) the long-range order in the B sublattice disappears - the I_{111}/I_{200} ratio changes gradually from 0.919 to 0.054 as x is changed from 0 to 0.3.

With the growth of the content of $PbTiO_3$, the angle β proceeds through a minimum in the pseudomonoclinic part of the phase diagram. At x=0.2 the structure becomes pseudocubic, the angle β differs from 90° within the accuracy of measurement $\pm$0.02° (Fig.1a, inset, the corresponding broadening of X-ray diffraction maxima 222 being at the limit of experimental resolution).

The morphotropic region (detected by appearance of the diffraction component of the relevant phase in the complex 200 profile) extends over the concentration interval x=0.38-0.49 (Fig.1). The MPB with the concentration ratio psd-M:T$\cong$1 corresponds to x=0.41. Within the MPB interval, compared to the PZT system, a rather strong distortion of the unit cell (c/a-1)$\geq$0.02, $\beta\geq$90.37°) is maintained (Fig.1b). For this reason the pure PLuNT is a "hard" piezoelectric.

Pure erbium niobate (PErN) has the orthorhombic Bmm2 lattice symmetry (Fig.1c,d). Its unit cell may be described as psd-M, similar to PLuN. The linear parameters of the pseudo-monoclinic cell were found to be a=c=4.2161 Å, b=4.0869 Å, the monoclinic angle β=90.55°. We have not found any previous reports on this composition.

The strong monoclinic distortion, rather weak superstructure reflections, and c $_{PErN}$ >>c $_{PT}$ suggest that crystallographic properties of the PErNT system are different from those of PLuNT. Indeed, solid solutions in the concentration interval x=0.1-0.2 do not have the proper perovskite structure. The absence of long-range ordering in the B sublattice of the perovskite solid solutions formed in concentration interval x=0.3-0.8 is another difference. The linear and angular lattice parameters are given in Fig.2c,d. There is a strong spontaneous deformation of the unit cell (both tetragonal and pseudomonoclinic) over the whole concentration interval (including a wide morphotropic region found to extend from x=0.4 to x=0.6) considered. The monoclinic angle β does not change much and tends to increase within x=0.3-0.4. To compare the unit cell distortion of different symmetry, the generalized parameter of deformation δ is used (Fig.2d); in case of tetragonal distortion δ=(2/3)(c/a-1), in case of monoclinic distortion δ=(a/b·sinβ)$^{1/3}$·(1-cosβ)$^{1/2}$. The MPB lies between x=0.48-0.52.

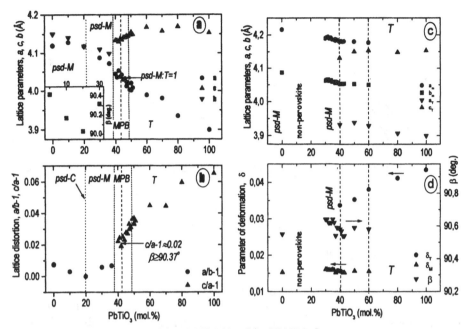

Figure 1. The phase diagrams of the PLuNT (a,b) and the PErNT (c,d) systems at room temperature. Lattice parameters *a,c,b* – (a,c); distortion of unit cell *a/b-1, c/a-1* - (b); pseudomonoclinic angle β - (d), inset (a); generalized parameter of deformation δ - (d) vs. concentration of PbTiO₃ (mol.%).

In this work attempts were made to synthesize $Pb(Tm_{1/2}Nb_{1/2})O_3$ and $Pb(Ho_{1/2}Nb_{1/2})O_3$, as well as the corresponding binary systems with PbTiO₃. $Pb(Tm_{1/2}Nb_{1/2})O_3$ ceramics has a pseudo-monoclinic unit cell with parameters a = c = 4.177 Å; b = 4.119 Å; β = 90.75° at room temperature. The unusually large monoclinic distortion with a,c > c $_{PT}$ and b>> a $_{PT}$ does not permit formation of the solid solution in the binary system $(1-x) Pb(Tm_{1/2}Nb_{1/2})O_3$-$xPbTiO_3$; the corresponding X-ray diffraction data show that a mechanical mixture of the two components is obtained.

The investigation of $(1-x)Pb(Ho_{1/2}Nb_{1/2})O_3$-x PbTiO₃ showed a failure of the synthesis. The nearly dense packing of the crystallographic structure in the $Pb(B_{1/2}^{3+}Nb_{1/2}^{5+})O_3$ accounts the geometric peculiarities in the composition line B=Lu, Er, Ho causing a gradual "hardening" of parameters of synthesis in the compositions. By use of the ion radia 0,93 Å, 0,96 Å and 0,97 Å of Lu^{3+}, Er^{3+} and Ho^{3+} accordingly, the tolerance factor $t = (R_A + R_O)/\sqrt{2}(R_B + R_O)$ is calculated. The gradually decreasing low values of t=0,83; 0,82; 0,81 became critical for the perovskite structure revealing a remarkable B-O bonds strain in BO₆ octahedra. Apparently, in the case of $Pb(Ho_{1/2}Nb_{1/2})O_3$ the overtension of bonds in the ABO₃ perovskite structure become thermodynamically unstable.

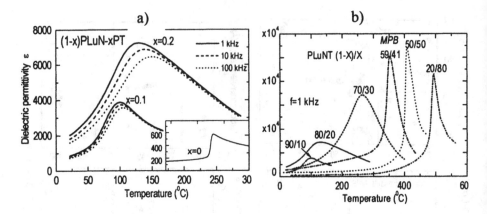

Figure 2. Dielectric permittivity ε vs. temperature for PLuNT ceramics.

3.3. DIELECTRIC PROPERTIES

The thermal dependence of the dielectric permittivity ε of (1-x)PLuN-xPT ceramics with different concentration x of lead titanate is shown in Fig.2. An antiferroelectric - paraelectric (AFE-PE) phase transition at ~258°C is observed in pure lutecium niobate (ε(T) data - Fig.2a, inset). Broadening of the phase transition and dielectric dispersion, characteristic features of ferroelectric relaxors are observed within the concentration interval of $0.1 \leq x \leq 0.3$ (Fig.2a).

The peak values of ε increase with the increase of PT content and exceed 25 000 in the MPB region. However, it still high (almost higher >26 500) even at x=0.5. The lowest T_m is observed at x=0.1. Near the MPB T_m is over 350°C and is the highest value among binary Pb(B',B'')O$_3$ - PT perovskites.

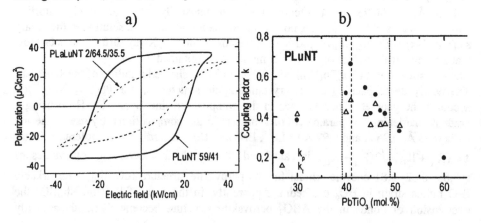

Figure 3. Dielectric hysteresis loop (a) and the electromechanical coupling factors k_p and k_t (b) for PLuNT ceramics.

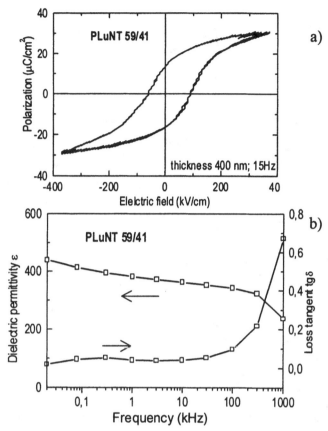

Figure 4. Dielectric hysteresis loop (a) and room temperature dielectric permittivity ε and loss tgδ (b) of PLuNT/LSCO/MgO thin film heterostructure.

The dielectric hysteresis of PLuNT 59/41 as shown in Fig.3a is characterized by rather high values of coercive field E_c=22 kV/cm and remanent polarization P_r = 37 μC/cm^2.

Highly oriented perovskite thin films of PLuNT with compositions near the morphotropic phase boundary were formed by pulsed laser deposition on La$_{1/2}$Sr$_{1/2}$CoO$_3$/(100)MgO [24]. The typical ferroelectric hysteresis loops measured at room temperature are presented in Fig.4a. The full polarization and coercive field are about 30 μC/cm^2 and 80 kV/cm, respectively; P_r=16 μC/cm^2. The room temperature dielectric permittivity is in the range 350-450 (Fig.4b).

Pure erbium niobate (PErN) and thulium niobate (PTmN) ceramics have a relatively low values of dielectric permittivity ε≈100 at room temperature reaching ca. 370 and 450 at AFE to PE transition at 305^{0}C and 315^{0}C, respectively (Fig.5).

46

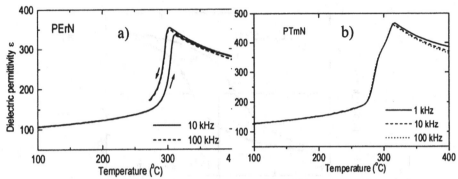

Figure 5. Dielectric permittivity eps vs. temperature for PErN (a) and PTmN (b) ceramics.

3.4. PIEZOELECTRIC PROPERTIES

Coefficients of electromechanical coupling (k_p - planar mode, k_t - thickness mode) as functions of the PT concentration x in the (1-x)PLuN-xPT ceramic series are shown in Fig.3b.

Both coefficients reach maxima at x=0.41 (MPB compound): k_p=0.66, k_t=0.48. Compositions obtained from stoichiometric mixtures have much lower electromechanical coupling coefficients compared to those containing 1 wt.% excess of PbO: k_p = 0.54, k_t=0.43 at x=0.38.

The properties of PLuNT ceramics are sensitive to the change of technological parameters – respectively, dielectric and piezoelectric characteristics are expected to be essentially improved by particular optimization of calcining and hot-pressing regimes, additional heat treatment of the hot-pressed ceramics at high temperatures, by use of proper amount of excess PbO, by selection of optimal doping material and concentration.

3.5. MODIFICATION OF PLuNT CERAMICS

In case of heterovalent substitution of Pb^{2+} by La^{3+} according to $2Pb^{2+} \rightarrow 2La^{3+} + \square_A$ (the corresponding formula of PLuNT solid solutions is $(1-x)Pb_{1-1.5y}La_yLu_{0.5}Nb_{0.5}O_3-xPb_{1-1.5y}La_yTiO_3$, where y=0.01; 0.02 is the molar concentration of La^{3+}), the following changes of the phase diagram near the morphotropic region are observed.

The morphotropic phase boundary shifts towards PLuN (the concentration of the monoclinic phase corresponding to the MPB increases). The MPB of initial compositions corresponds to x=0.41 while in case of modified compositions at y=0.01 it corresponds to x=0.355 and at y=0.02 to x=0.35. At the same time, the spontaneous deformation (c/a-1) of the unit cell decreases monotonously; outside region in the tetragonal part of the diagram c/a-1 diminishes at the rate 1.7×10^{-3} per 1 mol.% of La^{3+}; (c/a-1) corresponding to the MPB decreases to 0.015 at y=0.01 and to 0.009 at y=0.02, which means that modified composition are softer ferroelectrics compared to unmodified ones. The effect of La^{3+} on dielectric polarization is shown in Fig.3a. The relatively high values of induced and remanent polarization being maintained the coercive field E_c of composition PLaLuNT 2/64.5/35.5 has decreased to 15.8 kV/cm. Coercive field is still further reduced to E_c=11 kV/cm in PLaLuNT 2/65.5/34.5 which, regretfully, is accompanied by a decrease of the remanent polarization. The reason is the shift from morphotropic phase boundary (in case of composition PLaLuNT 2/65.5/34.5 the ratio psd-M:T≈4:1).

4. Conclusions

The phases of the $(1-x)Pb(Lu_{1/2} Nb_{1/2}) O_3 -xPbTiO_3$ (PLuNT) binary system at room temperature change from pseudomonoclinic (x<0.38) via pseudocubic at x≈0.2 to tetragonal (x>0.49). Pure PLuN has a pronounced long-range order in the B sublattice (the X-ray diffraction contains superstructure maxima of a pseudo-pseudomonoclinic symmetry) and an antiferroelectric-paraelectric phase transition at ~258°C. Dielectric dispersion and broadening of the phase transition - typical features of relaxors are observed within the concentration interval of 0.1≤x≤0.3. The morphotropic region extends over the interval x=0.38-0.49, the concentration ratio psd-M:T≅1 corresponding to x=0.41. Non-isovalent doping of PLuNT with La^{3+} in Pb sublattice shifts the MPB to lower values of x, diminishes the distortion of the unit cell (c/a-1=0.009 at 2mol.% of La) and reduces the coercive field.

The unit cell of erbium niobate (PErN) is described as pseudomonoclinic of orthorhombic Bmm2 symmetry: $a=c$=4.2161 Å; b=4.0869 Å; β=90.55° and composition is characterized with antiferroelectric phase transition at 305°C. The PErNT system has the morphotropic phase region extending over the x=0.4-0.6 interval.

The PLuNT system has a strong electromechanical coupling (k_p = 0.66, k_t = 0.48, k_{31}=0.36 in compositions near the MPB at x=0.41) and among binary $Pb(B',B'')O_3$-PT perovskites the highest T_m =353°C. It may be a favorable material for piezoelectric sensors and actuators, utilized at high temperatures, and may have an interest as promising basis for thin film and single crystal performance.

5. Acknowledgement

The research was supported by Council of Science of Latvia and Dr.K.Hayashi of Hayashi Chemical Ltd., Kyoto, Japan.

48

6. References

1. Jaffe, B., Roth, R.S., and Marzullo, S. (1955) Properties of piezoelectric ceramics in the solid solution series PT, PZ, PSn, and PHf, *J.Res. Natl.Bur.Standard* **55**, 239-254.
2. Zhang, Q.M., Zhao, J., and Cross, L.E. (1996) Aging of the dielectric and piezoelectric properties of relaxor lead magnesium niobate - lead titanate in the electric field biased state, *J.Appl.Phys.* **79** [6] 3181-3187 .
3. Noblanc, O., Gaucher, P., and Calvarin, G. (1996) Structural and dielectric studies of $Pb(Mg_{1/3} Nb_{2/3})O_3$-$PbTiO_3$ ferroelectric solid solutions around the morphotropic boundary, *J.Appl.Phys.* **79** [8] 4291-4297.
4. Yamashita, Y. (1993) Improved ferroelectric properties of niobium-doped $Pb[(Sc_{1/2} Nb_{1/2})Ti]O_3$ ceramic material, *Jpn.J.Appl.Phys.* **32**, 5036-5040.
5. Yamashita, Y. (1994) Large electromechanical coupling factors in perovskite binary material system, *Jpn.J.Appl.Phys.* **33**, 5328-5331.
6. Kalvane, A., Antonova, M., Livinsh, M., Kundzinsh, M., Spule, A., Shebanovs, L., and Sternberg, A. (1997) High electromechanical coupling in relaxor $Pb(Sc_{1/2} Nb_{1/2})O_3$ -$PbTiO_3$ system hot-pressed piezoelectric ceramics, *Key Engineering Materials* **132-136**, 1072-1075.
7. Giniewitcz, J.R., Bhalla, A.S., Cross, L.E. (1998) Variable structural ordering in lead scandium tantalate-lead titanate materials, *Ferroelectrics* **211**, 281-297.
8. Yamashita, Y., Harada, K., Tao, T., and Ichinose, N. (1996) Piezoelectric properties of the $Pb(Sc_{1/2}Nb_{1/2})O_3$-$Pb(Mg_{1/3}Nb_{2/3})O_3$-$PbTiO_3$ ternary ceramic materials near the morphotropic phase boundary, *Integrated Ferroelectrics* **13**, 9-16.
9. Dambekalne, M., Brante, I., Antonova, M., and Sternberg, A. (1992) Production and properties of ceramics of lead containing niobates, *Ferroelectrics* **131**, 67-73.
10. Zhu, H., Tan, Y., Wang, Q., Cai, Z., and Meng, Z. (1994) Dielectric and electrostrictive properties of $Pb(Zn_{1/3}Nb_{2/3})O_3$ - $Pb(Mg_{1/3}Nb_{2/3})O_3$ - $PbTiO_3$ ceramic solid solutions, *Jpn.J.Appl.Phys.* **33**, 6623-6625.
11. Yamamoto, T. and Ohashi, S. (1994) Dielectric and piezoelectric properties of $Pb(Yb_{1/2} Nb_{1/2})O_3$ - $PbTiO_3$ solid solution system, *Jpn.J.Appl.Phys.* **34**, 5349-5353.
12. Alberta, E.F., Bhalla, A.S. (1998) Piezoelectric properties of $Pb(In,Nb)_{1/2}O_3$-$PbTiO_3$ solid solution ceramics, *J.of the Korean Phys.Soc.* **32**, S1265-S1267.
13. Kondo, M., Hida, M., Tsukada, M., Kurihara, K., and Kamehara, N. (1997) Piezoelectric properties of $PbNi_{1/3}Nb_{2/3}$-$PbTiO_3$-$PbZrO_3$ ceramics *Jpn.J.Appl.Phys.* **36**, 6043-6045.
14. Yamamoto, T. and Yamashita, Y. (1996) Piezoelectric properties of hot-pressed $Pb(Sc_{0.5} Nb_{0.5})O_3$ - $PbTiO_3$ system," *Proc. ISAF'96, IEEE Catalog Number 96CH35948* **2**, 573-575.
15. Tai-Bor Wu, Ming-Jyx Shyu, Chia-Chi Chung, and Hsin-Yi Lee (1995) Phase transitions and ferroelectric characteristics of $Pb[(Mg_{1/3}Nb_{2/3})_{1-x}Ti_x]O_3$ ceramics modified with $La(Mg_{2/3}Nb_{1/3})O_3$," *J.Am.Ceram.Soc.* **78** [8] 2168-2174.
16. Zhao, J., Zhang, Q.M., Kim, N., and Shrout, T. (1995) Electromechanical properties of relaxor ferroelectric lead magnesium niobate - lead titanate ceramics, *Jpn.J.Appl.Phys.* **34**, 5658-5663.
17. Kuwata, J., Uchino, K., and Nomura, S. (1982) Dielectric and piezoelectric properties of $0.91Pb(Zn_{1/3}Nb_{2/3})O_3$ - $0.09 PbTiO_3$ single crystals, *Jpn.J.Appl.Phys.* **21**, 1298-1302.
18. Shimanuki, S., Saito, S., Yamashita, Y. (1998) Single crystal of the $Pb(Zn_{1/3}Nb_{2/3})O_3$-$PbTiO_3$ system grown by the vertical Bridgeman method and this cheracterization, *Jpn.J.Appl.Phys.* **37**, 3382-3385.
19. Park, S.E., and Shrout, T.R. (1997) Ultrahigh strain and piezolectric behaviour in relaxor based ferroelectric single crystals, *J.Appl.Phys.* **82** (4), 1804-1811.
20. Shrout, T., Zhang, Z.P., Kim, N., and Markgraf, S., (1990) Dielectric behaviour of single crystals near the $(1-x)Pb(Mg_{1/3}Nb_{2/3})O_3$ - $(x)PbTiO_3$ morphotropic phase boundary, *Ferroelectric Letters* **12**, 63-69.
21. Shebanovs, L.A. (1981) X-ray temperature study of crystallographic characteristics of barium titanate, *phys.stat.sol.(a)* **65**, 321-325.
22. Sternberg, A., Shebanovs, L., Yamashita, Y., Antonova, M., Livinsh, M., Tyunina, M., Kundzinsh, K., and Shorublako, I. (in press) Structure and properties of high piezoelectric coupling $Pb(B'_{1/2} Nb_{1/2})O_3$ binary systems, *Ferroelectrics*.
23. IEC Standard Publication 483, (1976) *Guide to Dynamic Measurements of Piezoelectric Ceramics with High Electromechanical Coupling.*
24. Tyunina, M., Levoska, J., Sternberg, A., Leppävuori, S. (in press) Dielectric properties of pulsed laser deposited films of $PbLu_{1/2}Nb_{1/2}O_3$-$PbTiO_3$.

NANO-SIZE FERROELECTRIC STRUCTURES

M. ALEXE, C. HARNAGEA, A. PIGNOLET,
D. HESSE, and U. GÖSELE
Max Planck Institute of Microstructure Physics
Weinberg 2, D-06120, Halle (Saale), Germany

Abstract

In the present work ferroelectric structures with lateral sizes as small as 100 nm were fabricated by electron beam direct writing. Switching of single 100 nm cells was achieved and piezoelectric hysteresis loops were recorded using scanning probe microscopy working in piezoresponse mode. The ability to scale down ferroelectric capacitors down to sizes below 1 μm^2 and the possibility to achieve densities of 64 Mbit to 4 Gbit for non-volatile ferroelectric memories is thus demonstrated.

1. Introduction

There is a growing interest in the physical properties of very small structures. Much attention has been directed towards semiconductors, but nowadays increasing interest in high-density non-volatile ferroelectric random access memories (FRAM) stress the importance of the size effect in ferroelectric materials. For a 1 Gbit memory, the lateral dimension of the entire memory cell should not exceed 150x150 nm^2 and this implies ferroelectric capacitors with lateral dimensions of about 100 nm. For ferroelectrics, many phenomena grouped under the generic name of "size effects" can be explained by: (i) monodomain configuration, which is energetically favorable for very small structures, (ii) depolarization fields causing the instability of the polar phase, and (iii) surface effects [1]. For free standing nano-size ferroelectric structures, the surface effects represented by a surface energy added to the total energy of the system are of great concern since it is proportional to the total surface area [2]. For a ferroelectric structure having lateral size of 100 nm or less the number of the unit cells seeing a bulk environment is extremely small whereas the number of the atoms associated with the surface becomes quite significant. Thus, the surface/interface effects like surface defects inducing domain pinning can easily dominate the ferroelectric properties.

Size effects in ferroelectrics were studied considering finite size modification in nanocrystalline bulk ferroelectrics [3] and mostly thin films in order to find the critical thickness under the ferroelectricity vanishes [4,5]. Three dimensional nano-size ferroelectric cells were only theoretically studied [1], whereas experiments involving three dimensional nano-size ferroelectric structures have not yet been performed since

49

C. Galassi et al. (eds.), Piezoelectric Materials: Advances in Science, Technology and Applications, 49–57.
© 2000 *Kluwer Academic Publishers. Printed in the Netherlands.*

such structures are beyond the resolution of the present photolithography techniques. The photolithography resolution nowadays is 0.18 μm and lateral size of the order of 100 nm needed for structures to be considered in the nano-size range can be patterned only using one of the next generation lithography methods. Potential candidates for the next lithography generation allowing features in the range of 100 nm lateral dimensions are extreme UV lithography, X-ray lithography, ion beam lithography and electron beam (EB) lithography [6,7].

The smallest stand-alone single ferroelectric structures exhibiting ferroelectric switching reported to date have sizes of 1.0 x 1.0 μm and 0.7 x 0.7 μm for cells obtained by conventional optical photolithography at Mitsubishi-Symetrix [8] and NEC [9], respectively, and 0.15 x 0.15 μm for cells obtained using focused ion beam at the University of Maryland [10]. All these cells were far from the nano-size dimensions needed to reveal size-effects in ferroelectrics.

We report here on the fabrication of regular $SrBi_2Ta_2O_9$ (SBT) and $Pb(Zr_{0.70}Ti_{0.30})O_3$ (PZT) structures with lateral sizes under 100 nm using electron beam direct writing (EBDW). We demonstrate the ferroelectric properties of these 100 nm cells by measuring their hysteresis loops *via* a scanning force microscope (SFM) working in piezoresponse mode.

2. Patterning of Nano-size Ferroelectric Cells

Patterning of ferroelectrics is a key step in the integration of ferroelectrics into silicon technology. The conventional patterning process based on usual resist lithography followed by dry etching of the oxide film exhibits unsolved problems like side wall redeposition and contamination [11,12]. Electron beam lithography (EBL) is one of the simplest method to produce features with lateral dimensions in the sub-100 nm range. However, it is known to be a low throughput process and considered too slow for IC manufacturing [13].

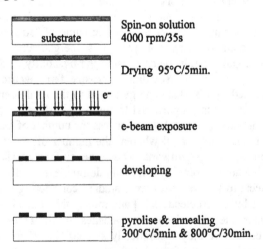

Figure 1. Schematic of the electron beam direct writing process

Electron beam direct writing (EBDW) was essentially used until now to write metallic nanostructures using metalorganic precursors or metal colloids [14,15].

It has been applied for the first time by Okamura et al. [16] to pattern ferroelectric structures. EBDW is based on chemical reactions that are locally induced in a metalorganic thin film by irradiation with an electron beam having sufficient energy and a sufficiently high dose. The desired pattern is written by scanning the electron beam over the sample and then the pattern is developed by dissolving the unexposed area in a specific solvent and further transformed into the desired metal or oxide by thermal annealing. EBDW alleviates the etching problems, such as side wall redeposition and contamination, and holds all the advantages of the EB lithography, *viz.* nanometer resolution and maskless process.

We used EBDW, outlined in Fig. 1, to pattern ferroelectric nano-size SBT and PZT structures with lateral dimension between 1 μm and 0.125 μm. The patterns were exposed into the corresponding metalorganic film using a commercial electron beam lithography system (ELPHY Plus) adapted to a JEOL JSM 6400 scanning electron microscope (SEM) working at an acceleration voltage of 40 kV. Exposure was performed at electron doses ranging from 1500 $\mu C/cm^2$ to 6000 $\mu C/cm^2$ for PZT and 600 $\mu C/cm^2$ to 1200 $\mu C/cm^2$ for SBT. These doses, needed to impress the PZT and SBT patterns into the metalorganic solution, are quite high compared with the doses of about 30 $\mu C/cm^2$ needed to impress normal e-beam resist.[17] After the e-beam exposure the metalorganic structures were developed by immersing the exposed sample 1 min. in toluene and dried by blowing with pure nitrogen. The developed metalorganic pattern is shown in Fig. 2. Note the overexposure due to the high spatial density of the cells (proximity effect) which occurs near the substrate.

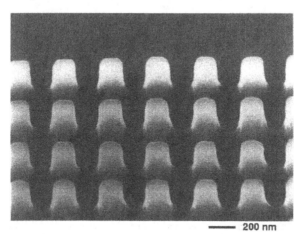

200 nm

Figure 2. SBT metalorganic mesas 200 nm in lateral size obtained after developing. The exposure dose was 1.0 mC/cm².

The metalorganic mesas were transformed into amorphous oxide mesas by subsequently annealing the samples in air for 5 min at 300°C. The ferroelectric crystalline phase was obtained by further annealing at temperatures ranging from 600°C to 850°C. During the pyrolysis the lateral dimensions of one mesa decrease and this shrinkage process continues during the crystallization anneal. For PZT 30% shrinkage is observed after the pyrolysis for a total shrinkage of about 45% after crystallization. Representative structures, consisting of periodic arrays of ferroelectric cells yielding ca. 1 Gbit/cm² density, are presented in Fig. 3.

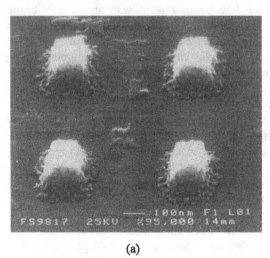

(a)

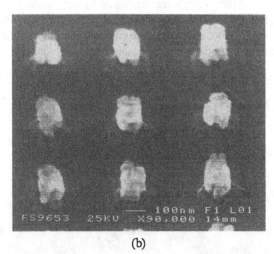

(b)

Figure 3. Scanning electron microscopy images of PZT nanostructures (a) after annealing at 650°C and SBT nanostructures (b) after 800°C annealing. Note the large grain structure in the SBT case compared to the PZT mesas that maintain the rectangular shape after annealing.

The cells are well defined in shape and are polycrystalline with grains of 20 nm or less (smaller for PZT than for SBT). During the crystallization process the SBT structures lose their rectangular shape while the PZT ones maintain their shape even after a shrinkage of 50%.

The smallest cell dimension ever fabricated using the above method is 100 nm, for nanostructures made of both PZT and SBT. These lateral sizes of 100 nm cells were the dimensions obtained after the final annealing. Fig. 4 shows such a 100 nm PZT cell.

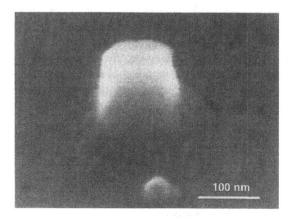

Figure 4. SEM micrograph of a PZT cell after annealing for 2 h at 700°C. The lateral size of the cell is less than 100nm.

3. Switching of Nano-size Ferroelectric Cells

As already mentioned, it is of high scientific interest to measure the ferroelectric properties of capacitors having lateral dimensions of 100 nm or less. It will not only give an answer to the practical question of the fabrication of high-density ferroelectric memories, viz. how small can be a switchable ferroelectric capacitor, but will also give the possibility to experimentally study size effects in ferroelectrics. In order to visualize and contact one single cell we used a scanning force microscope (SFM) with a conductive tip. Attempts to directly measure a ferroelectric hysteresis on a nano-size cell failed since the full switching of the polarization in such a 100 nm cell would lead to only approximately 2000 electrons on the electrodes of the nano-capacitor, provided that the cell is a single domain. Measuring this amount of charge without on-chip integration of the amplifiers is almost unfeasible. Therefore, we used the piezoelectric effect induced in the sample by SFM (piezoelectric response mode of SFM) in order to record the ferroelectric hysteresis and the switching of the nano-size cells.

SFM working in the piezoresponse mode is a powerful tool used in characterization of ferroelectric films at the nanometer scale [18,19]. Briefly, a conductive tip probes the sample surface in contact mode and a small testing AC signal with a frequency in the kHz range is applied between the tip and the bottom electrode of the sample. Due to the converse piezoelectric effect the sample underneath the tip will mechanically oscillate with the same frequency as the excitation signal and will generate oscillations of the cantilever. Using a lock-in technique, the piezoelectric signal is extracted from the total deflection signal of the SFM. Hysteresis loops of individual nano-capacitors were recorded by positioning the tip on top of a selected nano-cell and monitoring its piezoresponse signal as a function of a DC bias (switching) voltage superimposed on a small testing or probing (imaging) voltage.

54

Hysteresis in this strain-field loop is always associated with the ferroelectric properties of the sample [20]. Details on the method used here to measure nano-size cells are given elsewhere [21].

Fig. 5 shows the piezoelectric hysteresis loop of a 1 μm and of a 100 nm PZT cell, respectively. Each point of the hysteresis was acquired at zero DC bias after polarizing the sample with a bias pulse (100 ms duration) at the desired voltage. Both cell sizes exhibit well-defined hysteresis loops having exactly the same coercive voltages, which indicates that scaling the cell down to 100 nm does not modify the coercive field. The piezoelectric coefficient at zero field decreases as the size decreases, the value for a 1 μm cell being about two times larger than that of a 100 nm cell.

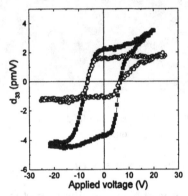

Figure 5. Piezoelectric hysteresis loops of a 1 μm PZT cell (–■–) and of a 100 nm PZT cell (–O–).

As the interaction between the tip and the sample is extremely complex [22] and still under investigation, neither the absolute polarization nor the true coercive field can be extracted from these measurements at the present time. Nevertheless, for ferroelectric materials with centrosymmetric paraelectric phase the polarization can be estimated from the piezoelectric coefficient using [23]:

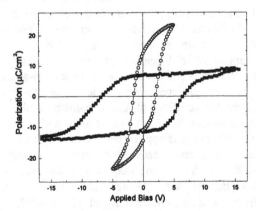

Figure 6. Polarization hysteresis loops of a 1 μm PZT cell (–O–) calculated from the piezoresponse signal and macroscopically measured for a 0.3 mm diameter PZT capacitor (–■–).

$$d_{33} = 2\varepsilon_{33}QP \tag{1}$$

where ε_{33} and Q represent the dielectric constant and electrostrictive coefficient, respectively. d_{33} can be measured from the first harmonic of the piezoresponse signal and Q can be estimated from the second harmonic.

The calculated polarization of a 1 µm cell and the measured polarization of a macroscopic capacitor having the same thin ferroelectric film processed in the same conditions as for the microscopic cells are presented in Fig. 6. A very good agreement between polarization values calculated from the piezoelectric hysteresis loop of the 1 µm cell and the values macroscopically measured can be seen, the remanent polarization of the microscopic cell being equal with the relaxed value of the macroscopic measurement.

The polarization switching and domain structure can be visualized using the piezoresponse mode of SFM. The image showing the piezoelectric signal can be acquired by scanning the tip over a certain area and simultaneously recording the topographic and piezoelectric signals. White and dark contrasts are associated with the two remnant polarization states, $\pm P_r$, and a gray contrast, indicating the absence of piezoelectric activity, is associated with zero polarization. Fig. 7 shows topographic and piezoelectric images, respectively, for four 1 µm PZT cells.

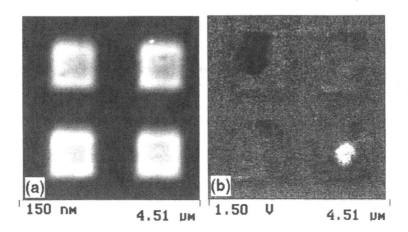

Figure 7. Topographic (a) and piezoresponse (b) image of a 1 µm cell assembling. The upper-left cell was polarized applying a –10V/10 ms poling pulse and the right cells were polarized with a +10V/10 ms pulse.

The positive and negative domains within the cells are easily revealed by the piezoresponse imaging. The negative domain (upper left cell) is larger than the positive domain (lower right cell) and extends to the side-wall of the cell, whereas the positive domain is confined to the center of the cell, indicating an asymmetry in the switching behavior.

The minimum diameter of the switched area under the tip is about 300 nm, one cell of 250 nm or less being fully switched by applying a single voltage pulse in the middle of the cell. Due to the tip shape (triangular with an apex radius ~ 50nm) piezoelectric images of structures under 250 nm lateral size are highly distorted but still show the piezoelectric contrast revealing the domain structure and switching process.

4. Conclusions

Three dimensional nano-size ferroelectric structures having lateral sizes down to 100 nm were patterned by electron-beam direct writing. Ferroelectric switching of these cells was achieved by scanning force microscopy working in piezoresponse mode.
It was shown that PZT cells with lateral sizes of 100 nm exhibit piezoelectric hysteresis loops. This shows that the problems of fringing fields which is important at aspect ratios lower than 5:1 and surface states on the side walls [24] are not preventing the switching of nano-size ferroelectric cells.

5. Acknowledgments

We acknowledge Dr. W. Erfurt for useful discussions and Mrs. U. Doss for performing e-beam exposures.

6. References

1. Li S., Eastman J. A., Li Z., Foster C. M., Newnham R. E., and Cross L. E. (1996) Size effects in nanostructured ferroelectrics, *Physics Letters A* **212**, 341-346.
2. Scott JF, Hartmann AJ. (1998) Effects of constrained geometries and fast access times in real ferroelectric memory devices, *Journal de Physique IV* **8**, 3-15.
3. Ishikawa, K., Yoshikawa, K., and Okada, N. (1988) Size effect on the ferroelectric phase transition in $PbTiO_3$ ultrafine particles, *Phys. Rev. B* **37**, 5852-5.
4. Bune, A. V. , Fridkin, V. M., Dusharme, S., Blinov, L. M. , Palto, S. P., Soronin, A. V., Yudin, S. G., and Zlatkin A. (1998) Two-dimensional ferroelectric films, *Nature* **391** 874-7.
5. Scott, J. F. (1991) Phase transitions in ferroelectric thin films, *Phase Trans.* **30**, 107-10.
6. Melngailis J., Mondelli A. A., Ivan L. Berry III, and Mohondro R (1998) A review of ion projection lithography, *J. Vac. Sci. Technol.* **B 16**, 927.
7. Glendinning W. B. and Cerrina F. (1991), in *Handbook of VLSI Microlithography*, ed. Glendinning W. B. and Herbert J. N., Noyce, Park Ridge, NJ.
8. Uchida H., Soyama N., Kageyama K., Ogi K., Scott M. C., Cuchiaro J. D., McMillan L. D., Paz de Araujo C. A. (1997) Characterization of self-patterned SBT/SBNT thin films from photo-sensitive solutions, *Integrated Ferroelectrics* **18**, 249-61.
9. Amanuma K. and Kunio T. (1996) Low-voltage switching characteristics of $SrBi_2Ta_2O_9$ capacitors. *Jpn. J. Appl. Phys. Part 1* **35**, 5229-31.
10. Stanishevsky A., Aggarwal S., Prakash A. S., Melngailis J., and Ramesh R. (1998) Focused ion-beam patterning of nanoscale ferroelectric capacitors, *J. Vac. Sci. Technol.B* **16** 3899-902.
11. Cofer A., Rajora P., DeOrnellas S., and Keil D. (1997) Plasma etch processing of advanced ferroelectric devices, *Intergrated Ferroelectrics* **16**, 53-61.
12. R. E. Jones (1997) Integration of ferroelectric nonvolatile memories, *Solid State Technology* **40**, 201-10.
13. Yamashita H., Tokunaga K., Kojima K., Nozue H., and Nomura E. (1995) *J. Vac. Sci. Technol. B* **13**, 2473.
14. Craighead H. G. and Schiavone L. M. (1986) Metal deposition by electron beam exposure of an organometallic film, *Appl. Phys. Lett.* **48**, 1748-50.

15. Lohau J., Friedrichowski S., Dumpich D., Wassermann E. F., Winter M., and Reetz M. T., (1998) Electron-beam lithography with metal colloids: direct writing of metallic nanostructures, *J. Vac. Sci. Technol.* **B 16**, 77-9.

16. Okamura S., Mori K., Tsukamoto T., and Shiosaki T. (1997) Fabrication of ferroelectric Bi/sub 4/Ti/sub 3/O/sub 12/ thin films and micropatterns by means of chemical solution decomposition and electron beam irradiation, *Integrated Ferroelectrics* **1**, 311-18.

17. McCord M. A. and Rooks M. J. (1997), in *"Handbook of Microlithography, Micromachining and Microfabrication"*, vol. 1, ed. P. Rai-Choudhury, SPIE, cap. 2.

18. Franke K., Besold J., Haessler W., Seefebarth C. (1994) Modification and detection of domains on ferroelectric PZT films by scanning force microscopy, *Surface Science* **302**, L283.

19. Gruverman A., Auciello O., and Tokumoto H. (1998) Scanning force microscopy: application to nanoscale studies of ferroelectric domains. *Integrated Ferroelectrics* **19**, 49-83.

20. Damjanovic D. (1998) *Rep. Prog. Phys.* **61**, 1267.

21. Harnagea C., Alexe M., Pignolet A., Satyalakhshmi K. M., and Hesse D. (1999) Switching of ferroelectric nanostructures, this volume.

22. Ahn C.H., Tybell T., Antognazza L., Char K., Hammond R.H., Beasley M.R., Fischer O., Triscone J-M. (1997) Local, nonvolatile electronic writing of epitaxial $Pb(Zr_{0.52}T_{0.48})O_3/SrRuO_3$ heterostructures, *Science* **276**, 1100-3.

23. Kholkin, A.L., Brooks, K.G. and Setter N. (1997) Electromechanical properties of $SrBi_2Ta_2O_9$ thin films, *Appl. Phys. Lett.* **71**, 2044.

24. Scott J. F. (1998) Limitations on ULSI-FeRAMs, *IECIE Trans. Electron.* **E81-C**, 477.

INFLUENCE OF Bi AND Y ADDITIONS ON THE ELECTROMECHANICAL ANISOTROPY OF LEAD TITANATE CERAMICS

L. AMARANDE, C. TANASOIU, C. MICLEA and C.F. MICLEA
National Institute for Materials Physics,
96700 Magurele-Bucharest, POBox MG-7, Romania

Abstract

Lead titanate (PT) is well known as a good piezoelectric material with a high Curie temperature (~490°C) and a low dielectric constant (~200). But pure lead titanate cannot be sintered as ceramic bodies, for practical purposes, due to its large crystal anisotropy (tetragonality c/a=1.064). However, dense PT ceramic bodies can be obtained if small additions of other elements are used. Ca and Sm modified PT were succesfully tried, but such dopings proved rather expensive for mass production. Therefore some other doping elements must be tried.

In this regard we investigated the effect of Bi and Y additives on the piezoelectric properties of PT ceramics. The general formula of our Bi, Y doped lead titanate ceramics was: $(Pb_{1-3x/2}Me_x)(Ti_{0.98}Mn_{0.02})O_3$, with Me=Bi,Y and x=0.04, 0.06, 0.08. The modified PT ceramics were prepared by conventional ceramic technique, using p.a.purity raw materials. The mixed powders were sintered at temperatures between 1000-1300°C. Poling was done in fields of about 70kV/cm. Density and coupling factors of the samples were determined as a function of sintering temperatures, and doping level. Temperature dependence of the main piezoelectric characteristics was also investigated for temperatures as high as 500°C.

It was found that Bi,Y modified lead titanate ceramics have a very high Curie point (T_C>450°C), a low dielectric constant (<140) and a large electromechanical anisotropy (the radial mode is nearly inexistent, at room temperature and the thickness coupling factor is about 0.4), which make them competitive to the rare-earth doped PT ceramics, for high temperature and high frequency transducers applications.

1. Introduction

Lead titanate ($PbTiO_3$) ceramics are very promising materials for high temperature and high frequency applications, due to their high Curie point and low dielectric constant [1]. Another interesting feature of lead titanate (PT) ceramics consists in the large anisotropy of their electromechanical coupling factors: the thickness coupling factor k_t is much larger than the planar coupling factor k_p, which makes them extremely useful

59

C. Galassi et al. (eds.), Piezoelectric Materials: Advances in Science, Technology and Applications, 59–66.
© 2000 *Kluwer Academic Publishers. Printed in the Netherlands.*

for special applications such as ultrasonic transducers and SAW (Surface Acoustic Waves) devices, at high frequency and hydrophones, at low frequency [2,4].

But as it is known [1,5], pure PT ceramics cannot be sintered as dense body, because they are fragile and break during cooling through their Curie point, as a result of the internal stresses generated by crystal anisotropy.

However, dense modified PT ceramics were possible to be obtained if small additions of dopants were used [1,2,8,9]. The influence of different substitutions, such as alkaline or rare earths elements for Pb and Mn for Ti, was investigated [1-3,6-9] and it was found that these dopants enhanced the mechanical strength, resistivity, dielectric and piezoelectric properties of such modified PT ceramics.

The present work reports the effect of Y and Bi additions on the dielectric and piezoelectric properties of modified PT ceramics, as well as the temperature behavior of their main piezoelectric, electromechanical and dielectric constants, from room temperature up to 500°C.

2. Materials preparation

Bi and Y doped lead titanate ceramics, with general formula $(Pb_{1-3x/2}Me_x)(Ti_{0.98}Mn_{0.02})O_3$, with Me=Bi,Y and x=0.04, 0.06, 0.08,were prepared by conventional ceramic technique, using p.a. purity raw materials. The starting oxides were wet (in methanol) mixed in a planetary ball mill, in agathe vessels, for 1h. After drying, the mixture was triple calcined at 900°C, for 2h, with intermediate millings, than finally milled for 5h. The resulting powders were pressed, in a steel die, into disks of 11.6 mm diameter and 1mm thickness. The disks were sintered for 3h, in air, at temperatures between 1000°C and 1300°C, than mechanically processed on a lapping machine, to about 0.6 mm thickness and silver electroded. Poling was done under d.c. fields of about 70 kV/cm, in silicon oil, at 150°C, for 5min.

3. Measurements

Dielectric, piezoelectric and electromechanical properties of Bi and Y modified PT ceramics were determined 24h after poling, using a HP 4194A Impedance/Gain-Phase Analyser.

The capacitance C_p and the dissipation factor D were measured at 1kHz, than dielectric constant K_{33}^T was calculated from the relationship $K = C_p t / \varepsilon_0 A$, where t is the sample thickness and A the electroded area.

Planar and thickness coupling factors, piezoelectric constants h_{33} and e_{33} , frequency constant of the thickness mode N_t , elastic stiffness c_{33}^D , dielectric permitivity ε_{33}^S were determined by resonance method [10]. k_p was given by the relationship (155) of IEEE Std.[10], after Poisson's ratio σ was calculated by a polynomial fit [11] of the data given by table 10 of IEEE Std.[10]. k_t was determined either by a polynomial fit [12] based on Onoe's method [13], or from ec.(163) of IEEE Std.[10],when series resonance frequencies of thickness mode overtones cannot be accurately measured.

N_t, c_{33}^D, ε_{33}^S, h_{33} and e_{33} were calculated using the following relationships:

$$N_t = t f_p \tag{1}$$

$$c_{33}^D = \rho(2tf_p)^2 \tag{2}$$

$$\varepsilon_{33}^S = \frac{2t(g_{max}/r_{max})^{1/2}}{\pi^2 d^2 f_p} \tag{3}$$

$$h_{33} = k_t \sqrt{c_{33}^D / \varepsilon_{33}^S} \tag{4}$$

$$e_{33} = k_t \sqrt{\varepsilon_{33}^S c_{33}^D}, \tag{5}$$

where ρ is material density (determined by geometric method), f_p the paralel resonance frequency of the thickness mode, g_{max} and r_{max} the maxima of conductance and resistance respectively of the thickness mode.

4. Results

The densities as a function of sintering temperatures, for 4%,6% and 8% Bi and Y doped samples, are shown in figs. 1 and 2, respectively. For both dopants, the density shows a maximum, situated around 1250°C for Bi and 1050°C for Y, respectively.

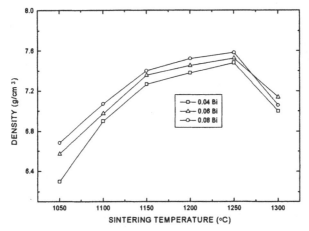

Figure 1. Density as a function of sintering temperature, for $(Pb_{1-3x/2}Bi_x)(Ti_{0.98}Mn_{0.02})O_3$ ceramics, with x=0.04, 0.06, 0.08.

The thickness coupling factor k_t shows a similar behavior with sintering temperature, as can be seen in figs. 3 and 4. One can notice that k_t reaches its maximum at the same temperature as the density does, for each type of additive, which means that the optima for the sintering temperatures are those stated above. Therefore, only samples sintered at these temperatures were selected for further investigations. Figs. 1-4 show that k_t and density increase with increasing amount of additive, which means an improvement of electromechanical and dielectric properties of such ceramic. k_t reaches a maximum value of about 0.4, while k_p is extremely low (it cannot be measured) at room

62

temperature. It proves that Bi and Y doped PT ceramics have a very large electromechanical anisotropy.

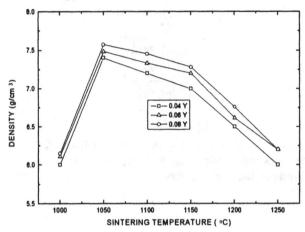

Figure 2. Density as a function of sintering temperature, for $(Pb_{1-3x/2}Y_x)(Ti_{0.98}Mn_{0.02})O_3$ ceramics, with x=0.04, 0.06, 0.08.

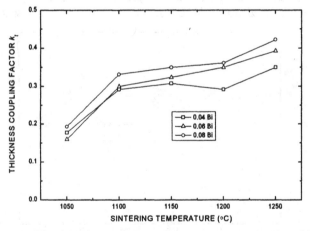

Figure 3. Thickness coupling factor k_t as a function of sintering temperature, for $(Pb_{1-3x/2}Bi_x)(Ti_{0.98}Mn_{0.02})O_3$ ceramics, with x=0.04, 0.06, 0.08.

Dielectric constant as a function of temperature is shown in fig.5, for Y modified PT samples. It increases with increasing temperature and reaches a maximum at the Curie point (T_c), above 450°C. T_c slightly decreases with increasing amount of additive, from 462°C to 457°C. The maximum value of the dielectric constant shows a similar behavior versus doping level, while dielectric constant slightly increases with increasing amount of additive, far below the transition temperature, similar to that of the rare-earths modified PT ceramics [7]. This behavior of the dielectric constant may be due to the decrease of lattice anisotropy which enhance the polarizability. One can notice the low dielectric constant (K_{33}^T <150) at room temperature and below 100°C. For Bi doping, dielctric constant shows a similar behavior with increasing temperature, but the Curie point is a little higher, of about 480°C.

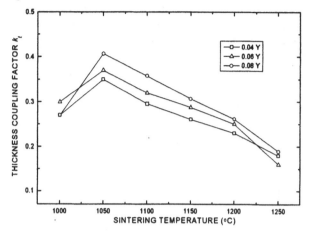

Figure 4. Thickness coupling factor k_t as a function of sintering temperature, for $(Pb_{1-3x/2}Y_x)(Ti_{0.98}Mn_{0.02})O_3$ ceramics, with x=0.04, 0.06, 0.08.

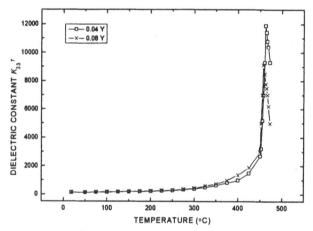

Figure 5. Dielectric constant $K_{33}{}^T$ as a function of temperature, for $(Pb_{1-3x/2}Y_x)(Ti_{0.98}Mn_{0.02})O_3$ ceramics, with x=0.04, 0.06, 0.08.

The influence of increasing temperature, on the electromechanical properties of Bi doped PT ceramics is shown in fig.6. k_t slightly increases with increasing temperature between room temperature and 125°C, than it is nearly constant up to 400°C, when it begins to decrease gradually up to 450°C, than suddenly above 450°C , corresponding to the irreversible degradation of piezoelectric properties by depoling. The increase of k_t may be due to a partial release of intergranular stress when domain walls mobility increases with increasing temperature. Higher values of k_t corresponding to higher amount of additive, can be explained by assuming an increasing degree of poling, when lattice tetragonality decreases [7].

The radial mode is extremely low between room temperature and 100°C. Above 100°C, it can be slightly observed, but k_p values are very low, with low corresponding

64

Poisson's ratio (σ =0.18-0.2). k_p slightly increases between 100°C and 300°C, than suddenly vanishes. The behavior of k_p is similar to that of rare-earths doped PT ceramics [6] and may be related to increasing 90° domain wall motion, with temperature and to the sign change of the real part of piezoelectric constant d_{31} [6]. Fig. 6 proves the large electromechanical anisotropy of such ceramics, which can be succesfully used for ultrasonic transducers.

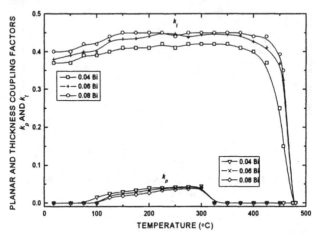

Figure 6. Thickness coupling factor k_t and planar coupling factor k_p as a function of temperature, for $(Pb_{1-3x/2}Bi_x)(Ti_{0.98}Mn_{0.02})O_3$ ceramics, with x=0.04, 0.06, 0.08.

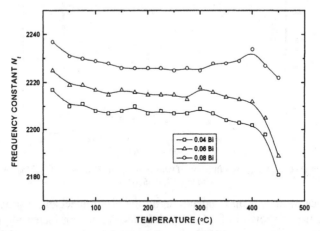

Figure 7. Frequency constant N_t as a function of temperature, for $(Pb_{1-3x/2}Bi_x)(Ti_{0.98}Mn_{0.02})O_3$ ceramics, with x=0.04, 0.06, 0.08.

Fig.7 shows the frequency constant N_t of the thickness mode, as a function of temperature, for Bi doped samples. N_t is nearly constant between room temperature and 400°C, than it decreases. Above 450°C, N_t decreases rapidly to zero, when thickness mode vanishes. The electromechanical coupling factors and the frequency constant, of Y modified PT samples, have a similar behavior.

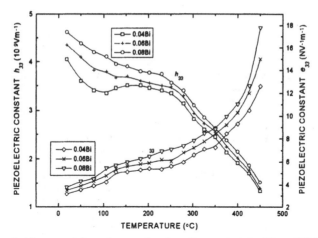

Figure 8. Piezoelectric constants h_{33} and e_{33} as a function of temperature, for $(Pb_{1-3x/2}Bi_x)(Ti_{0.98}Mn_{0.02})O_3$ ceramics, with x=0.04, 0.06, 0.08.

Figs. 6 and 7 shows a large thermal stability of thickness mode electromechanical behavior of Bi (and Y also) doped PT ceramics, which make them suitable for high-temperature applications (up to 400°C).

Temperature dependences of piezoelectric constants h_{33} and e_{33} are shown in fig.8, for Bi modified samples and they are similar for Y modified samples as well. h_{33} decreases, while e_{33} increases with increasing temperature, especially above 300°C.

5. Summary

Dense PT ceramics, doped with Bi and Y , having the chemical formula $(Pb_{1-3x/2}Me_x)(Ti_{0.98}Mn_{0.02})O_3$, where Me=Bi orY and x=0.04, 0.06, 0.08, were prepared by the usual ceramic technique and sintered at temperatures between 1000°C and 1300°C. Disc shaped samples, with 10mm diameter and 0.5mm thickness, were poled in fields of about 70kV/cm, high enough to assure a good poling degree.

The dielectirc and piezoelectric properties were determined as a function of composition and temperature, up to the Curie point (above 450°C), only for samples optimally sintered, at 1050°C, for Y doping and 1250°C, for Bi doping, respectively.

It was found that both additives, Bi and Y, are beneficial for PT ceramics, giving rise to an enhancement of the piezoelectric properties.

Thickness coupling factors of about 0.4 and nearly zero planar coupling factors were obtained between room temperature and 100°C, proving a large electromechanical anisotropy. These types of extremely anisotropic materials seems to be very promising for ultrasonic directional transducers.

66

6. References

1. Ikegami, S., Ueda, I. and Nagata, T. (1971) Electromechanical properties of PbTiO₃ ceramics containing La and Mn, *J.Acoust. Soc. Am.* **50**, 1060-66.
2. Takeuchi, H., Jyomura, S., Yamamoto, E. and Ito, Y. (1982) Electromechanical properties of (Pb,Ln)(Ti,Mn)O₃ ceramics (Ln=rare earth), *J.Acoust. Soc. Am.* **72**, 1114-1120.
3. Takeuchi, H., Jyomura, S. and Nakaya, C. (1985) New piezoelectric materials for ultrasonic transducers, *Jpn. J. Appl. Phys.* **24**, 36-40.
4. Takahashi T.(1990) Lead titanate ceramics with large piezoelectric anisotropy and their applications, *Am. Ceram. Soc. Bull.* **69**, 691-695.
5. Jaffe, B., Cook, W.R. and Jaffe, H. (1971) *Piezoelectric ceramics*, Acad. Press, London, N.Y.
6. Damjanovic, D., Gururaja, T.R. and Cross, L.E. (1987) Anisotropy in piezoelectric properties of modified lead titanate ceramics, *Am. Ceram. Soc. Bull.*, **66**[4] 699-703.
7. Suwannasiri, T. and Safari, A. (1993) Effect of rare-earth additives on electromechanical properties of modified lead titanate ceramics, *J. Am. Ceram. Soc.*, **76**[12], 3155-3158.
8. Mendiola, J., Jimenez, B., Alemany, C., Pardo, L. and del Olmo, L. (1989) Influence of calcium on the ferroelectricity of modified lead titanate ceramics, *Ferroelectrics*, **94**, 183-188.
9. Ichinose, N. and Fuse, Y. (1990) Anisotropy of piezoelectric properties in the modified PbTiO₃ ceramics, *Ferroelectrics* **106**, 369-374.
10. *IEEE Standard on piezoelectricity, 176-1987*, ANSI-IEEE Std.176 (1987).
11. Sherit, S., Gauthier, N., Wiederic, H.D. and Mukherjee, B.K. (1991) Accurate evaluation of the real and imaginary material constants for a piezoelectric resonator in the radial mode, *Ferroelectrics*, **119**, 17-32.
12. Sherit, S., Wiederic, H.D. and Mukherjee, B.K. (1992) A polynomial fit for calculating the electromechanical coupling constants of piezoelectric materials using the method described by Onoe et al.[J. Acoust. Soc. Am., 35, nr.1, 36-42 (1963)], *J. Acoust. Soc. Am.*, **91**(3), 1770-71.
13. Onoe, M., Tiersten, H.F. and Meitzler, A.H. (1963) Shift in the location of resonant frequencies caused by large electromechanical coupling in the thickness-mode resonators, *J. Acoust. Soc. Am.*, **35**, nr.1, 36-42.

$Bi_4Ti_3O_{12}$-BASED PIEZOELECTRIC CERAMICS FOR HIGH TEMPERATURE APPLICATIONS

M. VILLEGAS, A.C. CABALLERO, C. MOURE, J.F. FERNANDEZ,
P. DURAN
Electroceramics Department. Instituto de Cerámica y Vidrio. CSIC.
28500 Arganda del Rey, Madrid, Spain

Abstract

High temperature piezoelectric ceramics based on WO_3-doped $Bi_4Ti_3O_{12}$ (W-BIT) were prepared by two ways: chemical (hydroxide coprecipitation) and conventional oxides mixing. Sintering was carried out between 800°-1150°C in air. A rapid densification, >99% ρ_{th} at 900°C/2h, took place in the coprecipitated W^{6+}-doped BIT ceramics, whereas conventionally prepared BIT-based materials achieved only a maximum density of −94% ρ_{th}, at 1050°C. As expected, a plate-like morphology in both materials appeared, with the aspect ratio (l/t) of the grains increasing linearly with the sintering temperature. Plate-like grains were larger in the conventionally prepared W-BIT based materials. The sintering behaviour could be related both with the agglomeration state of the calcined powders and with the enlargement of the platelets at high temperature. The W^{6+}-doped BIT materials showed an electrical conductivity value 2-3 orders of magnitude lower than undoped samples. The most important result of the electrical characterisation of this type of materials was that the electrical conductivity increased exponentially with the aspect ratio (l/t) of the plate-like grains.

1. Introduction

Bismuth titanate $Bi_4Ti_3O_{12}$ (BIT), which belongs to the Aurivillius family, is perhaps one of the most studied compounds among the bismuth-based layered ceramics. BIT structure is formed by two unit cells of $BiTiO_3$ with perovskite-like structure interleaved with $(Bi_2O_2)^{2+}$ layers [1]. This structure promotes a plate-like morphology, with platelets growing preferentially in the basal plane (ab plane).

As many other compounds of the bismuth-based layered family, BIT shows ferroelectric properties. The ferroelectric-to-paraelectric phase transition temperature is T_c−675°C. Due to this high transition temperature, BIT ceramics are good candidates for high-temperature piezoelectric applications. At temperatures below T_c, BIT is monoclinic, but it can be represented as orthorhombic with the c-axis perpendicular to the $(Bi_2O_2)^{2+}$ layers [2].

Accepting this representation, the major component of spontaneous polarisation

67

C. Galassi et al. (eds.), Piezoelectric Materials: Advances in Science, Technology and Applications, 67–74.

$(P_s-50 \ \mu C/cm^2)$ lies in the ab plane of the perovskite [3].

Bismuth titanate and other bismuth-based layered compounds have a relatively low electrical resistivity. The electrical conductivity in BIT is highly anisotropic, with the maximum value in the same plane as the polarization [4-5]. As a consequence BIT ceramics are very difficult to pole. The reduction of electrical conductivity is, then, one of the main goals of BIT worldwide investigations. p-type conductivity has been reported for BIT materials [6], but there is a controversy about the origin of positively charged mobile carriers: hole compensation of cation vacancies, as in $Pb(Zr,Ti)O_3$ [7] or acceptor impurities introduced through the raw materials, as in $BaTiO_3$ [8]. Both kinds of defects could explain the increase of conductivity when BIT is annealed in an oxygen atmosphere [6]. As expected in a p-type material, acceptor dopants such as Fe^{3+} in Ti^{4+} positions or Sr^{2+} in Bi^{3+} positions increase the BIT electrical conductivity [6], whereas it has been shown that doping with donor cations such as Nb^{5+}, Sb^{5+} or Ta^{5+} in the Ti^{4+} positions decreases BIT conductivity [6,9].

The aim of the present work is to study both the influence of the processing method on the sinterability and the microstructure development and the effects of WO_3 doping and microstructure on electrical conductivity and piezoelectric properties of $Bi_4Ti_3O_{12}$ ceramics.

2. Experimental Procedure

Donor-doped (5 at% W^{6+}) $Bi_4Ti_3O_{12}$ materials ($Bi_4Ti_{2.95}W_{0.05}O_{12.05}$) were prepared by both the conventional and the chemical routes. For the conventional ceramic method (WBIT-O), stoichiometric amounts of TiO_2, α-Bi_2O_3 and WO_3 were mixed in a rotatory mill with ZrO_2 balls in isopropanol for 2h. After drying, this mixture was calcined at 800°C/2h in air and re-milled in an attrition mill under the same conditions. This calcined powder was granulated, uniaxially pressed at 50 MPa in disks and sintered in air 850°-1150°C for 2 hours.

The chemically prepared W-BIT ceramics were done by using the hydroxide coprecipitation method (WBIT-H). Titanium tetrabutoxide $Ti(C_4H_9O)_4AC_4H_9OH$, $Bi(NO_3)_3A5H_2O$ and WO_3 were used. Stoichiometric amounts of Ti^{4+} and Bi^{3+} precursors were dissolved in a slightly acid isopropyl alcohol solution. This solution was carefully added to an aqueous suspension of WO_3 by dropping while stirring. Then, with careful pH control, an aqueous solution of NH_4OH (50/50) was slowly added to achieve the complete precipitation of the hydroxides. The coprecipitated powders were washed and redispersed several times with isopropyl alcohol until a neutral pH was achieved. These coprecipitated powders were calcined at 650°C/1h and attrition milled for 2h in isopropanol. Then, the calcined powders were granulated, uniaxially pressed and sintered between 800°-1150°C for 2h.

WBIT-O and WBIT-H calcined powders were characterised by X-ray diffraction (D-5000 Siemens) and scanning electron microscopy (DSM950 Carl Zeiss). The apparent density of both the conventionally and the chemically processed W-doped BIT sintered ceramics was measured by the Archimedes' method in water. The microstructure of sintered compacts was studied on polished and thermally etched surfaces by SEM. The

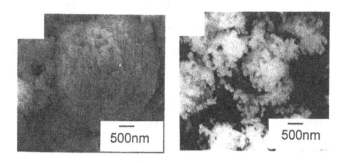

Figure 1. SEM micrographs of a) WBIT-O powders calcined at 800°C for 2 hours and
b) WBIT-H powders calcined at 650°C for1 hour

aspect ratio (length/thickness) of the plate-like grains was determined using an image analyser by measuring at least 300 grains.

Electrical conductivity measurements were done by complex impedance spectroscopy using a HP4192A impedance analyser on disks ground with 6 μm diamond paste and electroded with Ag-Pd (70/30) paste.

3. Results and Discussion

3.1. POWDER CHARACTERISATION

Coprecipitated WBIT-H powders were amorphous and after calcining WBIT-O (800°C/2h) and WBIT-H (650°C/1h) ceramic powders revealed the presence of only a $Bi_4Ti_3O_{12}$-type phase. The SEM observation of WBIT-O calcined powders (Fig. 1.a) showed the presence of dense spherical agglomerates with an average size of –2 μm, consisting of –200 nm primary particles. On the contrary, WBIT-H calcined powders consisted of smaller, more porous agglomerates of 1 μm formed by very small equiaxed particles of –100 nm (Fig. 1.b).

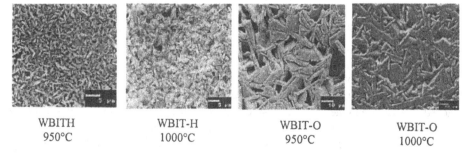

WBITH	WBIT-H	WBIT-O	WBIT-O
950°C	1000°C	950°C	1000°C

Figure 2. SEM micrographs of polished and thermally etched surfaces of WBIT-H and WBIT-O ceramics sintered at 950°C and 1000°C for 2 hours

TABLE I. Microstructure parameters of WBIT-H and WBIT-O ceramics as a function of sintering temperature

	Temperature (°C)	l (μm)	t (μm)	l/t
WBIT-H	850	0.90±0.10	0.30±0.02	3.0±0.7
	900	1.20±0.13	0.40±0.04	3.1±0.7
	950	1.40±0.15	0.40±0.05	3.5±0.7
	1000	2.70±0.26	0.80±0.08	4.0±0.8
	1050	3.40±0.30	0.80±0.10	4.3±0.9
	1100	10.0±1.1	1.80±0.20	5.6±0.9
WBIT-O	900	9.5±1.0	2.20±0.2	4.3±0.9
	950	10.6±1.0	2.30±0.20	4.6±0.9
	1000	11.5±1.1	2.40±0.20	4.8±0.8
	1050	12.5±1.3	2.50±0.25	5.0±1.0
	1100	18.8±2.0	3.30±0.35	5.7±1.2
	1150	21.7±1.9	3.50±0.30	6.2±1.1

3.2. SINTERING AND MICROSTRUCTURE DEVELOPMENT

Fig. 2 shows the microstructure of WBIT-H and WBIT-O ceramics, after sintering at 950° and 1000°C for 2h. The plate-like morphology expected in BIT-based ceramics was observed at any sintering temperature. The average length and thickness of the platelets observed in the different materials along with the aspect ratio data, are summarised in Table I.

The evolution of the sintered density and the average platelets length as a function of sintering temperature (soaking time, 2 hours) is plotted in Fig. 3. As it can be seen, WBIT-H ceramics densified at lower temperatures (>99% ρ_{th} at 900°C) than those of WBIT-O ones (94% ρ_{th} at 1050°C) , and the grain growth in conventionally processed ceramics was higher than in the chemically prepared ones.

The sintering behaviour of WBIT-O and WBIT-H strongly depended on the agglomeration state of the calcined powders. WBIT-O powders having hard spherical agglomerates that did not break during the pressing step promoted the existence of particles packed face-to-face in the green state which formed large plate-like grains at low sintering temperature and exaggerated platelet size with further sintering [10]. The existence of large plate like grains inhibited the densification of WBIT-O materials at low temperature and, although face-to-face aligned platelets (colonies) showed no pores between them, large pores between colonies appear at high temperature due to pore coalescence [11] and this interfered with further densification. In this manner, at temperatures higher than 1050°C,

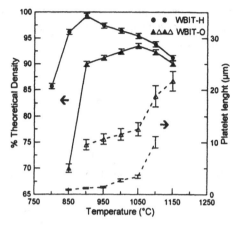

Figure 3. Densification and platelet length as a function of temperature
(soaking time, 2 h) for WBIT-H and WBIT-O ceramics.

the density of WBIT-O materials decreased along with a rapid enlargement of the platelets length (see Fig. 3). In WBIT-H powders the presence of more reactive particles and soft flowable agglomerates that were broken during the pressing step allowed to sinter this material at lower temperature than WBIT-O ceramics. In addition, WBIT-H powders would have a small number of face-to-face packed particles (powders less agglomerated) and the porosity removal was improved in the initial stage of sintering where the development of platelets was not obvious (at 800°C for 2 hours platelets were hard to distinguish). As occurred in WBIT-O ceramics, sintering at high temperatures decreased the apparent density coinciding with an enlargement of the platelet length (see Fig. 3)

3.3. ELECTRICAL CHARACTERISATION

The temperature dependence of the electrical conductivity in WBIT-O and WBIT-H materials sintered at 1050°C for 2h is plotted in Fig. 4, along with the data of the chemically prepared undoped BIT [12]. The inverse temperature dependence of the conductivity in a logarithmic scale was a straight line with an activation energy of 1.1±0.1 eV in the range 350°C-650°C for WBIT-O, WBIT-H and undoped BIT, in agreement with the E_a data given by Shulman et al [6] for Nb^{5+} doped and undoped BIT.

As can be seen in Fig. 4, W^{6+} doping decreased the electrical conductivity in BIT-based materials by 2-3 orders of magnitude when compared to the undoped BIT.

The electrical conductivity studies revealed that chemically prepared BIT-based ceramics had a lower electrical conductivity than the conventionally processed ones. The origin of this behaviour seems to be related to the growth of platelets. As is well known, the conductivity in the direction of the ab planes is higher than that parallel to the c-axis [13]. Therefore, the larger the aspect ratio (i.e. increasing the contribution of the ab planes), the higher the conductivity of the BIT ceramic. Note that differences in density had little effect on the conductivity values since the porosity observed in sintered specimens was closed.

The conductivity values measured on our WBIT samples (Fig. 5) (regardless of the processing method) increased exponentially with the aspect ratio of the platelets and can be fitted to the following expression, at any of the measurement temperatures plotted:

$$\ln \sigma = A\left(\frac{l}{t}\right) + B \qquad (1)$$

σ being the measured conductivity. A and B parameters are obtained from the fitted curves, and are different for each measurement temperature.

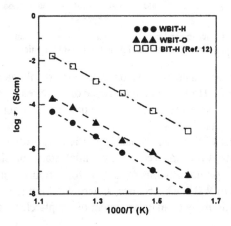

Figure 4. Arrhenius plots of the log of the conductivity as a function of the inverse of the temperature for the W-doped and undoped BIT ceramics sintered at 1050°C for 2 hours

The physical sense of the parameters A and B can be understood by taking into account the meaning of the aspect ratio l/t. If $l/t=1$ then no platelets would be observed and consequently the measured conductivity must be an average of the values for the ab and c directions. According to our experimental fitting this can be expressed as:

$$\ln \sigma_{av} = A + B \qquad (2)$$

with σ_{av} being the conductivity averaged on ab and c directions.
From equations (1) and (2) the following relationship can be written:

$$\sigma_m = \sigma_{av} \; e^{A\left(\frac{l}{t}-1\right)} \qquad (3)$$

Therefore, the measured conductivity is the averaged bulk conductivity on ab and c directions corrected by an exponential function which depends on the aspect ratio. The A parameter depends on the measurement temperature and on the material composition.

According to our fitting parameters, the σ_{av} of WBIT, i.e. the bulk conductivity of the ceramic material assuming that no platelets are formed, is $2.85A10^{-7}$ S/cm at 550°C. This value would be the minimum conductivity possible at 550°C in a BIT ceramic containing 5at% of W^{6+}. Lopatin et al. [9] have reported a conductivity value of $9.5A10^{-7}$ S/cm at 550°C for W^{6+}-doped BIT, however, the lack of microstructure data of their samples does not allow us to check our approach with their data. Note that the experimental fitting has been done for the l/t values obtained (ranging from 3 to 6). Different fitting parameters might be obtained for l/t values out of this range.

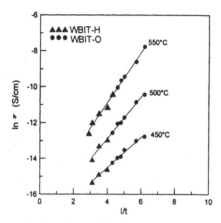

Figure 5. Exponential dependence of the conductivity (S/cm) at different temperatures with the aspect ratio (l/t) for the W-BIT doped ceramics

4. Conclusions

The preparation of $Bi_4Ti_3O_{12}$ (BIT) ceramics doped with WO_3 by a chemical (hydroxide coprecipitation) method produced better sintered materials than the conventional method.

The densification process in the chemically made WBIT ceramics was fast and a maximum density >99% ρ_{th} was obtained at a temperature as low as 900°C. In contrast, WBIT-O ceramics densified (ρ–94% ρ_{th}) at higher temperature (1050°C). This behaviour is attributed to an earlier formation of big plate-like grains and colonies in WBIT-O ceramics, whereas WBIT-H ceramics began to sinter prior to platelet development. The agglomeration state of the calcined powders influenced the platelet size of the sintered ceramics. In WBIT-O the face-to-face contacts between particles in the agglomerates promoted the formation of big plate-like grains and colonies in the early stages of sintering.

On the contrary, for WBIT-H powders, which were less agglomerated, colony formation and platelets growth was retarded.

As expected, doping with WO_3 reduced the electrical conductivity by about two orders of magnitude when compared with the undoped BIT. An exponential relation between bulk conductivity and the aspect ratio (l/t) of the platelets was found. According to the fitting parameters, σ_{av} of WBIT ceramics assuming $l/t=1$ (cubic shaped grains) is 2.85 A 10^{-7} S/cm at 550°C. However, to establish the validity of the approach, a wider range of l/t than studied here (3-6) needs to be examined.

5. Acknowledgements

The authors would like to thank to the Spanish CICYT for the financial support of this work under Project MAT-97-0694-C02-01.

6. References

1. Aurivillius, B. (1949). Mixed Bismuth Oxides with Layer Lattices. *Arkiv for Kemi*, 1, 499.
2. Dorrian, J.F., Newnham R.E. and Smith, D.K. (1971) Crystal Structure of $Bi_4Ti_3O_{12}$. *Ferroelectrics* 3, 17-27.
3. Cummins, S.E. and Cross, L.E.. (1968) Electrical and Optical Properties of Ferroelectric $Bi_4Ti_3O_{12}$ Single Crystals. *J. Appl. Phys.*, 39, 5, 2268-2274.
4. Fouskova, A. and Cross, L.E. (1970) Dielectric Properties of Bismuth Titanate. *J. Appl. Phys.*, 41, 2834-2838.
5. Takenaka, T. and Sakata, K. (1984) Grain Orientation Effects and Electrical Properties of Bismuth Layer Structured Ferroelectric $Pb_{1-x}(NaCe)_{x/2}Bi_4Ti_4O_{15}$ Solid Solution. *J. Appl. Phys.*, 55, 1092.
6. Shulman, H.S., Testorf, M., Damjanovic, D. and Setter, N. (1996) Microstructure, Electrical Conductivity and Piezoelectric Properties of Bismuth Titanate. *J. Am. Ceram. Soc.*, 79, 3124-3128.
7. Jaffe, B., Cook, W.R. and Jaffe, H. (1971) *Piezoelectric Ceramics*, Academic Press, New York.
8. Chan, N.H. and Smyth, D.M. (1976) Defect Chemistry of $BaTiO_3$. *J. Electrochem. Soc.*, 123, 1584-85.
9. Lopatin, S.S., Lupeiko, T.G., Vasiltsova, T.L., Basenko, N.I. and Berlizev, I.M. (1988) Properties of Bismuth Titanate Ceramics Modified with Group V and VI elements. *Inorg. Mater.* (Engl. Trans.) 24, 1328-1331.
10. Inoue, Y., Kimura, T. and Yamaguchi, T. (1983) Sintering of Plate-Like $Bi_4Ti_3O_{12}$ Powders. *Am. Ceram. Soc. Bull.*, 62, 704-711.
11. Kingery, W.D. and Francois, B. (1965) Grain Growth in Porous Compacts. *J. Am. Ceram. Soc.*, 48, 546-47.
12. Villegas, M., Moure, C., Fernández, J.F. and Durán, P. (1996) Low Temperature Sintering of Submicronic Randomly Oriented $Bi_4Ti_3O_{12}$ Materials. *Ceram. Int.*, 22, 15-22.
13. Swartz, S.L., Schulze, W.A. and Biggers, J.V. (1983) Fabrication and Electrical Properties of Grain Oriented $Bi_4Ti_3O_{12}$. *The Bulletin of the Bismuth Institute*, 40, 1-4 (1983).

INFLUENCE OF PROCESSING PARAMETERS ON THE PROPERTIES OF PZT MATERIALS

C. GALASSI, E. RONCARI, C. CAPIANI, A. COSTA
CNR-IRTEC Research Institute for Ceramics Technology
Via Granarolo, 64 – I-48018 Faenza, Italy

Abstract

Bulk PZT materials of composition near the morphotrophic phase boundary are obtained by conventional solid state reaction of the starting oxides. The combination of either donor or acceptor dopants allows the improvement of the electromechanical properties and tailoring the material to the specific application. Reliability and reproducibility are still critical points for these complex systems and careful control of every processing parameter from raw materials purity and granulometry to the intermediate steps of mixing and homogenisation must be assessed. The powder processing steps are examined and examples of the effects of different processing on the electromechanical properties are discussed.

An improvement on milling of agglomerates after solid state reaction of the oxides affects the liquid phase formation during sintering and hence the final microstructure with consequent substantially different final properties.

1. Introduction

Lead zirconate-lead titanate ceramics (PZT) show extremely strong piezoelectric effects for compositions near the morphotrophic phase boundary (MPB) [1]. Rhombohedral and tetragonal phases coexist and are related to the presence of a maximum in the dielectric constant, a larger number of orientable polarization directions and a maximum mechanical compliance preventing cracking during domain reorientation [2]. The MPB of the undoped material lies near the composition with $PbZrO_3$ 53.5 mol %. It is impurity sensitive and superior properties can be achieved through compositional modifications that can be summarized as donor and acceptor doping, isovalent substitutions and addition of low melting additives[1, 3]. Donor dopants of higher charge, like Nb^{5+} in the B sites or La^{3+} in the A sites, are compensated by cation vacancies in the crystal structure which enhance domain wall mobility and result in improved remanent polarization, coupling factors, dielectric constants, dielectric loss and reduced ageing and mechanical quality factor (electrically "soft" PZT) [4]. Acceptor dopants, of a lower charge, such as K^+ for A sites or $Mn^{3+/2+}$ or Fe^{3+} for B sites, are compensated by oxygen vacancies which pin the domain walls and result in lower dielectric constants and losses, a higher mechanical quality factor

C. Galassi et al. (eds.), Piezoelectric Materials: Advances in Science, Technology and Applications, 75–86.
© 2000 *Kluwer Academic Publishers. Printed in the Netherlands.*

and higher ageing rates (electrically "hard" PZT). Isovalent substitutions with ions of the same valence and similar size, such as Ba^{2+} or Sr^{2+} for A sites or Sn^{4+} for B sites, tend to inhibit domain reorientation and reduce the Curie temperature. Several other additives reduce the sintering temperature. These include low-melting additives [5] complex oxide additives [6], or by forming pseudoternary solid solutions [7].

Processing of PZT materials is mainly based on conventional ceramic powder technology, the perovskitic phase being obtained by the solid state reaction of the starting oxides. For most applications high density and homogeneous microstructure are among the most important features that must be achieved. Several limiting factors come into play, mostly related to the high volatility of PbO at elevated temperatures [8]: difficulty to achieve sintered densities close to the theoretical ones [9], nonstoichiometry in composition as well as compositional fluctuations [10] and poor microstructure. The use of ultrafine starting materials, produced by wet chemistry methods [11-13] or even mechanochemical synthesis [14] allows the lowering of the calcining temperature by hundreds of degrees, chemical homogeneity and reduction of agglomerate hardness. However it is not economical for most applications of bulk PZT. Critical steps are the milling routes prior to cold consolidation and the sintering conditions, because with a conventional sintering process, the full density is rarely achieved unless special techniques are used such as the use of a oxygen atmosphere, the addition of excess PbO to compensate for lead volatilization or hot pressing.

In the present study optimisation of the conventional processing steps was investigated starting from an evaluation of several raw materials, the effects of milling procedure and heating profiles. Two compositions were chosen, "soft" PZT $Pb(Zr_{0.52}, Ti_{0.48})O_3$ doped with niobium, widely used in the mass production of components, and "hard" PZT with a complex composition (PLN-PMN-PZT) showing excellent Q_m values while retaining high coupling factors.

2. Experimental

2.1. SOLID STATE REACTION: ROLE OF RAW MATERIALS

The optimisation of the performance of a "soft" PZT material of the composition $Pb_{0.988}(Zr_{0.52}, Ti_{0.48})_{0.976}Nb_{0.024}O_3$ was first carried out by investigating the role of raw material properties within a processing route defined as the "standard process":

- wet ball milling of the stoichiometric amount of the components for 48h,
- freeze drying of the suspension,
- heat treatment at 850°C for 4 h, and sieving at 250 µm,
- uniaxial pressing at 75 MPa (a few samples were isostatically pressed at 150 MPa) into disks (diameter 22 mm, thickness 2 mm),
- final sintering at 1200°C for 2 h.

One sample which showed higher green and sintered densities if heat treated at 800°C for 4 h and 1200°C for 1 h respectively, is reported for comparison. In order to avoid loss of PbO, which is fast at temperatures over 1000°C, all sintering processes were carried out with the samples placed on a ZrO_2 disk and covered with an Al_2O_3 crucible and sealed with ($PbZrO_3$ + 5wt% excess ZrO_2) powder to maintain a constant PbO

TABLE 1. Batches composition

| Batch | Oxide | | | | | | |
| | PbO | | ZrO$_2$ | | | | TiO$_2$ | Nb$_2$O$_5$ |
	Merck	Aldrich	Harshaw SP 103	Harshaw SP 102	Tosoh TZ-0	Mel SC 101	Degussa P 25	Fluka
1	✓		✓				✓	✓
2	✓			✓			✓	✓
3	✓				✓		✓	✓
4	✓					✓	✓	✓
5		✓			✓		✓	✓
6		✓				✓	✓	✓

activity at the sintering temperature [8]. The composition of the batches with trade names (labels) are reported in Table 1; purity and other properties of the raw materials are shown in Table 2. For some batches a milling step (wet ball milling for 100 h, freeze drying and sieving at 150 μm) was introduced after calcination. The freeze drying step to remove water from the wet milled suspensions was introduced because very soft powders are obtained without hard agglomerates so that the cold consolidation and the final microstructure are strongly enhanced. Specific surface area (S.S.A.) and mean particle size (d$_{50}$) of the powders, green and sintered density, linear shrinkage and weight loss of the samples are reported in Table 3. The phases present and microstructure were investigated by X-Ray Diffraction analysis (XRD) and Scanning Electron Microscopy (SEM) of the as calcined powders and polished fracture surfaces of the sintered samples respectively.

2.2. POWDER MILLING AS A CRITICAL INTERMEDIATE STEP

The development of the hard PZT material was extremely dependent on the intermediate milling steps especially when a very complex composition, such as [Pb(Li$_{0.25}$Nb$_{0.75}$)]$_{0.06}$[Pb(Mg$_{0.33}$Nb$_{0.67}$)]$_{0.06}$[Pb(Zr$_{0.50}$Ti$_{0.50}$)]$_{0.88}$O$_3$ (PLN-PMN-PZT) with the addition of 0.7 wt% MnO$_2$, was prepared. The raw material were the same as batch 6, above including MgO (Carlo Erba), Li$_2$CO$_3$ (Merck) and MnO$_2$ (Fluka) [15]. The processing route was the standard one defined previously except for: 72 h initial ball milling, dry milling in an agate mortar after calcination (sample A), wet milling for 100 h after calcination (sample B), or pre-grinding of the coarse MnO$_2$ powder (sample C).

TABLE 2. Characteristics of the raw materials

Oxide	Producer	Purity (%)	S.S.A. (m^2/g)	Grain size (μm)	d$_{50}$ (μm)	Crystalline Phase
PbO	Merck	99.0	0.2	8.11	===	orthorhombic
	Aldrich	99.9	0.8	===	3.9	orthor./tetrag.
ZrO$_2$	Harshaw SP 103	99.7 .	14.0	0.08	1.0	monoclinic
	Harshaw SP 102	99.7	28.0	0.04	0.6	monoclinic
	Tosoh TZ-0	99.8	14.0	0.07	0.2	monoclinic
	Mel SC 101	99.7	23.0	0.04	0.9	monoclinic
TiO$_2$	Degussa P 25	99.0	48.4	0.03	0.2	tetragonal
Nb$_2$O$_5$	Fluka	99.9	1.2	1.01	11.0	monoclinic

TABLE 3. Properties of the samples (calcination 850°C for 4 h, die pressing 75 MPa, sintering 1200°C for 2 h)

Batch	S.S.A. (m^2/g)	d_{50} (μm)	Green density (g/cm^3)	(%)	Sintered density (g/cm^3)	(%)	Shrinkage (%)	Weight loss (%)
1	=	1.25	4.59	57.3	7.50	93.7	13.7	1.1
1M	=	0.99	4.48	56.0	7.70	96.1	15.2	1.1
2	1.3	1.06	4.60	57.5	7.80	97.4	16.4	0.8
3	2.4	0.74	4.21	52.6	7.54	94.2	15.8	0.8
3c	2.4	0.74	4.67	58.3	7.64	95.4	13.6	0.8
4	1.8	1.02	4.58	57.2	7.70	96.2	15.1	1.0
4M	1.8	0.86	4.62	57.7	7.93	99.1	14.9	1.1
4Mc	1.8	0.86	4.88	61.9	7.87	98.3	12.0	0.8
5c*	=	0.58	4.58	57.2	8.00	100.0	15.6	0.8
5c	2.7	0.69	4.56	57.0	7.84	97.9	14.9	0.9
6c	2.4	0.68	4.92	61.5	7.93	99.1	13.0	0.9
6Mc	=	0.56	4.89	61.1	7.99	99.9	13.8	1.2

M: ball milling 100 h after calcination; c: cold isostatically pressed 150 MPa; (*): calcination at 800°C for 4 h and sintering at 1200°C for 1 h.

2.3. PIEZOELECTRIC PROPERTIES

In order to measure the piezoelectric properties of the sintered materials of the soft and hard compositions, disks with a length to diameter ratio >10 as recommended by the IEEE standards were prepared. The samples were ground to remove surface layers, screen printed with silver electrodes, fired at 700 °C and finally poled in silicon oil at 120°C, under a dc field of 3 kV/mm for 40 min. The electrical impedance was measured as a function of frequency by using an HP4194A impedance bridge. From the values of resonance and antiresonance frequencies of the planar mode, together with the minimum impedance, capacitance measured at 100 kHz, density and geometrical dimensions, the piezoelectric constants k_{31}, d_{31} and g_{31}, the elastic coefficients s_{11}^{E} and s_{12}^{E}, as well as the electromechanical coupling factor k_p and the mechanical quality factor Q_m were determined and are reported in Table 4 and Table 5.

TABLE 4. Piezoelectric properties of the soft PZT samples

Sample	ρ (g/cm^3)	σ_P	k_P	k_{31}	d_{31} $(10^{-12}m/V)$	g_{31} $(10^{-3}Vm/N)$	S_{11}^{E} $(10^{-12}m^2/N)$	Q_m
1	7.50	0.36	0.51	0.29	111.3	15.8	15.8	76.8
1M	7.70	0.36	0.54	0.31	117.8	12.1	15.3	92.3
2	7.80	0.34	0.62	0.35	161.5	11.0	14.2	99.9
3	7.54	0.38	0.60	0.33	144.1	12.7	16.6	90.4
3c	7.64	0.37	0.58	0.32	131.7	13.0	16.3	74.5
4	7.70	0.36	0.63	0.36	143.4	14.3	16.1	80.9
4M	7.93	0.37	0.65	0.37	148.7	14.6	16.2	83.7
4Mc	7.87	0.35	0.57	0.33	135.5	11.7	15.0	77.5
5c*	8.00	0.36	0.69	0.39	171.0	14.8	16.6	=
5c	7.84	0.36	0.70	0.40	182.7	14.9	17.2	88.5
6c	7.93	0.38	0.70	0.39	189.7	14.1	17.5	83.3
6Mc	7.99	0.33	0.70	0.40	183.6	14.6	16.5	83.2

TABLE 5. Piezoelectric and dielectric properties of the hard PLN-PMN-PZT samples

Sample	ρ (g/cm^3)	Weight loss (%)	σ_p	k_p	k_{31}	d_{31} (10^{-12}m/V)	g_{31} (10^{-3}Vm/N)	s_{11}^E (10^{-12}m^2/N)	Q_m
A	7.88	0.5	0.40	0.32	-0.18	-50.0	-8.2	13.0	380
B	7.90	0.4	0.31	0.54	-0.32	-64.7	-16.3	10.6	2132
C	7.86	0.7	0.30	0.54	-0.34	-69.0	-16.3	10.9	994

A: agate mortar milling after calcination, d_{50} = 1.22 μm. B: wet milling (100 h) after calcination; d_{50}=0.55μm.
C: sample A with MnO$_2$ pre-ground.

3. Results and Discussion

3.1 EFFECT OF RAW MATERIALS

The PZT batches are different only for the nature of the commercial ZrO$_2$ and PbO, TiO$_2$ and Nb$_2$O$_5$ being the same. The starting powder morphology greatly influences the extent of the solid state reaction together with the morphology of the calcined powder. The specific surface area of ZrO$_2$ SP102 is double that of SP103, while TZ-0 is finer and SC101 shows similar properties to SP102. PbO Aldrich powder is finer than Merk PbO and is partially in the tetragonal phase (see Table 2). The particle size distributions of the as calcined powders are reported in Figure1.

Batches 1-4 reflect the characteristic of the starting zirconia; in fact increasing the S.S.A. of ZrO$_2$ (batch 1 and 2) produces a significant reduction of the mean particle size of the calcined powder. The introduction of a powder with a smaller particle size

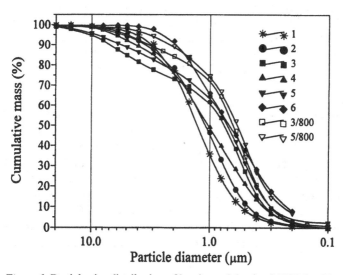

Figure 1. Particle size distribution of batches calcined at 850°C for 4 h (the two batches calcined at 800°C are mentioned in the graph).

Figure 2. SEM morphology of the as calcined powders

(Tosoh TZ0) allows a strong reduction of the mean particle size of the perovskitic phase (batches 3, 5) due to the improved reactivity that is further enhanced by the introduction of finer and purer PbO (batches 5 and 6). In these batches a very similar mean particle diameter is shown, but with a larger distribution for powder with Tosoh ZrO_2. Further tests verified that the optimal conditions of calcination, only for batch 3 and 5, is at 800°C, because the higher temperature (850°C) produces an enlargement and hardening of the agglomerates (Figure 1) with a low green density of the die pressed samples. SEM morphologies of selected batches are reported in Figure 2.

The XRD patterns of the as calcined powders show the absence of the pyrochlore phase and almost no sample is fully reacted to the perovskitic phase, with small quantities of PbO and $PbTiO_3$ still being present. Improvement of the cold consolidation behaviour of the powders was attempted by milling the as calcined powders and/or by cold isostatic pressing the die pressed samples (Table 3). The ball milling process reduces the mean particle size but does not produce significant effects on the green density which is often slightly reduced, whereas a strong increase in the green density is obtained by isostatically pressing the samples (compare 3 to 3c and 4M to 4Mc where a 6% increase of the green density is obtained).

The sintered densities of the samples reflect both the cold compactability of the calcined powders, i.e. agglomerate size distribution and hardness, and the intrinsic reactivity.

Powder 1 results in a lower final density despite its good green density (57.3%) moreover the milling step, introduced in the powder 1M, does not improve substantially the sinterability. Powder 3 sinters to 94% of the theoretical density, despite of its low green density. In the standard conditions (calcination 850°C for 4 h

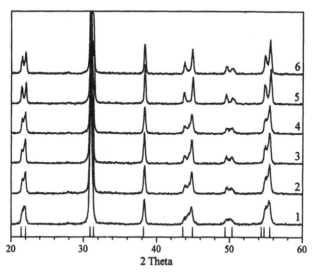

Figure 3. XRD patterns of samples 1-6 sintered at 1200°C for 2 h.

and sintering 1200°C for 2 h) only powder 4M, 6c and 6Mc reach final densities higher than 99%. One hour soaking time at the maximum temperature (1200°C) is enough to sinter powder 5 to the almost theoretical density (8.00 g/cm^3), while two hours soaking results in a significant decrease of the final density (7.84 g/cm^3). The XRD patterns of the external upper surface of sintered samples are shown in Figure 3; all the samples show the presence of pure perovskitic phase with different relative amounts of tetragonal and rhombohedral symmetry. The peaks of the tetragonal phase are all doublets except for the reflection 111 (2θ ≅ 38) while the rhombohedral one holds singlets almost in the same positions. Sample 1 shows the higher amount of rombohedral phase, samples 2 to 4 show different degrees of splitting, while samples 5 and 6 show well defined doublets.

SEM morphologies of the fracture surface of the sintered samples reveal a non homogeneous microstructure in sample 1 (Fig. 4 a) with the presence of a relatively quantity of small grains (mean diameter 0.3 µm) in a matrix of bigger grains (mean diameter 2 µm) with a partially transgranular fracture mode. As the purity and "reactivity" of the starting powders increase, the presence of small grains decreases (Fig. 4 b), so that a very homogeneous microstructure with well faceted grains is shown by samples 6 (Fig. 4 c) and sample 5 (Fig. 5 a); the application of more severe heat treatment conditions in sample 5 result in a slight increase in mean grain size and in the change in the fracture mode from intergranular to mainly transgranular (Fig. 5 b).

Figure 4. SEM morphologies of the fracture surface of the sintered batch 1 (a), batch 2 (b) and batch 6 (c).

Figure 5. SEM morphologies of the fracture surface of the sintered samples batch 5 (a) 800°C for 4h, 1200°C for 1h; (b) 850°C for 4h, 1200°C for 2h.

The differences in microstructure, as well as in the XRD spectra can be primarily discussed in terms of the reactivity of the starting powders that arises from their primary particle and agglomerate size. In fact, the more reactive is the ZrO_2 (compare samples 1 and 2) or PbO (compare samples 4 and 6 or 3 and 5), the lower is the mean particle diameter and the higher is the final density although the cold consolidation is enhanced by a broader particle size of the starting powders. In fact sample 2 shows higher final density than 1 despite of the same green density; ball milling enhances the final density while the green density remains almost the same.

The comparison of samples 3c and 5c shows that although the green density of sample 5c is lower its sintered density is significantly higher, due to the finer and purer starting lead oxide. Sample 6, in comparison with sample 5, shows the best performance if processed in our standard conditions, owing most probably to the better cold compaction behaviour (61.5% relative green density of sample 6c against 57.5% for sample 5c). Thanks to the contribution of the very fine and pure ZrO_2 powder the optimum calcining condition of sample 5 is at lower temperature (800 °C), that results in a strong reduction of the broadening of the particle size (see Fig.1); the green density remains the same, while the samples sinter to full density at 1200°C 1 h, with very soft processing parameters for a PZT material prepared with the conventional solid state reaction. From the XRD analysis of the reacted powders, although the composition is in the tetragonal side near the MPB, the rhombohedral phase seems to prevail and it can be attributed to the poor reactivity of ZrO_2 that slows down its reaction with lead titanate to the stoichiometric composition; so that a composition in the rhombohedral side is present together with lead titanate and residual PbO, until the reaction is almost complete. The presence of unreacted oxides probably promotes sintering by enhancing the formation of the transient liquid phase and their amount govern the differences in microstructure after sintering. SEM morphologies of the sintered samples shown in Fig. 4 and 5 reflect the improvement in the raw materials quality, the more homogeneous, being samples 5 and 6 which show well faceted and with a narrow size distribution grains.

Two conclusions can be drawn from the present investigation: 1) the rhombohedral phase persists as the temperature is increased and only in the final stage of sintering is the pure tetragonal phase formed. It is shown by the presence of the rhombohedral phase in a sintered sample 1(Fig. 3) made of the less reactive raw materials as well as in samples 2, 3, and 4 (or sample 6 soaked 1 h only at 1200 °C). Samples 5 and 6 show

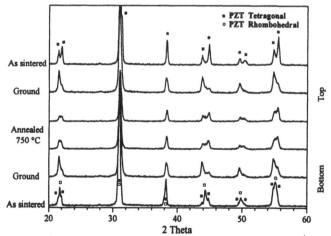

Figure 6. XRD patterns of bottom and top surfaces of sample 6: as sintered at 1200°C for 2h, after grinding and after further annealing at 750°C for 10 min.

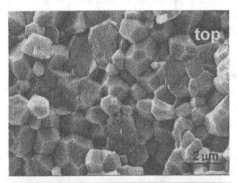

Figure 7. SEM morphology of the fracture surface of sample 6 investigated through the section.

almost pure tetragonal phase; 2) there is no homogeneity between the surface and the core of the samples. If XRD patterns are recorded on the bottom of the pellet (in contact with the zirconia substrate) top or core, it is evident that the core is homogeneous with prevailing tetragonal symmetry (but the presence of the rhombohedral phase can not be excluded), while the bottom surface is mostly rhombohedral (Fig. 6).

The non-homogeneity of the sample 6 is confirmed by the SEM morphology of the fracture surface (Fig. 7). It could be attributed to the different atmosphere surrounding the sample; the contact with the ZrO_2 setter in the bottom side prevents the interaction with the lead atmosphere, resulting in a slowed transformation to the pure tetragonal phase, in the soft processing conditions chosen. Grinding to a depth of about 30 μm, achieves identical upper and lower surfaces (Fig.6), but induces surface texture evidenced by the increase of the ratio of the intensities of the peaks I_{002} ($2\theta = 44°$) to I_{200} ($2\theta = 45°$) ratio that decreases again after annealing [17, 18].

The piezoelectric properties of the materials (Table 4) reflect the improvement of microstructural properties of the batches, showing that samples 5 and 6 have better properties than reported in literature for similar compositions [1].

3.2   POWDER MILLING AS A CRITICAL INTERMEDIATE STEP

As previously discussed, the ball milling process of the calcined powders improved the sintered density significantly while the final properties were only slightly enhanced (compare for example samples 1 and 1M or 4 and 4M).

Therefore, it is a critical step for the processing of a more complex composition. In fact the phase evolution towards the pure tetragonal phase, as expected for the chosen composition, and already discussed previously, is influenced depending on the efficiency of the milling procedures chosen. This results in significantly different final piezoelectric properties as can be seen in Table 5.

Figure 8. SEM morphology of the sintered PLN-PMN-PZT samples starting from crushed powder (sample A), 100 h ball milled powder (sample B), sample A with pre-ground MnO_2 (sample C).

Although not affecting significantly the sintered density of the samples the milling efficiency induces a different distribution of the dopants, in particular MnO_2, so that the sintering mechanism is modified therefore influencing, to a different extent, the microstructure (Fig. 8) and two important properties (electromechanical coupling factor K_p and mechanical quality factor Q_m) of the material. Sample A, poorly milled and with coarse MnO_2 shows low K_p and low Q_m, samples B and C show the same, higher, K_p and higher Q_m, but sample B presents really improved Q_m values, as expected. 100 h milling allows most probably a better distribution of Mn oxides in the perovskitic lattice so enhancing its acceptor effect with the formation of oxygen vacancies and an increase of Q_m. When MnO_2 is crushed but a only poorly efficient milling procedure is introduced after calcination, its finer subdivision allows it to participate on the formation of the intermediate liquid phase, so enhancing grain growth (sample C) (Fig.8).

4. Conclusion

The morphology and reactivity of the starting oxides strongly affects the microstructure and piezoelectric properties of PZT materials obtained by solid state reaction, as well as the processing steps, in particular intermediate ball milling, that were examined in this investigation.

Very high quality "soft" PZT can be obtained by calcining starting oxides at 800°C for 4 h and sintering at 1200°C for 1 h. Intermediate ball milling let us enhance substantially the mechanical quality factor of an "hard" PZT. The microstructure and inhomogeneities in the phase distribution were discussed in terms of the morphological and physical properties of the starting powder and the dopant distribution related to milling efficiency.

5. References

1. Jaffe, B., Cook, W.R., Jaffe, H. (1971) *Piezoelectric Ceramics*, Academic Press, London.
2. Wersing, W., Rossner, W., Eckstein, G., Tomandl, G. (1985) The Morphotropic Phase Boundary in PZT Ceramics Prepared by Spray Drying of Salt solutions and by the Mixed Oxide Method, *Silicates Industriels* **3-4**, 41-46.
3. Haertling, G.H. (1986) Piezoelectric and Electrooptic Ceramics, in R.C.Buchanan (ed.), *Ceramic Materials for Electronics*, Marcel Dekker Publisher, New York, pp.135-225.
4. Atkin, R.B., Holman, R.L., Fulrath, R.M. (1971) Substitution of Bi and Nb ions in Lead Zirconate-Titanate, *J. Am. Ceram. Soc.* **54** [2], 113-115.
5. Witter, D.E., Buchanan, R.C. (1981) Low-Temperature Densification of Lead Zirconate-Titanate with Vanadium Pentoxide Additive, *J. Am. Ceram. Soc.* **64** [8], 485-490.
6. Kaneko, S., Dong, D., Murakami, K. (1998) Effect of simultaneous Addition of $BiFeO_3$ and $Ba(Cu_{0.5}W_{0.5})O_3$ on lowering of Sintering Temperature of $Pb(ZrTi)O_3$ Ceramics, *J. Am. Ceram. Soc.* **81**[4], 1013-1018.
7. Moon, J.H., Jang, H.M. (1993) Densification behaviours and piezoelectric properties of MnO_2, SiO_2-doped $Pb(Ni_{1/3}Nb_{2/3})O_3$-$PbTiO_3$-$PbZrO_3$ ceramics, *J. Mat. Res.* **8** [12], 3184-3191.
8. Kingon, A.I., Clark, J.B., (1983) Sintering of PZT Ceramics:I, Atmosphere Control, *J. Am. Ceram. Soc.* **66** [4], 253-256.
9. Kingon, A.I., Clark, J.B., (1983) Sintering of PZT Ceramics:II, Effect of PbO Content on Densification Kinetics, *J. Am. Ceram. Soc.* **66** [4], 256-260.
10. Fernandez, J.F., Moure, C., Villegas, M., Duran, P., Kosec, M., Drazic, G. (1998) Compositional Fluctuations and Properties of Fine- Grained Acceptor-Doped PZT Ceramics, *J. Eur. Ceram. Soc.* **18**, 1695-1705.
11. Lencka, M.M., Anderko, A., Riman, R. (1995) Hydrothermal Precipitation of Lead Zirconate Titanate Solid Solutions: Thermodynamic Modeling and Experimental Synthesis, *J. Am. Ceram. Soc*, **78** [10], 2609-2618.
12. Choy, J-H., Han, Y.S., Kim, J.T. (1995) Hydroxide Coprecipitation Route to the Piezoelectric Oxide $Pb(ZrTi)O_3$ (PZT), *J. Mater. Chem.* **5** [1], 65-69.
13. Kim, H.B., Lee, J.-H., Park, S.J. (1995) Preparation of spherical $Pb(ZrTi)O_3$ powders by ultrasonic spray pyrolysis, *J. Mat. Sci.:Materials in Electronics* **6**, 84-89.
14. Xue, J., Wan,D., Lee, S.-E., Wang, J. (1999) Mechanochemical Synthesis of Lead Zirconate Titanate from Mixed Oxides, *J. Am. Ceram. Soc.* **82** [7], 1687-1692.
15. Galassi C., Roncari E., Capiani C., Craciun F. (1999) Processing and characterization of high Q_m ferroelectric ceramics, *J. Eur. Ceram. Soc.* **19** [6-7], 1237-1241.
16. Mehta, K.and Virkar, A. V. (1990) Fracture Mechanics in Ferroelectric-Ferroeleastic Lead Zirconate Titanate (Zr:Ti= 0.54:0.46) Ceramics, *J. Am. Ceram. Soc.* **73** [3], 567-574.
17. Cheng, S. and Lloyd, I.K. (1992) Modification of Surface Texture by Grinding and Polishing Lead Zirconate Titanate Ceramics, *J. Am. Ceram. Soc.* **75** [8], 2293-2296.

SMART STRUCTURES BY INTEGRATED PIEZOELECTRIC THIN FIBERS

D. SPORN[1], W. WATZKA[1], A. SCHÖNECKER[2] AND K. PANNKOKE[3]

[1] Fraunhofer-Institut fuer Silicatforschung ISC, Neunerplatz 2,
D-97082 Wuerzburg, Germany

[2] Fraunhofer-Institut fuer Keramische Technologien und Sinterwerkstoffe IKTS, Winterbergstrasse 28, D-01277Dresden, Germany

[3] Fraunhofer-Institut für Fertigungstechnik und Angewandte Material-forschung IFAM, Wienerstrasse 12, D-28359 Bremen, Germany

Abstract

A new type of multifunctional composite material has been established by integrating parallel arrays of tiny piezoelectric fibers with interdigital electrodes into polymer matrices. This follows the approach of structural conformity to maintain the mechanical properties of the composite materials. The access to highly integrated materials consisting of at least four solid phases is based on the combination of fiber technology, electrodizing, composite fabrication and adaptive control. The sensing as well as the actuating properties of the resulting materials are evident. This offers a wide range of possible applications like active and passive vibration damping, health monitoring, structural control and impact detection.

1. Introduction

Over the past 20 years, material science has been searching for novel, useful, highly integrated and efficiently working composite materials. In several approaches nature has become a model for engineers to develop a new type of materials, namely multifunctional composite materials (also called adaptive or smart materials). Effective biological structures have developed during the evolution of life. They are able to adapt efficiently to changing environmental conditions in their habitat. This ability is based on incorporated sensing as well as actuating functions, both connected by signal paths and controlled by e.g. a brain as processing unit. As soon as the sensors detect an impact, they generate a signal which is transmitted to the brain. This receiver in return sends a control signal which releases an adequate actuator reaction.

To copy biological structures is not possible at present because of the very high degree of integration, excellent efficiency of the sensor/actuator response and the capability of damage repair. Nevertheless, the idea to mimic biological principles for multifunctional materials, e.g. by combining active solid materials with polymers in well

87

C. Galassi et al. (eds.), Piezoelectric Materials: Advances in Science, Technology and Applications, 87–97.
© 2000 *Kluwer Academic Publishers. Printed in the Netherlands.*

organized architectures, came up in the early 80´s [1]. As active materials, electroceramics are the most promising candidates [2].

The first approach was to tailor the properties by combining active and passive materials in defined architectures. Then, it was suggested to integrate active materials into composite materials [3]. Such „smart materials" consist of 3 or more solid phases and can be potentially used in technical applications. They should contain sensors, actuators and controllers to make the material adaptive.

Amongst all known functional materials with useful sensor/actuator coupling the piezoceramics in the system $Pb(Zr,Ti)O_3$ (PZT) including substitutions and dopants occupy a special position. These materials offer both sensing as well as actuating behaviour although they exhibit a relatively small strain in the range from 1 to 1.5 ‰ and need high electric driving fields of 20 kV/cm. However, they are suitable for dynamic and precise positioning, their energy density is exceeded only by shape memory alloys and biological muscles [4] and they can develop stresses up to 3 kN/cm^2.

The development of tiny functional components like fibers opens new opportunities in material design in the sense of structural conformity. Structural conformity means that the properties of fiber-reinforced composite materials (high specific strength and stiffness as well as high strain to failure) should not be deteriorated by the integrated functional components and their wiring. The consequence is to minimize the geometric dimensions of the functional components as long as they can maintain their functionality as well as their controllability. According to this concept the integration of functional components in form of thin fibers seems to be a promising approach.

The development of highly integrated composite materials is a matter of fusion and extension of one new and at least 4 existing technologies: the PZT-fiber preparation is a new technology, whereas electroding, composite fabrication and adaptive control have to be adjusted to new requirements.

Using this approach PZT fibers can be applied as integrated compression as well as tension sensors. Active and passive vibration damping, health monitoring and control of geometry are even more important fields of application [5,6]. Last but not least, PZT fibers with small diameters will give access to new types of ultrasonic transducers [7].

2. Our approach: structural conformity

The developmental steps from passive structures towards highly integrated adaptive structures are the following: as soon as a passive structure is affected by disturbing forces vibrations and noise will be generated. If those structures are equipped with sensors and actuators, both connected with a controller, vibrations and noise can be actively damped. However, the structure becomes heavier due to the additional external components because piezoelectric structural actuator systems are traditionally based on monolithic planar materials in stacked architectures. Structures of this type are already available on the market [8]. The next consequent step is the integration of sensors and actuators within the structure, the adaptive controller still being external. Such systems can be designed to be free of vibrations, quiet and stable, but still being light weight! In the highest degree of integration the controller is also contained within the composite.

Structures of the last both types generate new requirements on the design of sensing as well as actuating components: if the mechanical properties of the composite material are to be affected as little as possible, then their geometric dimensions have to be minimized resulting in "structural conformity". What does this mean?

The mechanical properties of fiber-reinforced composite materials are determined by the type and content of reinforcing as well as functional fibers. In order to calculate qualitatively at which diameters the functional fibers do not interfere with the properties of fiber-reinforced composites, it makes sense to compare the elastic behaviour of fibers of different types and diameters. One useful measure are their bending moments. The bending moment M can be calculated qualitatively for cylindrical shapes - like fibers - by the formula:

$$M \approx E \cdot d^3 / 32 \qquad (1) [9]$$

where E = Young´s modulus, d = fiber diameter.

The formula tells that smaller fiber diameters result in smaller bending moments. To fulfill structural conformity, it can be concluded that the bending moments of reinforcing as well as piezoelectric fibers with different diameters and Young´s moduli should ideally be quite similar. As a realistic case a glass fiber-reinforced composite (GFC) can be considered: the Young´s moduli of glass and PZT are comparable, consequently their diameters should not differ. Another consequence can be drawn from this consideration: in the case of carbon fiber-reinforced composites with a distinctly higher Young´s modulus and a diameter of 6 μm of C-fibers PZT fibers with a diameter smaller than 20 μm should be used.

From this qualitative consideration the conclusion can be drawn that a technology is needed that allows to produce PZT fibers with a final diameter of at most 30 μm or smaller. By using such fibers structural conformity can be achieved with consequences for the improved lifespan and the reliability of adaptive composites.

The preparation and integration of PZT fibers with such small diameters require a technology which permits to handle the fibers carefully during all production steps, to align them precisely in an ordered architecture and to electrodize them. The electrodizing process as such requires microtechniques which are still under development.

3. Preparation of piezoelectric ceramic fibers

The preparation of piezoelectric fibers with diameters < 30 μm has been started in the 90´s [10], but is still a challenge for material technology. In the literature at least three different routes have been reported. They differ essentially in the precursors used for their preparation, the achievable diameters in the sintered state and in the sintering temperatures themselves.

Fiber preparation routes based on the extrusion of polymer-supported PZT powder suspensions cannot produce final fiber diameters smaller than 125 μm [11], but they are used to develop Active Fiber Composites (AFC) for an integratable actuator [12].

Another fiber preparation route has been reported which starts with a PZT powder suspension based on a cellulose xanthogenate solution (VSSP method = viscous solution spinning process). Final fiber diameters of 20 μm have been achieved, but the necessary sintering temperature is so high (> 1250 °C) that the fibers sinter together and cannot be handled as single fibers [13]. The result is a highly porous aggregate of PZT "fibers" which can be used only for the development of acoustically adapted materials, e.g. for hydrophones [14]. It was recently reported that single fiber diameters in the range of 10 to 25 μm have been achieved by the VSSP method [15].

The third route based on the sol-gel process is favourable, because in this way diameters < 30 μm fibers sinterable at lower temperatures have been achieved [15-17]. Due to the comparably low sintering temperatures (< 1000 °C) it is possible to fabricate fibers which can be handled in the sintered state as single filaments.

Sol-gel-derived PZT fibers have been used for the following steps [16]. The fiber stoichiometry presently used is adjusted to lead to the final composition $Pb(Zr_{0.53}Ti_{0.47})O_3$. This composition is not yet the optimum for the intended applications. A "hard" PZT material exhibiting a high coercitivity and low piezoelectric charge constant results. This means that its poling needs high electric fields up to 50 kV/cm at elevated temperatures, and a relatively small charge will be generated under compressive or tensile load.

Sol-gel-derived PZT fibers show an extraordinarily high sintering activity. They can be completely densified at temperatures < 950 °C (Figure 1) and can be handled as single fibers after sintering.

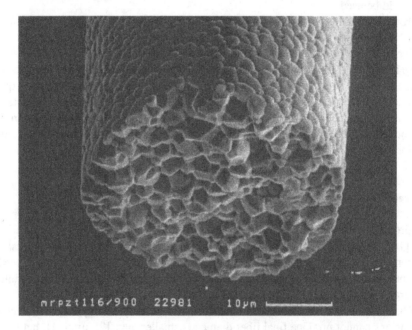

Figure 1. SEM micrograph of PZT fibers with the composition $Pb(Zr_{0.53}Ti_{0.47})O_3$
sintered at 925 °C [16].

The sintered fibers exhibit the targeted stoichiometry, the microstructure is pore free and consists of grains between 2 and 4 μm in diameter. The measured tensile strength is in the range of up to 300 MPa, Young´s modulus was estimated to be 70 GPa.

PZT fibers are not suitable to take over load bearing functions within the composite material. Their elastic properties cannot be improved for structural reasons, the tensile strength is - among other facts - a function of the failure probability within the microstructure, particularly on the fiber surface. For electromechanical reasons a coarse microstructure is preferred leading to a grooved fiber surface. These grooves are sources of local stress concentrations. Due to these reasons the fiber strength is limited.

Conclusions can be drawn concerning the structural arrangement of composite materials with integrated and electrodized PZT fibers: the resulting material will consist of at least four phases. The single components are the polymeric matrix, the reinforcing fiber (glass or carbon), PZT fibers and their electrodes.

4. Electrodizing and integration

The process route to active fiber laminates is shown in Figure 2. The structure of the electrodes has to allow that unidirectionally (UD) aligned PZT fibers can be poled and then addressed. The fiber integration should affect the structure of the composite as little as possible. The access to the internal electrodes is necessary to allow the in- and output of charges.

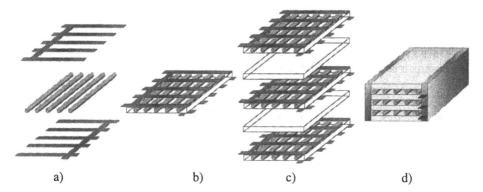

Figure 2. Scheme of a fiber-reinforced, laminated composite with integrated PZT fiber single sheets:
a) electrodizing of the UD piezofiber single sheet, b) electrodized piezofiber single sheet
c) piezofiber single sheets between GFC or CFC sheets, d) active laminate with electrodes.

The use of active single sheets can ensure a high degree of structural conformity but generates high demands on the PZT fiber technology. The gel fibers have to be precisely aligned in UD architectures; this geometry is not allowed to change over the whole process. It has to be taken into account that a linear shrinkage of the fibers of about 40 % occurs due to pyrolysis/oxidation and sintering. Sintered PZT fibers in aligned UD geometry can now be produced by careful process control. Only very small deviations within the fiber array can be tolerated, because the width and thickness of the electrodes as well as the distance between the electrodes must be kept with a tolerance below 5 %. In Figure 3 a real fiber/electrode assembly is shown.

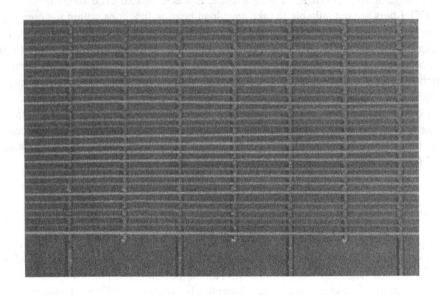

Figure 3. Real PZT fiber/electrode structure, electrode distance = 1 mm
(preparation: Fraunhofer IFAM).

The piezoelectric fibers in this array successfully have been poled. The sheets can be integrated on or between GFC sheets.

5. Properties of composites with integrated PZT fibers

5.1. PZT FIBERS AS INTEGRATED SENSOR

PZT fibers as described above were integrated within single sheets with a volume content of about 30 %. Such active sheets have been embedded in a GFC plate with a dimension of 20 x 20 cm^2. After that the PZT fibers can be poled with electric fields between 30 and 50 KV/cm at a temperature of 120 °C and behave anisotropically, as expected. As soon as the embedded PZT fiber array comes under a mechanical load charges are generated. The complete GFC plate is covered by a net of 81 measuring points in a regular manner. A mechanical load of 14.3 g has been dropped from a height of 15 cm on each of these points resulting in an impact energy of 0.02 Nm. The impacts generate a figure of measured charges as shown in Figure 4. This experimental evidence demonstrates that impact sensors can be built by tiny, integrated PZT fiber/electrode arrays.

A sensitivity of about 40 pC/N was calculated from the measurement of the generated voltage. The need for the change to a "soft" PZT composition as mentioned above is now evident. The charges generated are directly proportional to the piezoelectric charge constant d_{33} (d_{33} (hard) 220 pC/ N, d_{33} (soft) 600 pC/ N). With the change to soft PZT fibers the sensitivity of the sensor will be nearly tripled.

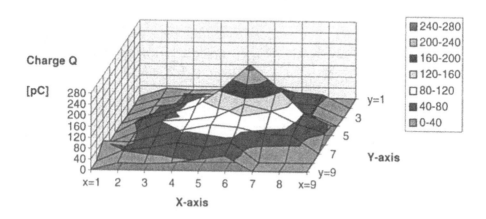

Figure 4. Charge diagram measured on a GFK sheet (20 x 20 cm^2) which contains poled PZT fibers after local impacts with an energy of 0,02 Nm (measurement: Fraunhofer IKTS).

5.2. PZT FIBER ARCHITECTURES FOR PASSIVE VIBRATION DAMPING

The same PZT fiber architecture as shown in Figure 2d also allows the damping of mechanical vibrations. As soon as poled PZT fibers in such arrays undergo compression or tension they generate charges by the direct piezoelectric effect. These electrical charges can be collected by the electrodes and can be transferred to Joule heat by shunting. By this mechanism mechanical energy is taken away from the system: it will be damped.

Derived from theoretical calculations of the possible energy transfer, comparably high damping values can be expected. Most probably damping of frequencies in the range of up to 2000 Hz will be reached. The maximum energy dissipation achievable, which is caused by a shunt after transformation of mechanical in electrical energy, can be calculated as follows:

$$\sin \delta = k^2 / 2 (1 - k^2)^{1/2} \text{ , with } \omega \varepsilon^\sigma \rho = 1 / (1-k)^{1/2} \qquad (2) \qquad [19]$$

with: k = electromechanical coupling coefficient
ω = frequency
ε = permittivity at constant mechanical stress
ρ = specific resistance

The achievable damping depends on the coupling factor of the piezoceramic (k_{33} = 0.6), the degree of coupling between the array and the source of vibration and the frequency used. At a frequency of 200 Hz the value varies between 8 and 22 % per single vibration (Figure 5). The experiment is in good agreement with the calculated form of the dissipation curve (Figure 5, cross marks).

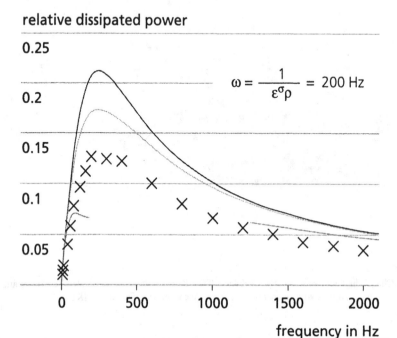

coupling of the damping element on the composite material (calculated)

— 100%, ⸱⸱⸱⸱ 10%, — 0%

Figure 5. Damping with integrated PZT fibers at a frequency of 200 Hz and different couplings between damping array and the vibrating exciter in comparison with a measurement (x) on a PZT fiber array (calculation: Dr. W. Kreher, TU Dresden, Germany, measurement: Fraunhofer IKTS).

The application of arrays, that can be integrated into composite materials in structural conformity, is now a matter of optimization of materials as well as the related technologies. This concerns the fibers themselves - here again the change of composition to a soft system is necessary - as well as their integration and control into devices.

5.3. INTEGRATED PZT FIBERS AS ACTUATOR

PZT fiber arrays as shown in Figure 3 can also be used as actuators. This has been demonstrated using a glass fiber-reinforced polymer (GFC) sheet with a laminated PZT fiber array. By applying an AC signal to the PZT fiber array the sheet is excited to vibrate audibly in its resonance frequency. The effective power P of the excited vibration mode was taken as the physical measure of electromechanical coupling. Figure 6 shows the measured function P (f) of a vibrating sheet (5 x 12 cm^2) driven by the integrated piezofiber transducer at an applied voltage of 2000 V (AC). The plotted function shows that at 400 Hz an effective amount of energy is exciting the vibration of the sheet in its characteristic vibration.

effective power [mW]

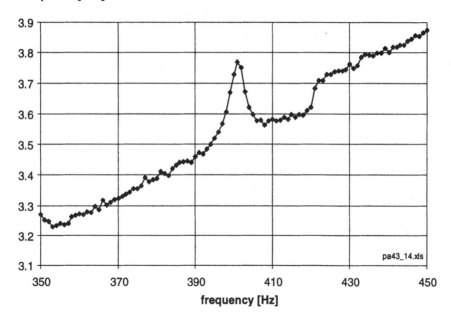

Figure 6. Plotted effective power in dependence of the applied frequency measured in a sheet which is driven by the integrated piezofiber transducer. At 400 Hz the characteristic vibration of the sheet is shown. (measurement: Fraunhofer IKTS).

This is the first experimental evidence of an actuatoric function of PZT fiber arrays. This can be seen as the starting point of material development opening wide fields of applications like active vibration damping as well as integration of actuatoric components into composites while maintaining their structural conformity.

6. Conclusions

The integration and control of PZT fibers with tiny diameters into composite materials leads to a new type of material. Tiny PZT fibers can be aligned in an array, electrodized with interdigital electrodes and then integrated. Evidence of sensoric and actuatoric properties of integrated PZT fiber architectures has been presented. The results achieved so far are a basis to explore the potentials of such materials by optimizing the fiber composition, the structural design of composite materials and - last but not least - the "intelligence" of the controller.

The fiber development is not yet finished. The next necessary steps are the optimization of the chemical composition and the microstructure with respect to the mentioned applications and the further development of the fiber preparation technology. Soft PZT formulations are now under investigation and will be available in the near future.

The fiber technology is still challenging. Our mid-term target is to develop and to establish a continuous process which can deliver larger quantities of fibers necessary for the development of components. At present the fibers are produced batchwise on a laboratory scale. We expect that "intelligent" composite materials with integrated PZT fibers can trigger the production of a large number of new products.

7. References

1. Newnham, R.E., Bowen, K.A., Klicker, K.A. and Cross, L.E. (1980) Composite piezoelectric transducers, *Mat. Eng.* **2**, 93-106.
2. Newnham, R.E. and Ruschau, G.R. (1991) Smart electroceramics, *J. Am. Ceram. Soc.* **74** 463-480.
3. Yoshikawa, S., Selvaraj, Brooks, K.G., and Kurtz, S.K. (1992) *IEEE, Proc. 8th Intern. Symp. Appl. Ferroelectr., Greenville, S.C. Aug. 30 - Sept. 2* , 269-272.
4. Hollerback, Hunter, Ballantyne (1998) A comparative analysis of actuator technologies for robotics, *MIT study*.
5. Hagood, N.W. and Bent, A.A. (1993) Development of Piezoelectric Fiber Composites for Structural Actuation", *Proc. 43th AIAA/ ASME, Adaptive Structures Forum, April 19-22, , La Jolla, CA* .
6. Schmidt, W. and Boller, Chr. (1992) Smart Structures - A Technology for Next Generation Aircraft, *75th Meeting AGARD - Structure and Materials Panel, Oct. 5.-7, , Lindau* .
7. Smith, W.A. (1989) The role of piezocomposites in ultrasonic transducers, *Proc. IEEE Ultras. Symp.*, 755-766.
8. Press release of ACX Inc., Cambridge, MA (1995).
9. Winter, G. (1972) Polykristalline, anorganische Fasern - Herstellung, Eigenschaften, Anwendung *Angew. Chemie* **84**, 866-875.
10. Selvaraj, U., Prasodarao, A.V., Komarneni, S., Brooks, K.G., and Kurtz, S., Sol-gel processing of $PbTiO_3$ and $Pb(Zr_{0.52}Ti_{0.48})O_3$ fibers, *J. Mat. Res.* **7/4**, 992-996.
11. Data sheet CeraNova Corp., Hopedale, MA.
12. Janos, B.Z. and Hagood, N.W., Overview of Active Fiber Composites Technologies, *Proc. ACTUATOR '98, Bremen, Germany*, in print.
13. Cass, R. B. (1993) Fabrication of Continuous Ceramic Fibers by Viscous Suspension Spinning Process", *Am. Ceram. Bull.* **70**, 424-429.
14. Safari, A. and Victor, V. F. (1997) *Ferroelectrics* **196 (1-4)**, 187-190.
15. French, J.D., Weitz, G.E., Luke, J.E., Cass, R.B., Jadidian, B., Bhargava, P., and Safari, A., Production of continuous piezoelectric ceramic fibers for smart materials and active control devices, *Proc. SPIE - Int. Soc. Opt. Eng.* **3044**, 406-412.

16. Glaubitt, W., Sporn, D. and Jahn, R. (1995) Sol-gel processing of PZT long fibers, *Advances in Science and Technology 10: Intelligent Materials and systems*, P. Vinzenzini, Techna Srl, Ed., 47-54.

17. Yoshikawa, S., Selveraj, U., Moses, P., Jiang, Q. and Shrout, T. (1994) *Ferroelectrics* **154**, 325-330.

18. Kitaoka, K., Kozuka, H., and Yoko, T. (1998) Preparation of lead lanthanum zirconate titanate (PLZT), $(Pb,La)(Zr,Ti)O_3$ fibers by sol-gel method, *J. Am. Cer. Soc.*, **81** [**5**], 1189-96.

19. W. Kreher, private communication

HYDROTHERMAL SYNTHESIS AND PROCESSING OF UNDOPED AND DOPED BATiO₃ FINE PARTICLES

JUAN YANG AND JOSÉ M. F. FERREIRA
Department of Ceramics and Glass Engineering,
UMIC University of Aveiro, 3810 Aveiro, Portugal

Abstract

A hydrothermal process was developed for the synthesis of fine powders of undoped BaTiO₃ and doped with 5mol% of Zr^{4+}, Ce^{3+} and La^{3+} at 150°C. Barium acetate, $Ba(OAc)_2$ and titanium butoxide, $Ti(OBu)_4$ were used as precursors. The results showed that hydrothermal treatment promoted the formation of solid solutions of doped BaTiO₃, and a reduction in particle sizes was observed in the doped powders, compared with the undoped BaTiO₃ one. All powders presented porous structures which retarded the densification of the final products. Compacts processed by slip casting and sintered at 1200°C showed higher values of density and dielectric permittivity than those shaped by dry pressing. Fine grains less than 1 □m were observed in the sintered bodies.

1. Introduction

Barium titanate ceramics have been widely used in many electronic devices, for example, multilayer capacitors [1] and PTC thermistors [2]. McNeal *et al* stated that high-permittivity in ferroelectric materials is also useful for the material selection for decoupling capacitors currently utilised in computer packaging [3]. It is generally accepted that the use of nanosized powders enables one to lower the sintering temperature and improves the properties of the final products due to the formation of finer sintered microstructures. However, there is a size effect on the dielectric properties of the BaTiO₃ ceramics [4-5]. It is reported that the relative permittivity increased with decreasing grain size, for sizes in the range of 400 and 1000nm, while for grain sizes below the above range, the permittivity decreased with decreasing the grain size [6]. But the size effect on the cubic-tetragonal transition still remains controversial [4, 7].

In order to obtain ceramics with fine microstructures, a critical fabrication step is the preparation of ultra-fine powders, which usually involves nonconventional processes including sol-gel [8], coprecipitation [9], citrate process [10] and hydrothermal technique [11]. Very serious agglomeration can occur in the dried nano-particles due to large specific surface areas. These agglomerates will cause inhomogeneites in the particle packing and the pore size distribution in the green bodies, which degrade the properties of the final ceramics. Colloidal forming processes, such as slip casting and pressure filtration, can minimise the agglomerates and possibly facilitate the formation of a network between particles by preparing well-dispersed suspensions with a suitable dispersant. These processing techniques, initially applied in the field of structural

C. Galassi et al. (eds.), Piezoelectric Materials: Advances in Science, Technology and Applications, 99–106.
© 2000 *Kluwer Academic Publishers. Printed in the Netherlands.*

ceramics [12], have received increasing attention by researchers in the field of electronic ceramics to optimise the properties of the green bodies [13].

The hydrothermal technique, in which chemical reactions are induced in aqueous or organic-aqueous media under the simultaneous application of heat and pressure, have used to prepare fine crystalline ceramic powders at relatively low temperatures. The avoidance of calcination at high temperature to obtain the desired crystalline phases allows the formation of fine particles with high specific surface areas.

The present work deals with the hydrothermal synthesis of undoped and doped $BaTiO_3$ powders with 5mol% La^{3+}, Ce^{3+} and Zr^{4+}, and with the processing of ceramic bodies by slip casting. The properties of samples shaped by slip casting and dry pressing techniques, are also compared.

2. Experimental Procedure

2.1 HYDROTHERMAL SYNTHESIS OF PARTICLES AND PROCESSING

Fine particles of $BaTiO_3$, $Ba_{0.95}Ce_{0.05}TiO_3$, $Ba_{0.95}La_{0.05}TiO_3$ and $BaZr_{0.05}Ti_{0.95}O_3$ were synthesised by a hydrothermal technique. $Ba(OAc)_2$ and $Ti(OBu)_4$ were used as precursors without further purification. Starting materials of Ce^{3+}, La^{3+}, and Zr^{4+} were $Ce(NO_3)_3 \cdot 6H_2O$, $La(NO_3)_3 \cdot 6H_2O$ and $ZrOCl_2 \cdot 8H_2O$ (A. R. Grade, Aldrich). Titania sols were prepared by hydrolysing 0.5M isoproponal solutions of $Ti(OBu)_4$ in the presence of tetraethyl-ammoniumhydroxide (TENOH) (Merck, A. R. Grade). Aqueous solutions containing Ba^{2+} with or without the dopants were mixed with the as-prepared titania sols. The molar ratio of TENOH/Ti was fixed at 2 in order to keep a pH value high enough to avoid dissolution of $BaTiO_3$ particles. The final concentration of metal ions was adjusted to 0.25M. The above mixtures were hydrothermally treated at 150°C for 2h and 6h, using a heating rate of 3°C/min. The resulting powders were then washed with distilled water to remove the remaining unwanted species and finally washed with absolute ethanol in an attempt to alleviate agglomeration among the particles. Table 1 shows the sample code, compositions and the time of hydrothermal treatments.

TABLE 1. The detailed information of the samples

Precursors	$Ba(OAc)_2$, $Ti(OBu)_4$,		$Ba(OAc)_2$, $Ti(OBu)_4$, $Ce(NO_3)_3 \cdot 6H_2O$		$Ba(OAc)_2$, $Ti(OBu)_4$, $La(NO_3)_3 \cdot 6H_2O$		$Ba(OAc)_2$, $Ti(OBu)_4$, $ZrOCl_2 \cdot 8H_2O$	
Time (h)	2	6	2	6	2	6	2	6
Sample code	BT-2	BT-6	CBT-2	CBT-6	LBT-2	LBT-6	ZBT-2	ZBT-6

Slip casting (SC) and dry pressing (DP) methods were used to consolidate green compacts. The suspensions were prepared by adding a suitable amount (4-wt% relative to the solid content) of a dispersant based on a polycarboxylic acid, Dolapix CE64 (Zschimmer & Schwarz, Germany) to distilled water. The powders were then added and dispersed by mechanical stirring. The solid loading was kept 37.5-wt% in all cases. TENOH was used again to adjust the pH value of the suspensions at 12. The as-obtained slurries were rolled in polyethylene bottles using Al_2O_3 balls for 24 hours. Subsequently, a de-airing and conditioning step was performed for further 24 h by rolling the slips in the milling container without balls. SC samples were consolidated by pouring the suspensions into plastic rings placed on plaster plates. The DP compacts were shaped by uniaxial pressing in a die (20MPa) and then isostatic pressing (about 200 MPa) to form green bodies. The compacts were first maintained at 300°C for 2 hours, and then sintered at 1200°C for 2 hours with a heating rate of 5°C/min.

2.2 CHARACTERIZATION

The crystalline phases of different samples were determined by X-ray diffraction (D/MAX-C, Rigaku), using Cu Kα radiation. The powder was characterised by thermogravimetric analysis (Linseis L18) in air at a heating rate of 10°C/min. The lattice parameters of cubic $BaTiO_3$ were determined by analysing the XRD patterns obtained at 2θ scanning step of 1°/min. The morphology of the powder was observed with a Transmission Electron Microscope (Hitachi H9000-NA). Micromeritics instrument (Gemini II 2370 Surface Area Analyzer) was used to measure the BET specific surface areas by nitrogen adsorption. The rheological behaviour of the suspensions was analysed with a controlled stress CSL rheometer (Carri 50-MED) at 20°C, using a cone-plate system (angle=0.02 rad, diameter=4cm). The microstructure of the polished and thermal etched surfaces and of the fracture surfaces of sintered bodies was observed by Scanning Electron Microscopy (Hitachi S-4100). The final densities were measured by the Archimedes methods in mercury. Au-Pd electrodes were deposited on the surfaces of the samples and dielectric measurements were carried on an automatic capacitance bridge (Solartron 1260 Impedance/Grain-Phase Analyser).

3. Results and Disscusion

3.1 PREPARATION AND CHARACTERIZATION OF THE POWDERS

Figure 1 shows the DTA-TG curves of the BT-2 powder. No remarkable peaks, especially for the transformation of cubic-tetragonal transition, were found. The weight loss started at about 60°C and was complete at about 850°C. The total weight loss reached ≈8.5% at about 1100°C. The main weight loss comes from the evaporation of the adsorbed molecular water on the surface at low temperature, and from the removal of structural (OH⁻ lattice) water and the decomposition of $BaCO_3$ at high temperature. The adsorbed molecular water and the structural (OH⁻ lattice) water usually occur in the hydrothermally prepared powders [14]. The lattice hydroxyl was believed to influence the defect chemistry and grain growth of the sintered $BaTiO_3$ ceramics.

Figure 2 shows the XRD patterns of various powders after hydrothermal treatment. Cubic $BaTiO_3$ phase and traces of $BaCO_3$ are found in all cases. No evidence of Ce_2O_3, La_2O_3 and ZrO_2 is detected in their corresponding doped-powders, although Ce_2O_3 and ZrO_2 could be hydrothermally synthesised alone [15-16]. The hydrothermal synthesis of La_2O_3 has rarely been reported. It seems reasonable to assume that most of the Ce^{3+}, La^{3+} enter in Ba^{2+} sites while Zr^{4+} could replace Ti^{4+} to form solid solutions during the hydrothermal synthesis, similarly to what usually occurs in the conventional calcined powders. To support such an assumption, the lattice parameters of the undoped and doped powders were measured, as shown in Table 2. It

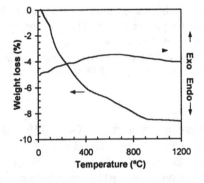 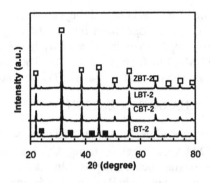

Figure 1. TG-DTA curves of BT-2 sample.   *Figure 2*. XRD patterns of powders, ■-BaCO₃, □- BaTiO₃.

can be seen that a reduction in lattice parameters occurred in La_2O_3 and Ce_2O_3-doped samples in comparison with the pure $BaTiO_3$, while a small increase was observed in the ZrO_2-doped samples. This clearly indicates the formation of solid solutions in the hydrothermally synthesised doped powders since the ionic radius of La^{3+} (1.016Å) and Ce^{3+} (1.034Å) are smaller than that of Ba^{2+} (1.34Å) and the ionic radius of Zr^{4+} (0.79Å) is larger than that of Ti^{4+} (0.68Å) [17]. The formation of these solid solutions occurs at a significantly lower temperature, compared with the conventional synthesis method. The specific surface areas of the powders are also presented in Table 2. All the doped powders show smaller particle sizes than the undoped ones. Solid solutions are more disturbed structures in which the chemical species added during crystal growth are more difficult to fit in. Furthermore, the synthesis time has no significant influence on particle size, although a slightly increasing trend with time can be noticed.

TABLE 2. Surface area, particle size and cubic-phase lattice parameter of the powders

Sample	BET (m²/g)	D_{BET} (nm)	Lattice parameter (nm)
BT-2	29.6	33.7	4.0309±0.0004
BT-6	24.8	40.1	4.0315±0.0006
CBT-2	52.8	18.9	4.0275±0.0004
CBT-6	45.6	21.9	4.0308±0.0003
LBT-2	40.3	24.7	4.0286±0.0006
LBT-6	34.3	29.1	4.0294±0.0004
ZBT-2	64.1	15.5	4.0314±0.0007
ZBT-6	56.0	17.8	4.0318±0.0005

Figure 3 presents typical TEM images and the corresponding electron diffraction patterns of BT-2 and ZBT-2 powders after hydrothermal treatment. The primary particle sizes of BT-2 and ZBT-2 are around 60nm and 40nm, respectively, as measured by SEM, which are larger than those calculated from BET. This is probably due to the existence of porosity in the particles, the dark parts together with the bright parts shown by the TEM morphology suggest that the particles are porous [6]. The TEM results also confirm that the particle size of the doped powder is smaller than that of the BaTiO₃ powder. Furthermore, both are well crystallized powders as indicated by ED and XRD patterns.

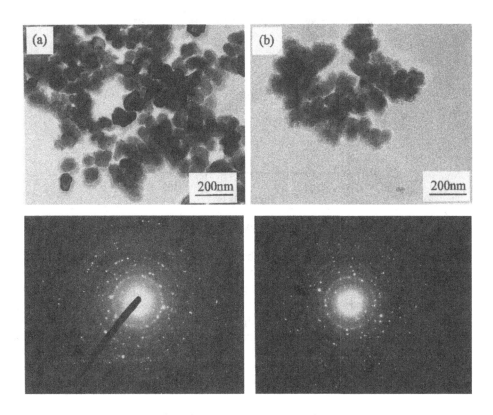

Figure 3. TEM images and ED patterns of powders, (a) BT-2; (b) ZBT-2

3.2 RHEOLOGICAL PROPERTIES

Figure 4 shows the flow curves of the suspensions of the various powders. The rheological behaviour is initially shear-thinning followed by a near Newtonian plateau at high shear rates in all powders hydrothermally treated for 6 hours. This suggests that better dispersions were obtained for powders prepared in these conditions. Contrarily, the suspensions of powders hydrothermally treated for 2 hours, exhibit a rheological behaviour more typical of an attractive particle network since the Newtonian plateau is hardly observed within all the shear rate range. This suggests that finer powders are more difficult to disperse than the coarser ones. This point can be well addressed by their larger specific surface areas, which allow the solid surfaces to consume larger amounts of water on hydration giving rise to more viscous suspensions.

3.3 CHARACTERIZATION OF THE SC AND DP SINTERED BODIES

Table 3 shows the density, dielectric permittivity (ε, at the frequency of 1kHz) and dielectric loss (tgδ, 1kHz) of the samples prepared from various powders by the different shaping methods. In all cases, the slip casting produced denser samples, with higher dielectric permittivity values than those obtained by dry pressing.

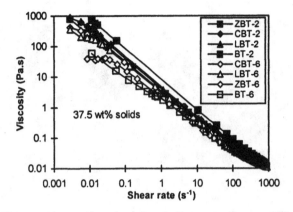

Figure 4. Viscosity versus shear rates for the various suspensions

Obviously, the slip casting seems to be a very promising process even for the fabrication of electronic ceramics. The advantage of slip casting lies in the formation of more homogeneous green bodies having a more uniform pore size distribution. However, the dielectric properties of the doped samples were not considerably improved compared with those of $BaTiO_3$. This can be attributed to a relatively low sintering temperature (1200°C), which is insufficient to fully densify the doped samples, especially the ZBT samples. The possible occurrence of structural defects due to the dopant may cause a transition from an insulating to a semiconducting behaviour, which is essential for the fabrication of PTC devices. This is still under investigation.

TABLE 3. Sintered and dielectric properties of the samples (the theoretical density of $BaTiO_3$ is 6.02g/cm^3)

Sample	BT-6		CBT-6		LBT-6		ZBT-6	
SC or DP	SC	DP	SC	DP	SC	DP	SC	DP
d (g/cm^3)	5.3	5.1	5.0	4.5	4.7	4.4	3.9	3.6
ε	2470	2210	2080	1512	2060	1470	626	468
tgδ	0.0247	0.0360	0.0246	0.0313	0.0033	0.1824	0.3310	0.3410

The sintered densities achieved are lower than usually obtained with powders prepared by conventional solid state reactions. Figures 5a and 5b present the SEM morphology of the polished and fracture surfaces of BT-6-SC sintered bodies. The average grain size is less than 1μm while a number of pores can be observed, which might derive from: (i), the porous and agglomerated nature of the fine particles; (ii), existence of $BaCO_3$ impurity and lattice hydroxyls remaining after calcination. It is expected that increasing the hydrothermal treatment temperature and using N_2 as protecting atmosphere might improve the sinterability.

The dielectric temperature curves of BT-6-SC sample at various frequencies are shown in Figure 6. A normal ferroelectric behaviour can be observed. The Curie temperatures ($T_C \approx 130$°C) remain unchanged with the frequency, while the relative permittivity decreases with increasing frequency. The permittivity at T_C is 4530 (1kHz), which is a slightly lower than that of $BaTiO_3$ ceramics fabricated by the conventional

Figure 5. SEM micrographs, (a) polished surface; (b) fracture surface of BT-6-SC sintered bodies.

Methods ($\varepsilon \approx 10000$ at T_C). This is due to the porosity of the sintered bodies. Further work will be done in the future to improve the sinterability.

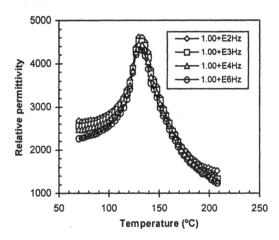

Figure 6. Variation of relative permittivity with temperature at different frequency for BT-6-SC samples.

4. Conclusions

Fine $BaTiO_3$ and $Ba_{0.95}Ce_{0.05}TiO_3$, $Ba_{0.95}La_{0.05}TiO_3$ and $BaZr_{0.05}Ti_{0.95}O_3$ powders were hydrothermally synthesised. Solid solutions were formed in the doped powders at a relatively low temperature. The particle sizes of the doped powders are smaller than that of the undoped $BaTiO_3$. Slip casting is a promising shaping method for manufacturing electronic cermics. The sintered bodies prepared by slip casting exhibit higher densities and better dielectric properties than those obtained by dry pressing. The porous and agglomerated nature of the hydrothermally synthesised powders combined with the existence of $BaCO_3$ impurity and lattice OH^- ions hinder their full densification. A normal ferroelectric behaviour was observed in the sample BT-6-SC.

106

5. Acknowledgement

This work was supported by Fundação para a Ciência e a Tecnologia of Portugal.

6. References

1. Selcuker, A.R. and Johnson, M.A. (1982) Termination sintering in multilayer ceramic capacitor: microstructural interpretation, *Am. Ceram. Soc. Bull.* **72**, 436-40.
2. Lewis, G.V., Catlow, C.R.A., and Casselton R.E.W. (1985) PTCR Effect in BaTiO₃, *J. Am. Ceram. Soc.* **68**, 555-58.
3. McNeal, M.P., Jang, S-J., and Newnham R.E. (1998) The effect of grain and particle size on the microwave properties of barium titanate (BaTiO₃), *J. App. Phy.* **83**, 3288-3297.
4. Li, X. and Shih, W.-H. (1997) Size effects in barium titanate particles and clusters, *J. Am. Ceram. Soc.* **80**, 2844-52.
5. Uchino, K., Sadanaga, E., and Hirose, T. (1989) Dependence of the crystal structure on particle size in barium titanate, *J. Am. Ceram. Soc.* **72**, 1555-58.
6. Wada, S., Suzuki, T., and Noma, T. (1995) Preparation of barium titanate ceramics from amorphous fine particles of the Ba-Ti-O system and its dielectric properties, *J. Mater. Res* **10**, 306-311.
7. Begg, B.D., Vance, E.R., and Nowotny, J. (1994) Effect of particle size on the room-temperature crystal structure of barium titanate, *J. Am. Ceram. Soc* **77** 3186-92.
8. Shih, W.-H. and Lu, Q. (1994) Ultrafine titanate powders produced via a precursors-modified sol-gel method, *Ferroelectrics* **154**, 241-46
9. Park, Z.H., Shin, H. S., Lee, B.K., et al (1997) Particle size control of barium titanate prepared from barium titanyl oxalate, *J. Am. Ceram. Soc.* **80**, 1599-604.
10. Tsay, J.-D., Fang, T.-T., Gubiotti, T. A, et al (1998) Evolution of the formation of barium titanate in the citrate process: the effect of the pH and the molar ratio of barium ion and citric acid, *J. Mater. Sci.* **33**, 3721-27.
11. Urek, S. and Drofenik, M. (1998) The hydrothermal synthesis of BaTiO3 fine particles from hydroxide-alkoxide precursors, J. Europ.Ceram. Soc. 18, 279-86.
12. Reed, J.S., (1995) *Principles of Ceramics Processing,* 2nd Edition, John Wiley & Sons, Inc., New York.
13. Jean, J.-H. and Wang, H.-R. (1998) Dispersion of aqueous barium titanate suspenisons with ammonium salt of poly(methacrylic acid), *J. Am. Ceram. Soc.* **81**, 1589-99.
14. Clark, I.J., Takeuchi, T., Ohtori, N., Sinclair, D.C. (1999) Hydrothermal synthesis and characterization of BaTiO3 fine powders: precursors, polymorphism and properties, *J. Mater. Chem.* 9, 83-91.
15. Li, G., Feng, S., and Li, L. (1996) Structural stability and valence characteristics in cerium hydrothermal system, *J. Solid State Chem.* 126, 74-9.
16. Segal, D. (1998) Soft chemistry routes to zirconia ceramics, *Key-Engineering-Materials* **153-154**, 241-50.
17. Weast, R. C. (1976-1977) *Handbook of Chemistry and Physics*, 57th Edition, CRC Press, Ohio.

CHEMICAL PROCESSING OF PbTiO₃/ORGANIC HYBRID AND Pb(Zr,Ti)O₃ THICK FILMS

S.HIRANO, T.YOGO, W. SAKAMOTO AND H.YAMAGUCHI[*]
*Dept. of Applied Chemistry, Graduate School of Engineering,
Nagoya University, 1, Furo-cho, Chikusa-ku, Nagoya, 464-8603,
Japan*
[*]*NGK Insulators Ltd. 2-56, Sada-cho, Mizuho-ku, Nagoya,
467-8530, Japan*

Abstract

Piezoelectric films have great potentials for applications in actuators, optoelectronics and integrated circuits for devices such as infrared detectors, surface acoustic wave devices, and so on. Inorganic/organic hybrid materials attract great attentions because of novel functionality as well as the combination of beneficial properties of each component. Tetragonal $PbTiO_3$ particles/organic hybrids were successfully synthesized by the reaction control of the specifically designed metal-organic precursor below 100°C. The demand for processing piezoelectric ceramics of sub-micron meter to 10 μm thick films has also been raised because of the functionality integration and improvement in performance. This paper reviews our recent results on the chemical processing of $PbTiO_3$/ organic hybrid and $Pb(Zr,Ti)O_3$ thick films.

1. Introduction

Inorganic/organic hybrids have been receiving great attention because they are expected to have novel features as well as the combined characters of each component. The conventional mixing of small particles into organics usually causes the particle agglomeration due to the van der Waals force, leading to the degradation of the properties. The authors have developed a novel method to synthesize $BaTiO_3$ particles/organic hybrids through the reaction control of the designed precursor from modified titanium and barium alkoxide below 100°C [1,2].

Tetragonal $PbTiO_3$ (PT) of perovskite structure is one of the ferroelectric and piezoelectric materials with attractive properties. Crystalline $PbTiO_3$ particles can be synthesized at 470°C from the precursor of lead acetate and titanium alkoxide. Usually, the heat treatment above 400°C is required to prepare crystalline perovskite $PbTiO_3$ particles. The authors succeeded to crystallize $PbTiO_3$ particles directly into perovskite phase below 100°C in organic matrix to form a hybrid [3].

In recent years, a demand for fabricating 1 to 10 μm thick ceramics has been increasing for applications in devices such as actuators, chip condensers and so on. Self-sustained film, i.e. films formed without substrate, of about 10 μm in thickness

C. Galassi et al. (eds.), Piezoelectric Materials: Advances in Science, Technology and Applications, 107–114.

108

can be produced by the modified doctor blade method. Question is still on whether this method can be commercially applicable for fabricating 1 to 10 μm thick film at one step. Several methods have been developing to prepare ceramic films. The screen printing method can not afford films of thinner than 10 μm thick regardless of its high productivity. The electrophoresis method can fabricate relatively thinner films, where substrates have to be electrically conductive. The sol-gel coating method provides efficiently thin films, by which repetition of dipping or spinning-firing process is required to reach several μm thickness. A dip-coating method using slurry gives a film thicker than hundreds μm per one dipping. However, a method for processing 1 to 10 μm thick film per one step has scarcely been developed.

A centrifugal method is considered to be applicable to fabricating thick films with desired thickness on substrates, where a mixture of starting powder and solvent in suspension centrifuges to form a thick film. In this method, the control of the

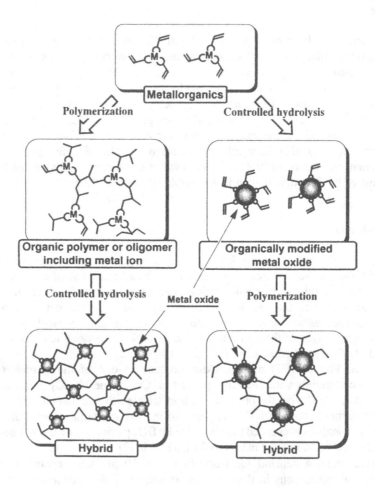

Figure 1. Chemical processing scheme of hybrids

dispersion in a suspension is an essential key factor to prepare dense thick films [4].

This paper reviews our recent results on processing perovskite PbTiO$_3$ particles/organic hybrid and Pb(Zr,Ti)O$_3$ (PZT) thick dense films free from cracks.

2. PbTiO$_3$ particles/organic Hybrid

Crystalline PbTiO$_3$ particles/organic hybrids were synthesized through a precursor derived from lead methacrylate and titanium isopropoxide, followed by the hydrolysis-polymerization reaction control below 100°C.

2.1. PROCESSING

Figure 1 shows the processing scheme for the preparation of hybrids. The fabrication of PbTiO$_3$/organic hybrids was carried out according to the scheme shown in Fig. 2.

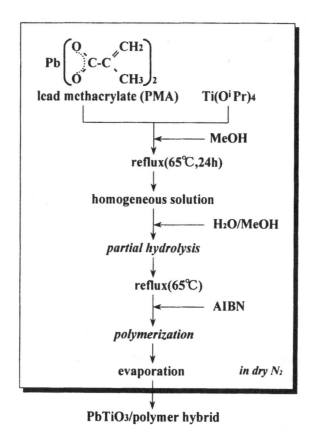

Figure 2. Processing of PbTiO$_3$/polymer hybrid

Lead methacrylate and titanium isopropoxide were weighed to be a molar ratio of 1.0 and dissolved in absolute methanol. The mixture was refluxed at 65°C for 24 h producing a clear PT precursor solution. The PT precursor is analyzed to comprise a complex alkoxide, Pb[Ti(OR)$_n$(methacryl)]. The solution was hydrolyzed with CO_2–free water, followed by the evaporation to afford a solid product, which was sealed with 2,2'–azobis (isobutyronitrile) (AIBN) and methanol under vacuum in a glass tube after freezing–melting treatments in order to remove oxygen and polymerized at 100°C. The solid hybrid was obtained after removal of solvent by drying under vacuum at room temperature.

The hybrid films were prepared from the hybrid powder by pressing between platens electrically heated at about 170°C.

2.2. CHARACTERIZATION OF HYBRIDS

Figure 3 shows a TEM photograph of a hybrid composed of fine particles dispersed in the organic matrix. The electron diffraction and EDX analysis of the black particles confirmed the particles to be the stoichiometric $PbTiO_3$. The increases both in amount of water and in reflux time result in the growth of $PbTiO_3$ (PT) particles, leading to the increase in size of the particles.

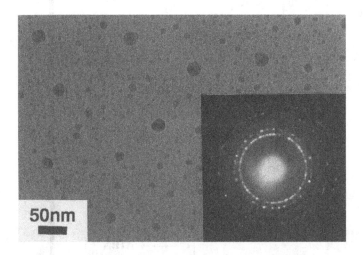

Figure 3. Microstructure and electron diffraction of PbTiO₃/polymer hybrid

Raman spectroscopy was studied on the hybrid. The scattering pattern of the hybrid was in good agreement with that of the crystalline tetragonal $PbTiO_3$ powder prepared by heating the precursor powder at 800°C. The results indicate that the crystalline tetragonal $PbTiO_3$ particles are formed in the organic matrix below 100°C.

The dielectric constant of the hybrid is 5.2 at 10 kHz at room temperature, which is higher than that of poly(methyl methacrylate)(PMMA) as the matrix. As the hybrid is composed of a continuous organic matrix and spherical PT particles, the dielectric constant of the hybrid can be calculated using an equation proposed for the model [5,6]. Approximately, the dielectric constant and density of the tetragonal PT particles were employed for ceramic dispersoid. The amount of the organic matrix was determined by TG. The calculated dielectric constant is 5.1, which is comparable with the measured value.

The hybrids suspended in silicone-oil exhibited the typical electro-rheological behavior on applying an electric field, which may lead to development of an emerging area.

3. PZT Thick Films

This work has been carried out to examine the applicability of the centrifugal sedimentation method which can afford 1 to 10 μm thick films at one step and to clarify the influencing factors in preparing the suspension on the densification in the piezoelectric PZT system.

3.1 PROCESSING

PZT powders used in this work were prepared by the conventional solid-state reaction method to be a composition of $(Pb_{0.95}Sr_{0.05})(Zr_{0.25}Ti_{0.375}(Mg_{1/3}Nb_{2/3})_{0.375})O_4$. An average grain size and specific surface area were 0.4 μm and 7.4 m^2/g, respectively.

A procedure for preparing a suspension to form a film is shown in Fig. 4. The concentration of the PZT powder was determined to 0.6 g in 100 cm^3 solvent to adjust the film thickness to 10 μm. A mixture of 75 vol. % xylene and 25 vol. % methylethylketone (MEK) was used as a solvent for suspension. Solbitan monoleate (Span 80^{TM}) and polyvinylbutyral (PVB, type BLS, Sekisuikasei, Japan) were selected

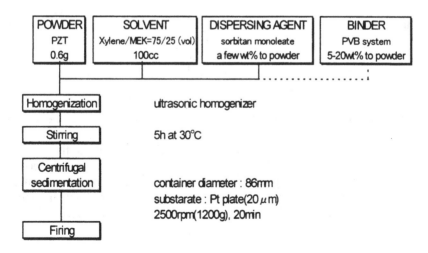

Figure 4. Flow chart for preparation of suspension and film forming

as a dispersing agent and binder, respectively. After mixing the PZT powder, solvent and organic agents with an ultrasonic homogenizer, the suspension was magnetically stirred for 5 h at 30°C., followed by centrifugation. A thick film of PZT was formed on Pt foil placed on flat bottom of a Teflon container. The sedimentation by centrifugation was carried out at the revolution of 2500 rpm (1200G) for 20 min. The film formed on Pt foil was fired in a MgO crucible at 1200°C for 3 h after removing organics by calcining at 500°C in air.

3.2. INFLUENCE OF PVB BINDER ON SUSPENSION

The apparent viscosity showed a maximum as a function of a shear rate for all suspensions with different amount of binders, which resembles thixotropy. This result suggests that the particles in suspension were not well dispersed but flocculated at some degree, and the flocculated particles may deform at higher shear rate. The measurement of the apparent viscosity revealed that the easiness of deformation of soft flocculates depends on the binder content. Although the shear force generated during the centrifugal treatment is not counted clearly, the relationship between the binder content and the densification is considered to depend upon the degree of the deformation adjustable by the binder content.

SEM photographs of the particles in suspension after freeze-dried indicate that the softly flocculated particles are about 1-2 μm in size and deform plastically to form dense films after deposited on substrates by centrifugal force.

3.3. INFLUENCE OF PVB BINDER ON MICROSTRUCTURE OF FILM

Both the PVB binder and the dispersing agent are effective for improving a degree of dispersion because the added PVB binder does bond on the hydrophobic side of the dispersing agent to cause the steric hindrance. Significantly dense packing structures of the particles were realized from the suspension with 10-15 wt % of the binder and 5 wt % of the dispersing agent. The increase in the binder content to 20 wt % yielded a porous structure.

Figure 5 shows the change of the fired film surfaces with the binder content in suspension. Dense films were prepared through the suspensions with 10 and 15 wt % of the binder content. It turns out from this result that the degree of dispersion and the amount of particles in suspension is a critical key to form the dense thick films by the centrifugal sedimentation, which can be controlled by adjusting both of the dispersion agent and the binder.

Consequently, the process for fabricating a thick film through the centrifugal sedimentation is proposed as follows (Fig. 6); (1) the formation of soft-flocculated particles in suspension, (2) the plastic deformation of flocculated particles on substrate and (3) the bonding between the deformed particles to form dense films. In each step, the binder plays an essential role for controlling the microstructure of green films.

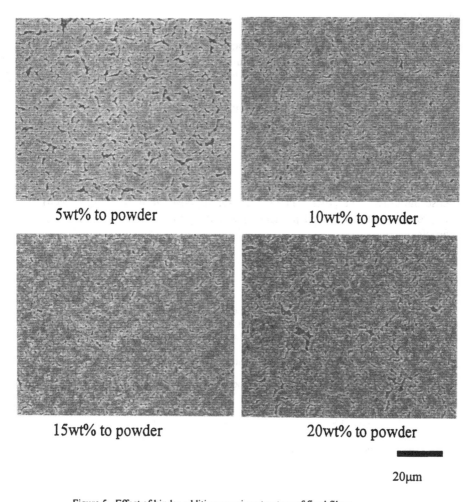

5wt% to powder

10wt% to powder

15wt% to powder

20wt% to powder

20μm

Figure 5. Effect of binder addition on microstructure of fired films

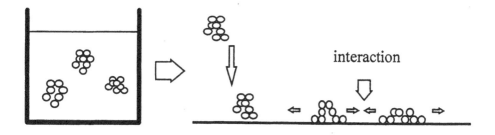

interaction

Figure 6. Film formation process in centrifugal sedimentation

114

4. Summary

Nanocrystalline perovskite PbTiO$_3$ particles/organic hybrids were successfully synthesized from designed metal-organics under controlled polymerization and hydrolysis conditions below 100°C. The PbTiO$_3$ precursor was designed to be a Pb-Ti complex alkoxide modified with methacryl group. The controlled hydrolysis of the precursor at 65°C affords uniform nano-sized crystalline tetragonal PbTiO$_3$ particles in organic matrix. The hybrid exhibits an interesting feature of electrorheological behavior.

The centrifugal sedimentation method was successfully applied as a simple method to process about 1 to 10 μm dense thick PZT films by one step operation. The proper selection of the dispersing agent and the binder was found to be the key factor to form dense films free from cracks. The binder was supposed to be selected so as to control the properties of flocculated particles, plasticity, and the bonding between deformed particles.

5. References

1. Yogo, T., Kikuta, K., Yamada, S., and Hirano, S. (1994) Synthesis of BaTiO$_3$/Polymer Composites from Metal Alkoxide, *J. Sol-Gel Sci. and Tech.* **2**, 175-179.
2. Hirano, S., Yogo, T., Kikuta, K., and and Yamada, S. (1996) Processing and Properties of Barium Titanate/Polymer Hybrid Materials by Sol-Gel Method, *Ceram. Trans.* **68**, 131-140.
3. Yogo, T., Ukai, H., Sakamoto, W. and Hirano, S. (1999) Synthesis of PbTiO$_3$/Organic Hybrid from Metal-Organics, *J. Mater. Res.* (in press).
4. Yamaguchi, H., Itoh, M. and Hirano, S. (1999) Fabrication of PZT Thick Film Through Centrifugal Sedimentation Method, *Ceram. Trans.* **83**, 219- 228.
5. Furukawa,T., Fujino,K., and Fukuda, E. (1976) Electromechanical Properties in the Composites of Epoxy Resin and PZT Ceramics, *Jpn. J. Appl. Phys.* **15**, 2119-2129.
6. Yamada, T., Ueda, T., and Kitayama, T. (1982) Piezoelectricity of a High-Content Lead Zirconate Titanate/Polymer Composite, *J. Appl. Phys.* **53**, 4328-4332.

PREPARATION AND CHARACTERIZATION OF THICK PIEZOELECTRIC FILMS ON ALUMINA SUBSTRATES

P. GONNARD, S. LE DREN, L. SIMON, M. TROCCAZ and L. LEBRUN
Laboratoire de Génie Electrique et Ferroélectricité – Bâtiment 504
INSA LYON – 20, av. Albert. Einstein – 69621 VILLEURBANNE Cedex France

1. Introduction

In the near future piezoelectric appplications in micro-devices or in high frequency sensors will require piezoceramic films with thickness up to 100 μm [1–2]. Most piezoelectric film research has been on thin film technologies but more recently studies on thick films have been increasing [3-4]. A typical piezoelectric thick film is a structure with an alumina substrate (or metallic one), a screen printed Pt bottom electrode, a screen-printed piezoelectric film and a top Ag or Pt electrode. Pt ink is inert during PZT sintering but is much more expensive than the Ag/Pd ink which was used in the present work. The high porosity of the PZT layers can be decreased by additions of glass phases in the inks but the piezoelectric and electrical properties are often degraded [5]. VECHEMBRE et al. [4] has obtained well densified PZT layers (14 μm), prepared by a double print method, with a piezoelectric constant d_{33} = 80 pC/N and a dielectric constant ε_r = 1040.

In the present work the realization and the characterization of 70 μm layers in thickness is reported and the reduced values of d_{33} and ε_r compared to the bulk are interpreted both by lateral clamping due to the substrate and by the porosity of the active layer.

2. Experimental procedure

2.1. POWDER

A modified PZT composition has been chosen because of its interesting ceramic piezo-electric characteristics combined with the ability to prepare the powder by a solid state process : $Pb_{0.93}$ $Sr_{0.07}$ ($Zr_{0.455}$ $Ti_{0.445}$ $W_{0.036}$ $Sn_{0.036}$ $Mn_{0.028}$)O_3 (sintering 1250° C, 3 h) d_{33} = 410 pC/N, ε_r = 1950, k_p = 0.59, tgδ = 0.8 %

The mixed oxides are fired at 950°C to 6h and are then ground to obtain the desired granulometry (Figure 1a-full line).Powders obtained from a coprecipitation process are unsuitable for the preparation of an ink because agglomerates exist even after several hours milling(Figure 1a-dotted line).

The specific area of powders made by the solid process is around 0.93 m²/g. The theoretical diameter (deduced from BET) is 0.86 μm whereas the mean diameter measured with the granulometer is 0.76 μm in number (Figure 1b).

115

C. Galassi et al. (eds.), Piezoelectric Materials: Advances in Science, Technology and Applications, 115–122.
© 2000 *Kluwer Academic Publishers. Printed in the Netherlands.*

116

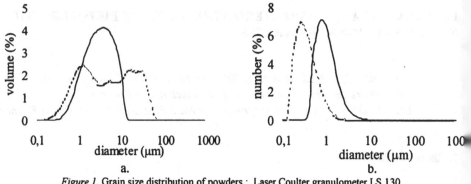

Figure 1. Grain size distribution of powders : Laser Coulter granulometer LS 130

    a. % in volume ⎫ ——— solid process
    b. % in number ⎭ ------ coprecipitation process

2.2. INK

The components of the ink are the active material (85 % PZT in weight) and the organic vehicle. The relative proportions in weight of the organic part are adjusted to obtain the necessary rheological characteristics :

- solvent 50 % (α-terpineol)
- binder 10 % (PVB : polyvinylbutyraldehyde)
- dispersing agent 25 % (BEEA : butoxy-ethoxy-ethyl-acetate)
- plasticiser : 15 % (PEG : polyethylene glycol)

       This ink which presents a Newtonian behaviour must be used immediately after its preparation because the ink loses its homogeneity within a few hours. The origin of this inhomogeneity is powder sedimentation which could be due to the progressive destruction of the complex made with PVB-PZT [6].

2.3. ELABORATION OF THE THICK FILMS ON SUBSTRATES

96 % Alumina substrates plates ($25 \times 12 \times 0.250$ mm^3) are electroded by screen-printing of a Ag/Pd paste diluted at 15 %, dried at 120° C and fired at 850°C during 15 min. The resulting electrode thickness is around 10 μm.

       The PZT ink is screen printed on these substrates. The final thickness of the PZT layers depends on the mesh of the sieve. A thickness of 70 to 90 μm has been chosen.

       The different firing stages of the paste deposited on the substrate are given in Table 1.

TABLE 1. Firing stages of thick films

Stage	Solvent phase evaporation	Organic part removal	Sintering Normal	Sintering Rapid
Heating rate	Free	1° C/min	3° C/min	Free*
Stage temperature (°C)	60	500	1050	900-1100
Stage time	20-30 min	5 h	1 h	2-30 min

*sample put into a furnace at the sintering temperature

top Ag electrode is applied by screen printing (firing 650°C – 15 min). Poling is carried t in an oil bath at 120°C by applying a 6 kV/ mm dc field

Results and discussion

1. OPTIMIZATION OF THE EXPERIMENTAL PROCEDURE

1.1. *Optimization of the rapid sintering (RS)*

ie best results are obtained for a 7 min sintering time at 1050° C (Figure 2) and for a 4 in sintering time at 1080° C (Figure 3).

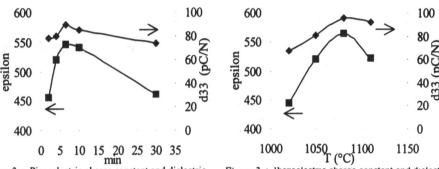

gure 2. : Piezoelectric charge constant and dielectric constant versus sintering time (1050° C)

Figure 3. : Piezoelectric charge constant and dielectric constant versus sintering temperature (4 min)

1.2. *Poling optimization*

igure 4 shows that 6 kV/mm for 2 min are sufficient to obtain the maximum polarization 120° C. An increase of poling time up to 25 min does not improve the piezoelectric narge constant.

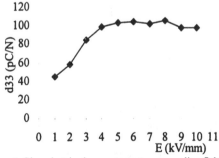

Figure 4. Piezoelectric charge constant versus poling field (2min – 120° C)

.2. SEM OBSERVATION OF THE FILMS

EM observation on polished cross sections of samples shows that the PZT films obtained y rapid sintering (RS) have a porous structure with closed pores. Such a structure is imilar to a so called 0-3 composite. However the interfaces "Al_2O_3 –Ag/Pd" and "Ag/Pd – ZT" are well defined indicating reduced diffusion effects (Figure 5).

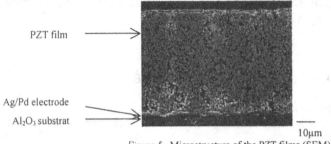

PZT film ————————→

Ag/Pd electrode ——————
Al₂O₃ substrat ——————————→

10μm

Figure 5. Microstructure of the PZT films (SEM)

3.3. INTERFACE EFFECTS

As previously reported the properties of PZT films are greatly improved when substituting the normal sintering NS (1050° C during 60 min) by a Rapid Thermal Annealing RTA (1150° C during 30 s with a ramp rate of 100° C/s) [7] (Table 2).

TABLE 2. Comparison between NS and RTA (Hard PZT – Bulk properties ε_r = 940 d_{33} = 245 pC/N)

	Thickness μm	ε_r	d_{33} pC/N
Normal sintering	50	130	35
RTA	50	340	90

Moreover for NS samples the Curie point (280° C for the bulk) is completely erased whereas it is preserved for RTA or RS samples (Figure 6)

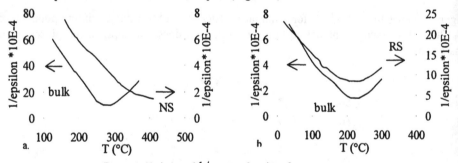

Figure 6. Variation of $1 / \varepsilon_r$ as a function of temperature

a. Hard type material   b. Material used in the present paper

Such a behaviour can be explained as being due to diffusion effects at the interfaces "substrate-electrode-PZT film" which generate a low permittivity non ferroelectric layer in series with the active PZT layer. These diffusion effects are greatly reduced by using RTA or RS sintering. In order to limit the reaction between the substrate and the active film KOSEC has proposed to use an intermediate PLZT barrier layer between the alumina substrate and the electrode [3]. Such a method experimented for our RS sintered films leads to a negligible increase of the permittivity and of the piezoelectric coefficient. We conclude that the diffusion of the aluminium and titanium ions is negligible when using RS sintering.

4. CLAMPING EFFECTS OF THE SUBSTRATE

4.1. *Effects on the hysteresis loop*

The hysteresis loop of the film (Figure 7b) is different from the bulk (Figure 7a). The slanted look of the film could originate from a lateral clamping effect of the substrate during the poling treatment. As a consequence saturation is not obtained at the maximum applied field and the remanent polarization P_r is smaller than that in the bulk ceramic. However the coercive fields are nearly the same in the two cases.

An increasing poling field slightly increases the remanent polarization but at the expense of the breakdown risk.

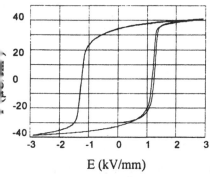
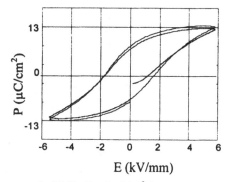

a. bulk ceramic $P_r = 33 \ \mu C/cm^2$ $E_c = 1.25$ kV/mm

b. RS film $P_r = 8 \ \mu C/cm^2$ $E_c = 1.7$ kV/mm

Figure 7 : Hysteresis loops

Let us consider for a bulk material the schematic model of loop of Figure 8a. and the corresponding strains S_3 and S_1 during poling (Figure 8b.). The dipole orientation (zone B) at zero stresses ($T_1 = T_2 = T_3 = 0$) can be linearized as follows :

$$S_3 = a \ (E_3 - E_c) \text{ and } P_3 \sim S_3 \qquad S_1 = S_2 = b \ (E_3 - E_c)$$

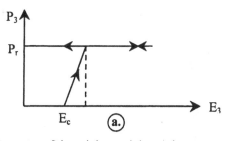
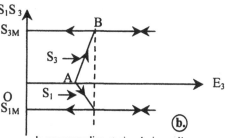

Figure 8. a. Schematic hysteresis loop and

b. corresponding strains during poling

Assuming the poling process to be isochoric :

$$S_1 + S_2 + S_3 = 0 \quad \text{so} \quad b = -a/2 \quad \text{and} \quad S_{1M} = S_{2M} = -S_{3M}/2$$

For a film deposited on a rigid substrate the lateral strains are negligible during poling and consequently important lateral stresses T_1 and T_2 are induced and can be estimated. In this model the lateral thermal stresses that originate from sintering and during cooling because of the different thermal expansion coefficients of alumina, Ag/Pd and PZT have not been taken into account.

$$S_1 = S_2 = 0 = b \ (E_3 - E_c) + s_{11} \ T_1 + s_{12} \ T_2$$
$$S_3 = a \ (E_3 - E_c) + s_{13} \ (T_1 + T_2) \text{ with } T_3 = 0 \text{ and } T_1 = T_2$$

One obtains : $\quad S_3 = a\left(1 + \dfrac{s_{13}}{s_{11}+s_{12}}\right)(E_3 - E_c) = a_f\,(E_3 - E_c)$

With $a_f < a$ typically : $0.25 < a_f/a < 0.4$

      As a result the hysteresis loop of a film deposited on a rigid substrate is slanted and poling remains incomplete for normal applied electric fields E_M (Figure 9). Moreover the lateral stresses are at the origin of a partial depoling when the electric field is removed (CD).

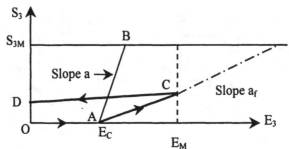

Figure 9. Hysteresis bulk loop (OABS$_{3M}$) and film loop (OACD)

At last residual depolarizing lateral stresses persist after poling and they could be at the origin of an important ageing of the films : the decrease in d_{33} reaches 15 to 25 % after one month.

3.4.2. *Permittivity and piezoelectric coefficient*

Let us consider as previously a film whose lateral strains S_1 and S_2 parallel to the substrate are negligible. The classical piezoelectric equations give the effective values of the piezo-electric coefficient d_{33eff} [8], the permittivity ε^T_{33eff} and the compliance s^E_{33eff} of a clamped film :

$$d_{33eff} = d_{33}\left[1 - \frac{2s^E_{31}}{\left(s^E_{11}+s^E_{12}\right)}\frac{d_{31}}{d_{33}}\right] \qquad \varepsilon^T_{33eff} = \varepsilon^T_{33}\left[1 - kp^2\right] = \frac{\varepsilon^S_{33}}{1-kt^2} \qquad s^E_{33eff} = s^E_{33} - \frac{2\,s^{E2}_{31}}{s^E_{11}+s^E_{12}}$$

where kp and kt are respectively the planar and thickness coupling factor of the bulk.

Some typical values of the reduction ratios for PZT are :

$$0.36 < \frac{d_{33eff}}{d_{33}} < 0.4 \qquad 0.5 < \frac{\varepsilon^T_{33eff}}{\varepsilon^T_{33}} < 0.75 \qquad 0.4 < \frac{s^E_{33eff}}{s^E_{33}} < 0.6$$

      For lead titanate type materials, lateral effects are negligible so these ratios are # 1. As a conclusion the rigidity of the substrate reduces the permittivity of the films and even further the piezoelectric coefficient d_{33} which is doubly reduced by the decrease of the remanent polarization and by the lateral clamping described in this paragraph.

3.4.3. *Comparison with bulk ceramics glued on substrates*

Table 3. shows the d_{33} values of ceramics with various thicknesses measured under the following conditions :

Free bulk ceramic : Polished □ Electroded □ Poled at 6 kV/mm, 115° C, 1 min
Poled samples used in ⤙ and glued on the alumina substrates
Bulk ceramics poled on substrates : Electroded on one side □ Glued on alumina
substrate □ Polished and electroded □ Poled at 6 kV/mm, 115° C, 1 min

TABLE 3. Values of d_{33} in pC/N

	Thickness μm	290	225	180	135	90	65
d_{33} (pC/N)	Condition ⤙	416	390	404	380	-	-
	Condition ✸	375	335	315	280	-	-
	Condition +	293	263	238	194	160	127

It clearly appears that d_{33} of free samples ⤙ is independent of the thickness
whereas d_{33} of the same samples glued on substrates ✸ is all the more reduced because
they are thin. At last when ceramics are poled after gluing on substrates d_{33} is drastically
reduced (+) probably because of an incomplete poling. Globally these results are in good
agreement with those obtained on films. However the permittivity of thin samples poled on
substrates is higher than the permittivity obtained on the films (ε_r # 1140 +, ε_{rf} # 600).
As discussed in the next paragraph such differences are due to the porosity of the films.

5. INFLUENCE OF THE FILM POROSITY

The volumic mass of the films ρ_f is much smaller than the bulk one ρ_o. The porosity p is
given by the relation $\rho_f = \rho_o (1 - p)$. Let us consider the modified cubes model used by H.
BANNO [9] for 3-0 or 0-3 composites and applied to the porous structures

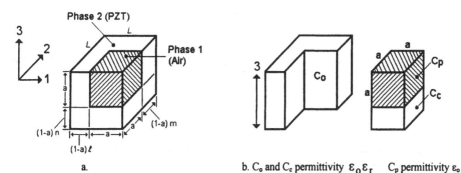

Figure 10. Schematic representation of :
a. unit cell of a porous film - b. decomposition into three capacitors in 3- direction

For a unit volume : $[a + (1-a)\,\ell]\,[a + (1-a)\,m]\,[a + (1-a)\,n] = 1$ and $p = a^3$.

Because of symmetry along 1 and 2 axis we have $\ell = m$. According to the poling
direction "3" the model can be divided into 3 capacitors (Figure 10b). Finally one obtains
the permittiviy ε_{33f}^T of the porous film :

$$\varepsilon_{33f}^T = \varepsilon_{33eff}^T \left(1 - p - p^{2/3}\left(1 - p^{1/3}\right)n\right)$$

The Banno model shows also that the porosity does not modify the piezoelectric coefficient $d_{33f} = d_{33eff}$. As a summary Table 4 gives the theoretical values which could be obtained for five material configurations.

TABLE 4. Theoretical values for five material configurations x = poling percentage and $d_{33} \sim P_r$

Material Configuration	ε_r	d_{33} pC/N	
Free ceramic plate (reference)	1950	410	
Laterally clamped ceramic plate	1290	164	
Non porous and fully poled film on substrate	1290	164	
Porous and fully poled film p = 0.4	$590 < \varepsilon_r < 774$	164	
Porous (p = 0.4) and partially poled films	$590 < \varepsilon_r < 774$	x = 75 %	123
		x = 50 %	82

4. Conclusion

Porous PZT films (70 µm) on alumina substrates have been obtained by Rapid Sintering. Despite their porosity these films withstand high electric fields and can be poled. The dielectric and piezoelectric properties are attractive but they are strongly limited by the high mechanical stiffness of the substrates which laterally clamps the films. Such a clamping effect also limits the film poling and slants the hysteresis loop. The porosity reduces mainly the permittivity. New investigations will be made on thinner Al_2O_3 substrates, on metallic substrates and on modified PT films. Futhermore a $PbO-PbF_2$ flux addition could be used to improve the quality of the films.

5. References

1. M. Kurosawa,T. Morita, T. Higuchi (1994), A cylindrical ultrasonic micro motor based on PZT thin film – *Proc. Ultrasonic Sym.*, 549-552, IEEE Piscataway USA
2. M. Prudenziati (1994) *Thick films Sensors*, Elsevier Sciences BV, Amsterdam
3. M. Kosec, J. Holc, B. Malic, V. Bobnar (1999), Processing of high performance lead lanthanum zirconate titanate thick films, *J. Europ. Ceram. Soc. 19, 949-54*
4. J.P. Vechembre, G.R. Fox, N. Setter (1999) Parameters influencing PZT thick films densification, *Ferroelectrics vol 225 (1-4), 775- 782*
5. T. Yamaguchi, S. Takeota, T. Iizuka, T. Nakano (1994), *Scr Metal. Mater. 31 (8), 1013-18*
6. P. Grandou, P. Pastour (1962), *Peintures et Vernis*, Ed. Hermann Paris
7. P. Drogui, P. Gonnard, L. Lebrun, M. Troccaz, D. Barbier (1996) Lead ziconate titanate thick film Electrical properties and characterization by a LDV technique, *Proc. ISAF X 96, 153-56. IEEE : Piscataway USA*
8. D. Damjanovic (1998). Ferroelectric, dielectric and piezoelectric properties of ferroelectric thin films and ceramics, *Rep. Prog. Phys. 61, 1267-1327*
9. H. Banno (1990) Recent progress in science and technology of flexible piezoelectric composite in Japan, *Proc. ISAF 7th, 67-72. IEEE : Piscataway USA*

NONLINEAR PIEZOELECTRIC RESPONSE IN FERROELECTRIC CERAMICS

DRAGAN DAMJANOVIC
Ceramics Laboratory, Materials Department,
EPFL - Swiss Federal Institute of Technology,
1015 Lausanne, Switzerland

ABSTRACT. The piezoelectric nonlinearity and electro-mechanical hysteresis in ferroelectric ceramics are reviewed in terms of recently proposed models. Examples of reduction and control of the nonlinearity and hysteresis by material engineering are presented.

1. Introduction

The term "nonlinear piezoelectric effect" is used here to describe relationship between alternating ("ac") mechanical and electrical fields (charge density D vs. stress X, strain x vs. electric field E) in which the proportionality constant, d, known as the piezoelectric charge or strain coefficient, is dependent on the driving field (Figure 1). Thus, for the direct piezoelectric effect one may write $D=d(X)X$ and for the converse effect $x=d(E)E$. Similar relationships may be defined for other piezoelectric coefficients (g, h, and e) and combination of electro-mechanical variables. The piezoelectric nonlinearity is usually accompanied by electro-mechanical (D vs. X or x vs. E) hysteresis, as shown in Figure 2.

The nonlinearity and hysteresis have a profound influence on application of piezoelectric sensors and actuators, particularly in high precision devices. Take as an example a piezoelectric actuator. In an actuator based on a linear piezoelectric material, for example lithium niobate, $LiNbO_3$ [1], the expected strain may be easily determined from the applied voltage using relationship $x=dE$ (where d is a known material coefficient). In actuators based on nonlinear materials one must calibrate actuator's output by determining experimentally $d(E)$ dependence. Moreover, because of the hysteresis, $x(E)$ relationship is a multiple-valued function. Nonlinearity and hysteresis may be functions of external static fields, temperature, and time. Several methods have been developed to reduce hysteresis and nonlinearity by using special driving circuits or compensating elements [2].

C. Galassi et al. (eds.), Piezoelectric Materials: Advances in Science, Technology and Applications, 123–135.
© 2000 *Kluwer Academic Publishers. Printed in the Netherlands.*

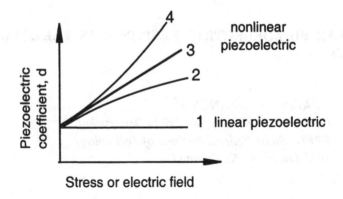

Figure 1. Schematic presentation of the field dependence of piezoelectric coefficients observed in ferroelectric ceramics: (1) in $SrBi_4Ti_4O_{15}$ [3], (2) in tetragonal $Pb(Zr,Ti)O_3$ [4], (3) in coarse grained $BaTiO_3$, Nb-doped $Bi_4Ti_3O_{12}$ and in soft $Pb(Zr,Ti)O_3$ [5] and (4) in soft $Pb(Zr,Ti)O_3$ [6].

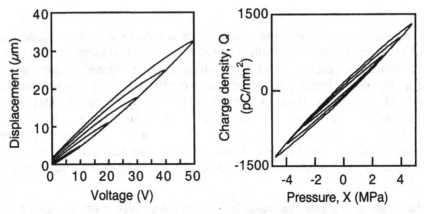

Figure 2. Electro-mechanical hysteresis (subswitching conditions) for the converse piezoelectric effect in a $Pb(Zr,Ti)O_3$–based stack-actuator (left) and for the direct effect in a $Pb(Zr,Ti)O_3$ ceramic single element sensor (right).

Many ferroelectric materials, including the most widely used $Pb(Zr,Ti)O_3$ (PZT) compositions, exhibit piezoelectric nonlinearity and hysteresis, even at relatively low driving fields. There are some notable exceptions, such as strontium bismuth titanate, $SrBi_4Ti_4O_{15}$ [3] and already mentioned $LiNbO_3$ [1]. Both materials show a nearly perfect linear and hysteresis-free piezoelectric behavior in the range of fields that are of interest for most applications.

In this paper, the origins of the piezoelectric nonlinearity in ferroelectric ceramics and recent models proposed to describe it are discussed. Using three examples, it is then shown how nonlinearity and hysteresis may be controlled and reduced by material engineering.

2. Origin and description of the nonlinear piezoelectricity

The origin of electro-mechanical nonlinearity in ferroelectric materials is usually attributed to displacement of domain walls. It is interesting that in most studies this assumption is made without being directly verified experimentally. There are only few reports in which a more or less direct observation of domain-walls displacement has been correlated to the nonlinearity in dielectric [7] and piezoelectric [8] properties. While assumption that domain-wall related mechanisms are dominant contribution to the nonlinearity is certainly reasonable for most ferroelectrics, other process may contribute to the piezoelectric response [9] and should not be neglected *a priori*. Effects of interphase boundary motion have been proposed, for example, to describe piezoelectric nonlinearity in PZT compositions near the morphotropic phase boundary [10] where tetragonal and rhombohedral phases coexist.

In this paper, so-called extrinsic nonlinearity, which is due to non-lattice related effects, is discussed. Under experimental conditions considered, the lattice nonlinearities (higher order material coefficients [11]) are taken to be negligible compared to extrinsic effects. Furthermore, the nonlinearity and hysteresis are considered in the regime of weak to moderate fields (subswitching conditions).

As already mentioned, the piezoelectric nonlinearity in most ferroelectrics is accompanied by electro-mechanical hysteresis. The opposite is not true, i.e. a large electro-mechanical hysteresis may exist in the field range where material behaves (nearly) linearly. Examples of such behavior are $(Pb,Sm)TiO_3$ and related modified lead titanate compositions [12]. Origins of the piezoelectric hysteresis in these materials are presently not clear. To the best of author's knowledge the nonlinear nonhysteretic extrinsic piezoelectric behavior has not yet been reported. Clearly, in a given material, more than one nonlinear or hysteretic mechanism may operate simultaneously, which, as will be seen later, complicates considerably attempts to describe their physical origins.

Two approaches to description of nonlinear piezoelectricity can be identified. In the first approach, the main goal is to find an algorithm to describe the hysteresis loop and nonlinearity, without necessarily considering underlying physical mechanisms. This approach is used when there is a need to control devices, for example actuators in active control systems. The most often used mathematical description is based on Preisach model [13]. Preisach originally developed his model to treat magnetic hysteresis. However, it was soon recognized that his model is very general and can be applied for many types of hysteretic behavior (ferromagnetic, mechanical, ferroelectric, piezoelectric, in superconductors) [14-16]. Other general and formal approaches, such as Maxwell slip model [17], have also been used to treat electro-mechanical nonlinearity and hysteresis.

The second approach comprises attempts to describe physical mechanisms that govern electro-mechanical hysteresis and nonlinearity. It should be emphasized that all presently available descriptions of the piezoelectric nonlinearity and hysteresis are purely phenomenological. One description is based on Néel-Rayleigh (N-R) approach originally developed for treatment of weak-field magnetic hysteresis and nonlinear magnetic susceptibility [12]. By considering displacement of magnetic domain-walls in a medium with randomly distributed pinning centers (Figure 3) Néel [18] derived expressions for magnetic hysteresis loop and field dependence of magnetic susceptibility. In the first approximation, Néel's theoretical relations are identical to empirical Rayleigh equations [19]. Recently, it has been demonstrated experimentally that the piezoelectric effect [5,12] in ferroelectric ceramics and piezoelectric and dielectric response in ferroelectric thin films [20,21] may be described, at least in the first approximation, by Rayleigh relations. Thus, it appears that Néel's theoretical approach can serve as a good starting point for description of electro-mechanical nonlinearity and hysteresis in ferroelectrics. It is interesting that Rayleigh relations can be derived as a special case of the Preisach model [22-25]. Other attempts to describe the dielectric and piezoelectric nonlinearity in ferroelectrics include polynomial expansion [26-28] and scaling behavior [6].

To illustrate application of the Néel-Rayleigh approach to ferroelectric materials, we take as an example the direct piezoelectric effect. In the framework of N-R model, the field dependence of the direct piezoelectric coefficient and associated charge-stress hysteresis may be written as:

$$d(X_0) = d_{init} + \alpha X_0 + ... \tag{1}$$

$$D(X) = (d_{init} + \alpha X_0)X \pm \frac{\alpha}{2}(X_0^2 - X^2) \tag{2}$$

Equations (1) and (2) are known as Rayleigh equations. X_0 is the amplitude of ac stress $X(\omega,t)$, and α and d_{init} are so-called irreversible and reversible Rayleigh parameters (see Figure 3 and for details Ref. [12]). Equation (1) describes the field dependence of the piezoelectric coefficient whereas equation (2) describes charge density versus pressure hysteresis loop (Figure 4). Note that hysteresis loop can be calculated from nonlinear parameter α and d_{init}. Analogues relationships are valid for the converse piezoelectric effect [20] and dielectric permittivity [21,29].

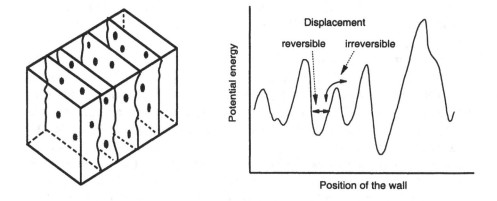

Figure 3. Schematic illustration of interaction of domain walls with randomly distributed pinning centers (left) and corresponding potential energy profile seen by a domain wall (right).

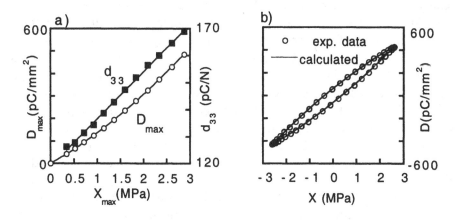

Figure 4. Illustration of the nonlinear and hysteretic piezoelectric response in a Pb(Zr,Ti)O$_3$ ceramic. Symbols present experimental data. Data for d_{33} in (a) are fitted with equation (1) and thus obtained α and d_{init} are then used, together with equation (2), to calculate hysteresis that is shown in (b) with full line. For details see [4, 12]. X$_{max}$ and D$_{max}$ in the figure have the same meaning as D$_0$ and X$_0$ in the text.

Since N-R relations are valid for the direct piezoelectric effect where piezoelectric nonlinearity is assumed to be due to displacement of ferroelastic-ferroelectric non-180° domain walls, it is tempting to propose extension of their validity to the nonlinear elastic compliance and strain-stress hysteresis in purely ferroelastic materials. If this were indeed the case, N-R description of domain-wall pinning would be valid in all ferroic materials. That, however, does not mean that each nonlinearity and hysteresis in any given ferroic must be of N-R type.

An important result of Néel's theoretical treatment is that empirical Rayleigh relations (1) and (2) are truncated forms of more complex expressions. Néel's derivation of equation (1), for example, includes higher order odd-power terms which are due to irreversible jumps of domain walls, and higher-order even-power terms which describe reversible displacement of domain walls [18] (Figure 3). Thus, term Néel-Rayleigh behavior is not limited only to piezoelectric response that can be expressed by the simplest Rayleigh relations presented by equations (1) and (2). Rather, it is used for the most general relations that can be derived by considering displacement of domain walls in a potential similar to that shown in Figure 3. Such general relations are expected to reduce, as their special case, to Rayleigh equations (1) and (2).

The treatment of piezoelectric nonlinearity in terms of N-R approach is interesting for several reasons. First, the electro-mechanical hysteresis follows naturally from the Néel's model (in fact, the nonlinearity is the consequence of the hysteresis). This is not the case for other models that will be presented briefly at the end of this section. Second, Rayleigh parameters can been correlated to concentration of pinning centers and to various microstructural parameters. This has been made extensively in magnetic materials [19,30]. Analysis [29] of dielectric nonlinearity in Fe-doped $BaTiO_3$ ceramics showed that a simple correlation between pinning centers and nonlinear parameters ($\alpha \propto 1/n$, where n is concentration of dopant) also holds for displacement of ferroelectric domain walls. Third, the fact that Néel-Rayleigh relations can be derived from the general Preisach model offers powerful formal background for N-R approach.

The nonlinear Néel-Rayleigh type behavior has been observed in dielectric and piezoelectric properties in several ferroelectric ceramics ($BaTiO_3$, $Bi_4Ti_3O_{12}$-family, $Pb(ZrTi)O_3$) and thin films ($Pb(Zr,Ti)O_3$) [5,20,21,29]. It has been shown that N-R approach can qualitatively account for the most important features of the dielectric and piezoelectric nonlinearity and hysteresis-loops (for details see Ref. [21]). Experiments made on a large number of different compositions also showed features of nonlinear behavior which cannot be interpreted by the simplest N-R relations (1) and (2). This should not be surprising, considering the fact that derivation of relations (1) and (2) assumes a rather idealized behavior of domain walls and the medium in which they are moving: (i) the height of energy barriers is distributed uniformly to zero, (ii) the displacement of domain walls is symmetrical for ascending and descending fields (symmetrical hysteresis), (iii) no other loss mechanisms are present, and (iv) irreversible parameter α is independent of field. Each of these conditions is difficult to meet in a real system. For example, experiments [21,31] show that both dielectric and piezoelectric nonlinearity exhibit a threshold field, below which permittivity and piezoelectric coefficient are field independent. In terms of Néel's description, a threshold field would suggest existence of a minimum height of potential barriers. At very low fields, domains walls would then be confined within the potential wells and only when the threshold field is overcome, irreversible jumps over barriers are possible (Figure 3). Second, a nonsymmetrical hysteresis is often observed experimentally. The asymmetry may be due to experimentally imposed bias fields [15] or may suggest a presence of internal fields. Such fields could favor movement of domain walls in a certain direction, leading to a difference between ascending and descending branches of the hysteresis loop. The asymmetry in the hysteresis could in turn lead to appearance of even-power harmonics in the Fourier expansion of the hysteresis-loop signal, which are forbidden by the fully symmetrical equation (2), but which have been observed experimentally [32]. Third, a presence of other types of loss mechanisms could explain why areas of measured hysteresis loops are often larger than the areas of loops calculated from Rayleigh parameters [12]. The additional loss mechanism may be either field independent [12] or could contribute to nonlinearity through a non-N-R type mechanism. Fourth, the assumption that irreversible Rayleigh parameter α is field independent may not hold in many real systems. A field dependent α could explain, in principle, appearance of even-power harmonics forbidden by equation (2), asymmetry of hysteresis loops, and higher order terms in equation (1) [10]. A detailed investigation of these issues is presently underway.

Two other approaches have been suggested to describe nonlinearity of piezoelectric and dielectric properties of ferroelectrics. The classical polynomial expansion [26-28]:

$$x = x_0 + dE + a_1 E^2 + a_2 E^3 + ... \qquad (3)$$

is shown in (3) for the strain $vs.$ electric field $[E=E_0\sin(\omega t)]$ response in a noncentrosymmetric material. This approach cannot account for the presence of linear term in Fourier expansion of $d(E_0)$ [21], nor it leads naturally to hysteresis loop that can be described by relation of type (2), as observed experimentally. Another recently suggested description [6] of the nonlinearity is based on scaling behavior:

$$m(E_0) = m_0 + A[(E_0 - E_c)/E_c]^{\phi} \qquad (4)$$

where m is dielectric or piezoelectric coefficient, E_0 driving field amplitude, E_c a threshold field, and ϕ is an exponent. In soft PZT, for the dielectric and piezoelectric coefficients measured in direction parallel to polarization, ϕ was found to be ~1, which effectively reduces equation (4) to Rayleigh expression (1). In direction perpendicular to polarization, ϕ was ~1.2 for the dielectric and the shear piezoelectric coefficient. However, when the dielectric and piezoelectric data measured on undoped PZT thin films were analyzed [32] by the same equation, ϕ was found to depend on film history, and varied between 0.8 and ~1 for the piezoelectric and between ~1 and 1.3 for the dielectric coefficient. It has been suggested [6] that the scaling behavior is a result of "dynamics of domain walls in a randomly pinned medium". Thus, the scaling behavior and N-R approach appear to describe a similar physical mechanism. A disadvantage of the description based on equation (4) is that it has not been shown how hysteresis, the essential feature of the piezoelectric nonlinearity in many ferroelectrics, follows from the model.

3. Control of piezoelectric nonlinearity and hysteresis

In this section, three examples are chosen to demonstrate how piezoelectric nonlinearity and hysteresis may be controlled and reduced by material engineering. It must be pointed out that exact microscopic mechanisms responsible for the decrease in the nonlinearity and hysteresis in the three cases are presently not known. It is assumed, however, that the experimental results can be interpreted in terms of domain-wall contributions to the piezoelectric properties.

Figure 5 shows the direct longitudinal d_{33} piezoelectric coefficient in coarse and fine grained BaTiO$_3$ ceramics as a function of the driving pressure amplitude. Domain wall contribution to d_{33} appears much stronger in the coarse

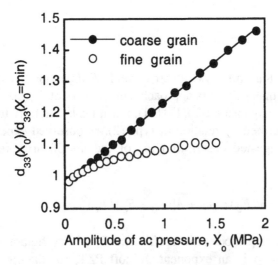

Figure 5. The relative direct d_{33} piezoelectric coefficient in fine and coarse grained BaTiO$_3$ ceramics as a function of the amplitude of driving pressure. For details see [31].

grained ceramic. This is consistent with recent theoretical predictions that the force constant of vibrating domain walls is stiffer for thin domains found in smaller grains than for thicker domains [33]. It is interesting that in the coarse grained material d_{33} can be well described by relation (1).

Apparent dependence of mechanisms responsible for the piezoelectric nonlinearity and hysteresis on microstructure of ceramics has been observed in other ferroelectrics. Figure 6 shows piezoelectric charge *vs.* driving pressure hysteresis in coarse grained and fine-grained ceramics of $0.95Bi_4Ti_3O_{12} - 0.05Bi_3TiNbO_9$ solid solution. Hysteresis and nonlinearity are significantly reduced in fine-grained samples, which is consistent with results observed in $BaTiO_3$. It is interesting to speculate on possible mechanism of domain wall pinning in this system. Bismuth titanate belongs to layer-structured ferroelectrics and consists of alternating bismuth-oxide and perovskite-like bismuth titanate layers. In small concentrations, and under normal processing conditions, Bi_3TiNbO_9 is completely soluble in $Bi_4Ti_3O_{12}$ [34,35]. To obtain samples with a small grain size, ceramics are fabricated at a low temperature and sintered for a short time. It is possible that under such conditions the crystal structure is not completely homogenized and that regular sequence of $Bi_4Ti_3O_{12}$ layers becomes broken by inclusions of Bi_3TiNbO_9 layers [36]. The resulting stacking faults (or inter-growth layers) could create local stress concentrations effectively pinning domain walls. This situation would thus be somewhat analogous to the well-known clamping effect that acceptor dopant-oxygen vacancy dipole defects (e.g. $Fe_{Ti}-V_O$) have on domain walls in acceptor-doped $BaTiO_3$ and PZT [37].

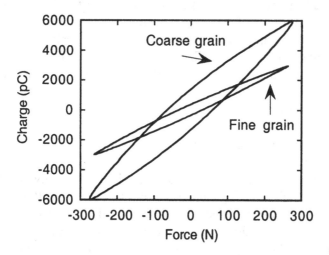

Figure 6. Piezoelectric charge *vs.* driving force hysteresis for a coarse and fine grained $0.95Bi_4Ti_3O_{12}-0.05Bi_3TiNbO_9$ ceramic (Courtesy of Fan Chu)

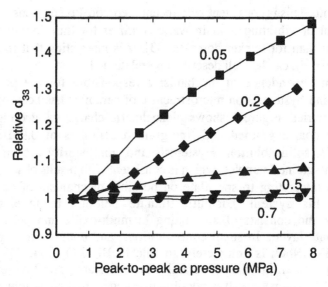

Figure 7. Relative piezoelectric d$_{33}$ coefficient in Nb-doped Bi$_4$Ti$_3$O$_{12}$ ceramics as a function of driving pressure for different concentrations of Nb. Numbers designate atom% concentration of Nb which replaces partially Ti.

Finally, the third example shows how piezoelectric hysteresis and nonlinearity in Bi$_4$Ti$_3$O$_{12}$ ceramics can be completely suppressed by doping. Figure 7 presents evolution of the nonlinearity in Nb-doped Bi$_4$Ti$_3$O$_{12}$ as a function of Nb concentration. The direct piezoelectric effect becomes linear (field independent and nonhysteretic) at large concentrations of Nb. As in the previous case, addition of Nb might create structural defects, which act as pinning centers for domain walls. Another origin of domain-wall pinning could be related to the donor nature of Nb^{+5} that replaces Ti^{+4}. The extra electrons may diffuse into charged domain walls making their irreversible displacement energetically unfavorable [38]. Several other mechanisms have been considered to interpret piezoelectric properties of Nb-doped Bi$_4$Ti$_3$O$_{12}$ [36].

4. Future developments and conclusions

A satisfactory description of the piezoelectric nonlinearity in ferroelectric materials (under subswitching fields) still does not exist. N-R approach which describes nonlinearity and hysteresis in terms of domain-wall pinning in a medium with randomly distributed pinning centers, appears to be a promising starting point. This approach clearly must be extended to describe experimentally observed features that cannot be accounted for by simple relations (1) and (2). More important, the future experimental and theoretical research should focus on physical origins of nonlinearity and hysteresis (pinning of domain walls *vs.* other mechanisms) and development of microscopic models rather than on their formal description.

Acknowledgement
This work was financially supported by The Swiss Federal Office for Education and Science (OFES) in the framework of the Brite-Euram Project "RANNTAC" (OFES Contract No. 97.0515-2). Author would like to thank his colleagues M. Demartin, F. Chu, D. Taylor, G. Robert, L. Sagalowitz, and P. Duran for help and many useful and challenging discussions.

5. References

1. Nakamura, K. and Shimizu, H. (1989) Hysteresis-free piezoelectric actuators using LiNbO$_3$ plates with a ferroelectric inversion layer, *Ferroelectrics* **93**, 211-216.

2. Uchino, K., (1997) *Piezoelectric actuators and ultrasonic motors*, Kluwer, Norwell, MA.

3. Reaney, I. M. and Damjanovic, D. (1996) Crystal structure and domain-wall contributions to the piezoelectric properties of strontium bismuth titanate ceramics, *J. Appl. Phys.* **80**, 4223-4225.

4. Damjanovic, D. and Demartin, M. (1997) Contribution of the irreversible displacement of domain walls to the piezoelectric effect in barium titanate and lead zirconate titanate ceramics, *J. Phys.: Condens. Matter* **9**, 4943-4953.

5. Damjanovic, D. and Demartin, M. (1996) The Rayleigh law in piezoelectric ceramics, *J. Phys. D: Appl. Phys.* **29**, 2057-2060.

6. Mueller, V. and Zhang, Q. M. (1998) Nonlinearity and scaling behavior in donor-doped lead zirconate titanate piezoceramic, *Appl. Phys. Lett.* **72**, 2692-2694.

7. Tikhomirov, O. A. (1997) Vibrations of domain walls in ac field and the low frequency permittivity of ferroelectrics, *Ferroelectrics* **190**, 37-42.

8. Tsurumi, T., Kumano, Y., Ohashi, N., Takenaka, T., and Fukunaga, O. (1997) 90° Domain Reorientation and Electric-Field-Induced strain of Tetragonal lead Zirconate Titanate Ceramics, *Jpn. J. Appl. Phys.* **36**, 5970-5975.

9. Arlt, G. (1982) Piezoelectric relaxation, *Ferroelectrics* **40**, 149-157.

10. Kugel, V. D. and Cross, L. E. (1998) Behavior of soft piezoelectric ceramics under high sinusoidal electric fields, *J. Appl. Phys.* **84**, 2815-2830.

11. Zaitseva, M. P., Kokorin, Y. I., Sandler, Y. M., Zrazhevskii, V. M., Sorokin, B. P., and Si'soev, A. M. (1986) *Nonlinear electro-mechanical properties of noncentrosymmetric crystals (in Russian)*, Nauka, Novosibirsk.

12. Damjanovic, D. (1997) Stress and frequency dependence of the direct piezoelectric effect in ferroelectric ceramics, *J. Appl. Phys.* **82**, 1788-1797.

13. Preisach, F. (1935) Über die magnetische Nachwirkung, *Z. Physik* **94**, 277-302.

14. Mayergoyz, I. D., (1991) *Mathematical Models of Hysteresis*, Springer-Verlag, New York.

15. Ge, P. and Jouaneh, M. (1995) Modeling hysteresis in piezoceramic actuators, *Precission Engineering* **17**, 211-221.

134

16. Ge, P. and Jouaneh, M. (1997) Generalized Preisach model for hysteresis nonlinearity of piezoceramic actuators, *Precision Engineering* **20**, 99-111.
17. Royston, T. J. and Houston, B. H. (1998) Modeling and measurement of nonlinear dynamic behavior in piezoelectric ceramics with applications to 1-3 composites, *J. Acoust. Soc. Am.* **104**, 2814-2827.
18. Néel, L. (1942) Théories des lois d'aimantation de Lord Rayleigh, *Cahiers Phys.* **12**, 1-20.
19. Jiles, D., (1991) *Introduction to magnetism and magnetic materials*, Chapman and Hall, London.
20. Taylor, D. V., Damjanovic, D., and Setter, N. (1999) Nonlinear contributions to dielectric and piezoelectric properties in lead zirconate titanate thin films, *Ferroelectrics* **224**, 299-306.
21. Taylor, D. V. and Damjanovic, D. (1998) Domain wall pinning contribution to the nonlinear dielectric permittivity in $Pb(Zr,Ti)O_3$ thin films, *Appl. Phys. Lett.* **73**, 2045-2047.
22. Rammel, R. and Souletie, J. (1982) Spin Glasses in *Magnetism of metals and alloys*, edited by Cyrot M. (North-Holland, Amsterdam, 1982), p. 379-486.
23. Turik, A. V. (1963) Theory of polarization and hysteresis of ferroelectrics, *Soviet Phys.-Solid State* **5**, 885-887.
24. Turik, A. V. (1964) A statistical method for the investigation of repolarization processes in ferroelectric ceramics, *Soviet Phys.-Solid State* **5**, 1751-1753.
25. Turik, A. V. (1964) Experimental investigation of the statistical distribution of domains in a ferroelectric ceramic, *Soviet Phys.-Solid State* **5**, 2141-2143.
26. Saito, Y. (1995) Measurements of complex piezoelectric d_{33} constant in ferroelectric ceramics under high electric field driving, *Jpn. J. Appl. Phys.* **34**, 5313-5319.
27. Li, S., Cao, W., and Cross, L. E. (1991) The extrinsic nature of nonlinear behavior observed in lead zirconate titanate ferroelectric ceramic, *J. Appl. Phys.* **69**, 7219-7224.
28. Robels, U., Zadon, C., and Arlt, G. (1992) Linearization of dielectric nonlinearity by internal bias field, *Ferroelectrics* **133**, 163-168.
29. Boser, O. (1987) Statistical theory of hysteresis in ferroelectric materials, *J. Appl. Phys.* **62**, 1344-1348.
30. Kronmüller, H. (1970) Statistical theory of Rayleigh's law, *Z. Angew. Phys.* **30**, 9-13.
31. Demartin, M. and Damjanovic, D. (1996) Dependence of the direct piezoelectric effect in coarse and fine grain barium titanate ceramics on dynamic and static pressure, *Appl. Phys. Lett.* **68**, 3046-3048.
32. Taylor, D. V. (1999), Dr. Sci. Thesis, Swiss Federal Institute of Technology, .
33. Arlt, G. and Pertsev, N. A. (1991) Force constant and effective mass of 90° domain walls in ferroelectric ceramics, *J. Appl. Physics.* **70**, 2283-2289.
34. Chu, F., Damjanovic, D., and Setter, N. (1995) *An investigation of dielectric and piezoelectric properties of $Bi_4Ti_3O_{12}+Bi_3TiNbO_9$ ceramics*, Proceedings of Fourth Euroceramics (Gruppo Editoriale Feanza Editrice), p. 197-202.
35. Chu, F., Damjanovic, D., Steiner, O., and Setter, N. (1995) Piezoelectricity and phase transitions of the mixed-layer bismuth titanate niobate $Bi_7Ti_4NbO_{21}$, *J. Am. Ceram. Soc.* **78**, 3142-3144.

36. Damjanovic, D., Chu, F., Taylor, D. V., Demartin Maeder, M., Sagalowicz, L., and Duran Martin, P. (1999) Engineering of piezoelectric properties in ferroelectric ceramics and thin films, *Bol. Soc. Esp. Cer. Vid.*, in print.

37. Carl, K. and Haerdtl, K. H. (1978) Electrical after-effects in $Pb(Ti,Zr)O_3$ ceramics, *Ferroelectrics* **17,** 473-486.

38. Al-Shareef, H. N., Dimos, D., Warren, W. L., and Tuttle, B. A. (1997) A model for optical and electrical polarization fatigue in $SrBi_2Ta_2O_9$ and $Pb(Zr,Ti)O_3$, *Integrated Ferroelectrics* **15,** 53-67.

INFLUENCE OF MICROSTRUCTURE ON MICROSCOPIC AND MACROSCOPIC STRAIN BEHAVIOR OF SOFT PZT CERAMICS

MARIANNE HAMMER[*1)], AXEL ENDRISS[*1)], DORU C. LUPASCU[2)], MICHAEL J. HOFFMANN[1)]

[1)] University of Karlsruhe, Institute of Ceramics in Mechanical Engineering, Germany
[2)] Darmstadt University of Technology, Department of Materials Science, Germany

Abstract

The influence of grain size and domain size/configuration on the electromechanical microscopic and macroscopic properties of soft lead zirconate titanate (PZT) ceramics were studied. Fine- and coarse-grained PZT were prepared using the conventional mixed oxide route and different sintering conditions. The experiments were performed on PZT with different Zr/Ti-ratios in order to obtain pure tetragonal (45/55), pure rhombohedral (60/40), and morphotropic (54/46) compositions. To distinguish between the intrinsic and extrinsic effect microscopic measurements (Synchrotron X-ray diffraction) were compared with macroscopic measurements of the mechanical hysteresis loop.

1. Introduction

Lead zirconate titanate solid solution ceramics (PZT) with compositions near the morphotropic phase boundary (MPB) reveal excellent electromechanical properties due to the coexistence of the rhombohedral and tetragonal modifications at room temperature [1,2,3]. Beside the Zr/Ti-ratio, acceptor and donor dopants show a strong influence on the electromechanical properties. In general it is observed that dopants lower the lattice distortion [1,2,4-6] which should lead to a decrease of the intrinsic contribution (shift of the ions in the unit cell). Furthermore dopants influence the domain wall motion (extrinsic mechanism) which could be seen clearly on the different nonlinear behavior of the electrical and mechanical hysteresis loops of undoped, "soft", and "hard" PZT ceramics. The influence of the grain size on the electromechanical

---

[1] Now with Robert Bosch GmbH, Stuttgart, Germany

C. Galassi et al. (eds.), Piezoelectric Materials: Advances in Science, Technology and Applications, 137–147.
© 2000 *Kluwer Academic Publishers. Printed in the Netherlands.*

properties has been intensively investigated by Randall *et al.* [5]. The purpose of this paper is to study the intrinsic and extrinsic contribution to strain in fine- and coarse-grained tetragonal, rhombohedral and morphotropic soft PZT ceramics. The analysis is based on microscopic high resolution X-ray diffraction and macroscopic strain measurements. The microstructure was investigated using SEM.

2. Experimental Procedure

Processing Stoichiometric soft $(Pb_{1-3/2x}La_x(Zr_{1-y}Ti_y)O_3)$ PZT ceramics were prepared by the conventional mixed oxide process. In order to obtain the morphotropic, tetragonal and, rhombohedral phase content the Zr/Ti-ratio was fixed to 54/46, 45/55, and 60/40, respectively. It was assumed that the La-ions will substitute A-site ions only. The processing and sintering conditions are described in detail by Hammer [4]. Different grain sizes were obtained by the variation of the sintering temperature and time. For the fine- (coarse-) grained PZT compositions the sintering temperature was 1225°C (1300°C) and the isothermal sintering time was kept constant for 2h (10h).

Microstructure Microstructures of polished and etched surfaces were studied by SEM. The relative densities and grain sizes were determined by the Archimedes method and quantitative image analysis (SIS, Germany) based on more than 500 grains [4], respectively.

Synchrotron X-ray Measurements High resolution X-ray diffraction measurements were performed using synchrotron radiation with the powder diffractometer. The experimental setup and sample geometry is described in detail by Arnold et al. and Endriss et al. [7-9]. In-situ measurements with electric field strengths in the range of -25 to 25 kV/cm were performed on tetragonal and rhombohedral PZT ceramics. The lattice distortion for the tetragonal and rhombohedral samples were determined by calculating the c/a-ratio and the $111/11\bar{1}$ -ratio, respectively. The intrinsic effect, i.e. the relative shift of the ions under bipolar loading (±25kV/cm and ±32kV/cm) was determined in the direction of the maximal strain for both modifications. For the extrinsic effect, i.e. the domain wall motion the intensity-ratio $I_{(111)}/I_{(11-1)}$ (rhombohedral) and $I_{(002)}/I_{(200)}$ (tetragonal) was determined.

Macroscopic Strain The strain was monitored using an inductive linear variable displacive transducer (LVDT) in a 5kHz AC-bridge. The resolution was about 10 nm at stabilized temperature. The samples were disk shaped (10 mm diameter, 2 mm thick), polished and subsequently sputtered with Au/Pd. The electrical wiring was connected using silver epoxy. The charge flow was determined from the voltage reading on an electrometer (Keithley 6127) measuring the voltage as a capacitive voltage divider. The high voltage (HV) was applied at 20 mHz triangular bipolar voltage. For electrical insulation the samples were immersed in silicone oil.

3. Results and Discussion

The microstructural features for the different compositions and sintering conditions are listed in Table1. All compositions could be sintered close to their theoretical densities (for F_T ρ_{th}=7,74g/cm^3 and for F_R ρ_{th}=7,90g/cm^3) which were calculated considering the volume of a tetragonal and rhombohedral unit cell [4]. The 100% ρ_{rel} densities of the tetragonal compositions could be explained by the existence of secondary phases. In comparison to the Zr-rich samples the Ti-rich ones exhibit a lower PbO-vapour pressure which leads during sintering to a reaction with the PbO-rich atmosphere and subsequently to the development of PbO-rich secondary phases.

Independent of the Zr/Ti-ratio a higher sintering temperature of 1300°C and holding time of 10h lead to mean grain sizes nearly twice of those which were sintered at 1225°C for 2h (Table 1). Furthermore a broadening of the grain size distribution could be observed for all coarse-grained PZT ceramics. The influence of the Zr/Ti-ratio and the grain size on the lattice distortion is listed in Table 1. It could be seen that a decreasing grain size and an increasing Zr-content lead to a smaller lattice distortion. Considering only the influence of the grain size fine-grained tetragonal ceramics reveal an approx. 1.0% smaller lattice distortion (c/a-ratio) than coarse-grained tetragonal PZT, whereas the difference between fine- and coarse-grained rhombohedral ceramics is approx. 4.6% ($111/11\bar{1}$ -ratio). A comparison of the tetragonal and rhombohedral modification clearly shows the higher distortion of the Ti-rich tetragonal phase (45/55) which could be observed for fine- and coarse-grained PZT. This phenomenon has been explained in terms of stereochemistry [10,11].

TABLE 1. Relative density ρ_{rel}, mean grain size, and lattice distortion of 2 mole-% La-doped tetragonal (F_T), rhombohedral (F_R), and morphotropic PZT compositions sintered at 1225°C for 2h and at 1300°C for 10h.

Zr/Ti-ratio 2 mole-% La	ρ_{rel} 1225°C/2h	ρ_{rel} 1300°C/10h	grain size 1225°C/2h (μm)	grain size 1300°C/10h (μm)	distortion [%] 1225°C/2h	distortion [%] 1300°C/10h
45/55 F_T	100	100	1.6	3.0	2.77	2.80
54/46 morph	98.7	98.9	1.9	2.9	-	-
60/40 F_R	98.2	97	1.8	3.3	0.62	0.65

Figure 1 shows the bipolar strain of the rhombohedral, morphotropic, and tetragonal compositions for coarse-grained PZT. The highest macroscopic strain could be observed for the morphotropic composition (Fig. 1). The lowest strains were found in the tetragonal modification, whereas the rhombohedral modifications reached values in between. No general tendency of the influence of the grain size on the maximum bipolar strain could be stated.

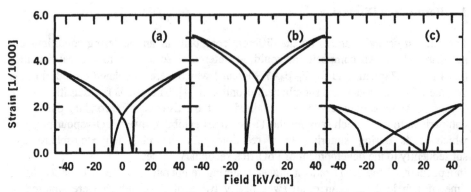

Figure 1. Bipolar strain as a function of the electric field for (a) rhombohedral, (b) morphotropic, and (c) tetragonal coarse-grained PZT.

The coercive field strongly depends on the Zr/Ti-ratio and to a much lower extent on the grain size (Fig.2b). The highest coercive field was determined for the Ti-rich tetragonal samples, which could be explained by the high lattice distortion of 2.77% and 2.80% (Fig.2a) for fine- and coarse-grained PZT, respectively. The high distortion leads to clamped domains, which are more difficult to be switched. From the shapes of the butterfly curves it can be seen that the coercive fields for the tetragonal ceramics do not show a definite value, but a range in which the domains start to align with the electric field. The coercive fields of the rhombohedral and the morphotropic PZT are well defined and only slightly different. The similar behavior between the rhombohedral and the morphotropic composition can be related to the coexistence of both phases (F_T and F_R) in the morphotropic composition yielding E_c in between the tetragonal and rhombohedral values. The fact that morphotropic PZT reacts more similar to rhombohedral PZT is due to the Zr/Ti-ratio of 54/46 in the rhombohedral-rich region relative to the morphotropic phase boundary [4].

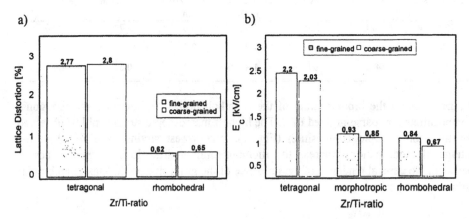

Figure 2. Influence of composition on a) the lattice distortion and on b) coercive fields (minimum values of the hysteresis).

Taking into account that the number of grain boundaries per unit volume is much larger for fine-grained ceramics, it can be deduced that the clamping of the grains will increase. Due to the overall increasing internal stress a higher amount of domain walls will be created [4,5,12-14] which can be pinned to grain boundaries. Furthermore the observed decrease in lattice distortion with the decrease in grain size should lead to an increase of internal stress in the ceramic. Consistent results were obtained by Randall et al. [5] who additionally investigated the unit cell distortion of free PZT powder which revealed a higher lattice distortion in comparison to the clamped unit cells in ceramics. Consequently, higher coercive fields are required for poling fine-grained ceramics.

To investigate the contribution of the intrinsic and extrinsic effect on the strain the tetragonal and rhombohedral compositions were investigated by high resolution X-ray diffraction measurements. Figure 3a and b schematically shows the measured effect described in detail by Endriss [9].

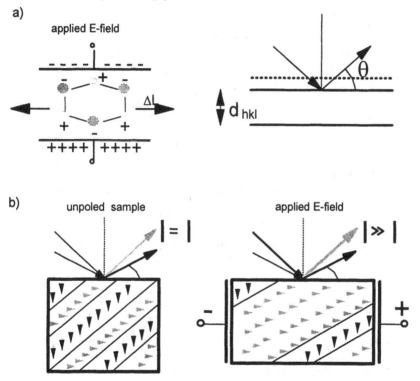

Figure 3. In-situ X-ray measurement. a) Intrinsic effect. Applied electric field leads to a shift of the ions which results in a change of the d-spacing and the corresponding deflection angle. b) Extrinsic effect. Applied electric field leads to domain wall motion which results in an increase or decrease of distinct crystal orientations and therefore in an intensity shift.

For the intrinsic effect of the fine- and coarse-grained rhombohedral PZT the change of the d_{100}-spacing under electrical loading of ±25 kV/cm was studied. Figure 4 shows the

142

typical strain behavior of a ferroelectric material. For both compositions butterfly curves with coercive fields of 6.2 and 5.2 kV/cm (fine- and coarse-grained, respectively) were obtained (Fig.4). One possible explanation of the difference to the macroscopic measurements, which reveal higher E_c's (8.4 and 6.7kV/cm, Fig.2) can be kinetic effects. The macroscopic strain measurements were obtained under much faster cycling (20mHz ≈ 50s/hysteresis-cycle) than the microscopic ones. During the X-ray diffraction measurements the domain orientation and the subsequently occuring intrinsic effect have more time to align with a slowly increasing electric field (80 min/hysteresis cycle).

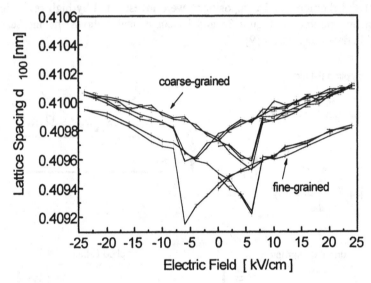

Figure 4. High resolution X-ray diffraction investigations of the intrinsic effect of fine- and coarse-grained rhombohedral PZT ceramics measured under bipolar field in direction of the maximum strain [100].

A smaller lattice distortion and a higher strain (0.07%) could be calculated from the measurements for fine-grained rhombohedral PZT ceramics.

Similar to the rhombohedral PZT the tetragonal PZT shows the intrinsic effect under electrical loading (Fig.5). Due to the already mentioned higher internal stress of the tetragonal ceramics (higher lattice distortion) and the related difficulty in poling, the measurements were performed up to ±32kV/cm. The maximum strain for fine- and coarse-grained F_T-ceramics was 1.1% at 24kV/cm.

A comparison between the microscopically and macro-scopically measured grain size dependent E_c's shows the same tendency. Independent of the Zr/Ti-ratio fine-grained PZT ceramics reveal higher E_c's which is consistent with the above made assumption that the internal stress increases with decreasing grain size. The domain wall motion occurs under higher electric fields due to the clamped microstructure.

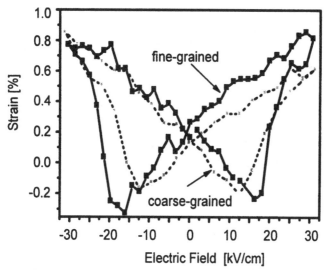

Figure 5. High resolution X-ray diffraction investigation of the intrinsic effect of fine- and coarse-grained tetragonal PZT ceramics measured in direction of the maximum strain [111].

The observed smaller strains for the microscopic measurements could be explained due to the interaction of neighboring grains [12-14]. Especially in a fine-grained ceramic, which reveals a higher volume fraction of grain boundaries than a coarse-grained one, domains could be clamped on the grain boundaries due to transgranular forces of adjacent domains. The clamping leads to a smaller d-spacing, i.e. a smaller lattice distortion (measured strain) of the observed domains with distinct orientation (parallel) to the X-ray (electric field). However, for the macroscopic measurements the decrease in strain due to clamped domains of distinct direction is compensated by domains which contribute to the overall strain even so if they are not completely aligned with the electric field. Furthermore the macroscopic measurements were obtained under much higher field strength (25 / 32kV/cm and 46kV/cm, respectively). It could be assumed that the higher loading used by the macroscopical investigations will align the domains more completely in direction of the applied electric field resulting in the observed higher maximum strains.

Figure 6a shows the intensity-ratio $I_{(111)}/I_{(11-1)}$ of the X-ray diffraction peaks from undoped rhombohedral PZT as a function of a unipolar applied electric field (up to 30 kV/cm). For the rhombohedral ceramic a considerable part of the domains align in direction of the electric field after reaching the E_c (approx. 16.5 kV/cm) during the first increase of the electric field. When the field is turned off, the intensity-ratio $I_{(111)}/I_{(11-1)}$ nearly keeps constant. Practically no reverse 71°(109°) domain wall motion occurs. Thus, a decrease in the macroscopic strain during the decrease of the electric field from its maximum value to zero is due to the intrinsic effect (Fig. 6b). Interestingly in subsequent cycles the rhombohedral composition reveals no further changes in the intensity-ratio for unipolar driving (Fig 6 c).

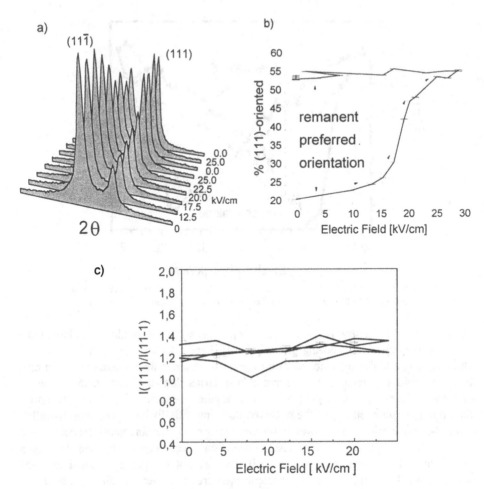

Figure 6. Extrinsic effect of rhombohedral PZT ceramics measured by X-ray diffraction under unipolar poling electric field: a) Change in intensity during poling, b) intensity ratios during poling and c) intensity change under unipolar driving after poling.

Figure 7 shows SEM images of etched surfaces of tetragonal and rhombohedral ceramics with a typical domain configuration for both modifications (90°/180° and 71°, respectively). A comparison of the tetragonal and rhombohedral ceramic shows that the density of domains per unit volume is much higher for the tetragonal ceramics. The microstructural observations are in good agreement with the assumed higher internal stress in the tetragonal modification, which develops due to the higher lattice distortion during the phase transformation.

A microstructural investigation performed by transmission electron microscopy of the grain and domain sizes [4] shows a correlation following a parabolic relationship. For grain sizes in the range of approx. 1 to 10 μm the domain sizes are proportional to the square root of the grain sizes (Fig. 8).

Figure 7. Scanning electron microscopy image of 90°/180° domain configuration in tetragonal PZT (left) and 71°/180° domain configuration in rhombohedral PZT (right).

The observation is consistent with results from Arlt for $BaTiO_3$ [12,13] and Randall *et al.* for PZT ceramics [5,14]. However, it was stated that there is a deviation to smaller domain widths for submicrometer grains which could be explained due to clamping of domain walls at the grain boundaries which increase the internal stress and thereby the domain density [5,12-14]. Grain sizes beneath 1 μm were not reached in our experiments.

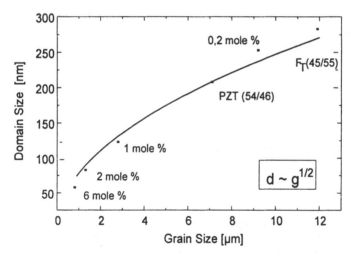

Figure 8. Parabolic relationship of grain and domain sizes in 2 mole-% la containing PZT (in the range of approx. 1 μm. to 12 μm).

4. Conclusions

Chemically and microstructurally tailored fine- and coarse-grained soft-PZT ceramics could be synthesized using different sintering conditions and Zr/Ti-ratios. The fine- and coarse-grained morphotropic, rhombohedral, and tetragonal ceramics revealed dense

146

microstructures and grain sizes of approx. 1.5 μm and 3 μm, respectively. Both, the Zr/Ti-ratio and the grain size reveal an influence on the strain behavior and piezoelectric properties under uni- and bipolar electrical loading.

Considering the grain size effect it could be stated that independent of the Zr/Ti-ratio:
- fine-grained ceramics reveal smaller lattice distortions
- fine-grained ceramics reveal higher coercive fields
- a higher domain density per unit volume with decreasing grain size

Considering the influence of the Zr/Ti-ratio it could be stated that:
- an increasing Zr-content leads to a decreasing lattice distortion
- an increasing Ti-content leads to increasing coercive fields
- a higher domain density per unit volume develops with increasing Ti-content
- the highest strain is observed for morphotropic PZT
- after poling, the strain in rhombohedral PZT is dominated by the intrinsic effect
- the strain in tetragonal PZT is due to intrinsic and extrinsic contributions

According to the microscopic and macroscopic results it can be summarized that an increase of internal stress in PZT ceramics leads to higher coercive fields and stronger contributions of ferroelastic domain switching. However, different origins of the increase of internal stress have to be considered. First the influence of grain size and second the influence of the Zr/Ti-ratio.

Acknowledgment   The authors would like to thank the DFG for financial support under contract No. Ho 1156/3 and Lu729/1.

5. References

1. Jaffe, B. Roth, R.S. and Marzullo, S. (1954) Piezoelectric Properties of Lead Zirconate-Lead Titanate Solid-Solution Ceramics, *J. Appl. Phys.*, **25**, 809.
2. Xu, Y. (1991) *Ferroelectric Materials and Their Applications*, Elsevier Sci. Pub, Amsterdam, Netherlands.
3. Endriss, A. (1995) *Reorientierungsverhalten von Domänen und mikroskopische Gitterdeformation in piezoelektrischen PZT-Keramiken*, Ph.D. Thesis, University of Tübingen, Germany.
4. Hammer, M. (1996) *Herstellung und Gefüge-Eigenschaftskorrelationen von PZT-Keramiken*, Ph.D. Thesis, University of Karlsruhe, Germany.
5. Randall, C.A. Kim, N. Kucera, J.-P. Cao, W. Shrout, T.R. (1998) Intrinsic and Extrinsic Size Effects in Fine-Grained Morphotropic-Phase-Boundary Lead Zirconate Titanate Ceramics, *J. Am. Ceram. Soc.* **81**, 677-88.
6. Hammer, M. Monty, C. and Hoffmann, M.J. (1998) Correlation of Surface Texture and Chemical Composition in Undoped, Hard and Soft Piezoelectric PZT Ceramics, *J. Am. Ceram. Soc.* **81**, 721-24.
7. Arnold, H. Bartl, H. Fuess, H. Ihringer, J. Kosten, K. Löchner, U. Pennartz, P.U. Prandl, W. Wroblewski, T. (1989) New Powder Diffractometer at HASYLAB/DESY, *Rev. Sci. Instrum.* **60**, 2380-2381.
8. Endriss, A. Ehrenberg, H. Fischer, B. Hammer, M. Hoffmann, M.J. Knapp, M. (1997) Influence of grain size, Ag- and La-doping on piezoelectric strain in PZT ceramics, *HASYLAB Jahresbericht* **551**.

9. Endriss, A. Hammer, M. Hoffmann, M.J. Kolleck, A. Schneider, G.A. (1999) Microscopic and Macroscopic Ferroelectric-Ferroelastic and Piezoelectric Behavior of PZT Ceramics, *J. European Ceram. Soc.* **19**, 1229-1231.

10. Thomann, H. (1990) Piezoelectric Ceramics, *Adv. Mater.* **2**, 458-463.

11. Heywang, W. and Schöfer, R. (1965) Aufbau ferroelektrischer Phasensysteme in Perowskiten, *Z. Angew. Phys.* **20**, 10-15.

12. Arlt, G. Perttsev, A. (1991) Force Constant and Effective Mass of 90° Domain Walls in Ferroelectric Ceramics, *J. Appl. Phys.* **70**, 2283-2289.

13. Arlt, G. (1990) Twinning in Ferroelectric and Ferroelastic Ceramics: Stress Relief, *J. Mater. Sci.* **25**, 2655-2666.

14. Cao, W. Randall, C.A. (1996) Grain Size and Domain Size Relations in Bulk Ceramic Ferroelectric Materials, *J. Phys. Chem. Solids* **57**, 1499-1505.

THE EFFECT OF STATIC COMPRESSIVE STRESS ON THE HIGH FIELD DIELECTRIC PROPERTIES OF HARD PZT CERAMICS

D A HALL, P J STEVENSON AND S W MAHON*

Materials Science Centre, University of Manchester and UMIST, Grosvenor St., Manchester M1 7HS, UK.
* Mechanical Sciences Sector, Building A7, Room 1079, DERA Farnborough, Hampshire GU14 0LX, UK.

Abstract

The application of a static compressive stress to hard PZT ceramics was found to cause significant increases in dielectric permittivity and loss, which were attributed mainly to a stress-induced deageing mechanism caused by ferroelastic 90 ° domain switching. The $\varepsilon_r^{'}$, $\varepsilon_r^{''}$ and tan δ values all decreased gradually during subsequent ageing under stress. Comparable increases in the dielectric coefficients, and subsequent ageing behaviour, were observed on releasing the stress, which were related to recovery of the original domain configuration. The results can be understood in terms of a time-dependent domain wall stabilisation process caused by the reorientation of dipolar defect associates.

1. Introduction

Piezoelectric ceramics represent a well-proven technology for the fabrication of electromechanical transducers. However, despite their widespread applications there are still many aspects of their behaviour that are poorly understood. This is particularly true of the nonlinear effects that characterise their dielectric, elastic, and piezoelectric properties. Only relatively recently, good progress is being made in the characterisation and understanding of nonlinearity in piezoelectric ceramics; the nonlinear properties are usually attributed to a ferroelectric domain wall contribution [1]. A current trend in this research is a drive towards a more-complete understanding of the materials' behaviour under a wider range of measurement conditions (e.g. high field, high stress), which serve to mimic the environmental conditions within practical devices. This is an important step forward, since the functional properties of piezoelectric ceramics often vary considerably with the external loading conditions.

149

C. Galassi et al. (eds.), Piezoelectric Materials: Advances in Science, Technology and Applications, 149–157.

The focus of the present study is on the dielectric properties of hard PZT ceramics, measured under a high AC electric field and static uniaxial compressive stress. The choice of these particular conditions is motivated by the use of such materials in high power acoustic transducers, where the application of a static compressive load is necessary in order to obviate any potential problems caused by mechanical failure. Several papers on this topic have been published previously, notably the detailed experimental study of Krueger [2] and the review of early work by Butler et al. [3]. Earlier studies tended to focus on the practical implications of increased loss and irreversible depoling, which could result from the application of a high stress. More recently, a number of papers have appearing concerned with more fundamental aspects of the problem [4,5]. The present study was initiated following observations at DERA (Farnborough) that the dielectric loss of a hard PZT ceramic increased considerably when a static uniaxial compressive stress was applied. This effect could potentially have an important influence on the performance and stability of high power acoustic transducers. The aims of the work were to characterise the effect and to identify the underlying mechanisms responsible for it.

2. Experimental Procedures

The ceramic material used in the study was a hard PZT, type PC8, manufactured by Morgan Matroc (Unilator Division, Ruabon, UK). Specimens were provided in the form of electroded and poled disks, with a thickness of approximately 1 mm and a diameter of 10 mm. Previous studies indicated that the material had a tetragonally-distorted perovskite crystal structure with a c/a ratio of 1.019. The grain size was determined by the line-intercept method as 4.2 ± 0.5 μm and the Curie point was approximately 290 °C.

A modified mechanical loading device (Hounsfield Tensometer) was used to apply a static uniaxial compressive stress to the specimen through 2 brass electrodes. Alumina spacers were employed to isolate the frame of the mechanical testing device from the high voltage supply. Stresses up to 100 MPa were readily achievable using this arrangement. For the high field measurements, a test signal was generated using a computer-controlled Hewlett Packard HP 33120A function generator in combination with a Trek 609D-6 high voltage amplifier. By using an intermittent 'burst-mode' signal comprising two complete sinusoidal cycles, and with due regard to the stable specimen polarity, it was possible to obtain a series of high field measurements up to a field amplitude of 3.5 kV mm^{-1} without causing any field-induced 'deageing' effects. The current induced by the applied field was measured using a Stanford Research Systems SR570 current amplifier; the current and voltage signals were then downloaded to a PC via a Tektronix TDS 420 DSO. Numeric integration of the current signal yielded the change in polarisation, thereby enabling construction of the P-E (polarisation-electric field) relationship. The effective high field dielectric coefficients ε_r', ε_r'' and tan δ were then calculated from the measured P-E curves according to the method described previously [6].

Preliminary measurements indicated that the dielectric permittivity and loss of hard PZT ceramics increased considerably in response to an applied compressive stress, confirming the observations made by previous authors. However, it was also immediately obvious that the initial high values then decayed gradually during ageing under constant stress. A rapid increase in dielectric permittivity was again obtained as the stress was released at a given ageing time; this was followed by a further reduction in values during ageing at zero stress.

Therefore, a series of experiments was devised in which a relatively low stress of 20 MPa was applied to a well-aged, poled specimen and then held constant for a period of 24 hours. The change in dielectric properties during ageing was monitored periodically over a period of 17 hours using a field amplitude E_0 of 1.0 kV mm^{-1}, and then the field-dependence of dielectric behaviour was determined at a fixed ageing time of 20 hours. The stress was released after the 24 hour period and the subsequent change in dielectric properties during ageing at zero stress was monitored over a period of 17 hours. Higher stresses of 60 MPa and then 100 MPa were then investigated according to the same sequence of stress application, ageing, and release. All measurements were conducted at a frequency of 1 Hz and a temperature of 20 ± 2 °C.

3. Results and Discussion

3.1. AGEING OF HIGH FIELD DIELECTRIC COEFFICIENTS AFTER APPLICATION OF COMPRESSIVE STRESS

As noted above, the dielectric coefficients increased remarkably on the first application of a compressive stress. For example, the value of ε_r' (for $E_0 = 1.0$ kV mm^{-1}) increased from 1080 to 1590 due to an applied stress of 100 MPa. The variation in dielectric loss was even more dramatic, with ε_r'' increasing from 20 to 157 and tan δ from 0.018 to 0.099 for the same stress. Similar observations were made at the lower stress levels of 20 MPa and 60 MPa, although the variations in values were somewhat less pronounced. However, it was observed that all of the dielectric coefficients exhibited a pronounced ageing effect and began to reduce with time under load, as shown in fig. 1. These observations can be attributed to a stress-induced deageing effect, since the applied compressive stress must induce some 90 ° ferroelectric domain switching, the extent of which increases with the applied stress level.

152

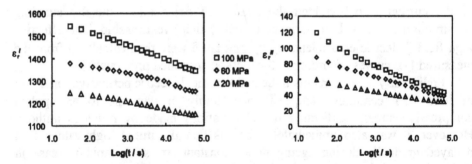

Figure 1. Variation in high field dielectric coefficients during ageing under various levels of applied compressive stress.

3.2. FIELD-DEPENDENCE OF DIELECTRIC COEFFICIENTS UNDER COMPRESSIVE STRESS

The dielectric coefficients measured after ageing for 20 hours under various compressive stresses all showed significant increases as a function of the applied field amplitude E_0, as shown in fig. 2. In contrast to previous results for hard PZT ceramics [7], the 'threshold field' E_t for dielectric nonlinearity and the linear portion in the $\varepsilon_r'-E_0$ relationship (the so-called Rayleigh behaviour) were not well-defined. Instead, a gradual increase in the gradient of the $\varepsilon_r'-E_0$ curves was evident over the field range up to 2.0 kV mm^{-1}. This curvature in the relationship may be related to the relatively fine grain size of the material and an overlap of the contributions from different domain-related polarisation mechanisms [7].

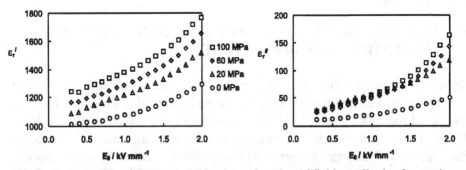

Figure 2. Dielectric coefficients obtained as a function of field amplitude after ageing for 20 hours under various levels of applied compressive stress.

There was a clear separation of the $\varepsilon_r'-t$ and $\varepsilon_r'-E_0$ curves as a function of the applied stress level, indicating that the stress may have had a permanent effect on the stable permittivity value. In contrast, the stress level appeared to have little effect on the ε_r'' values for field amplitudes below 1.5 kV mm^{-1}, although the curve obtained for the well-aged 'initial' condition lay distinctly below the other ones. It is likely that the ε_r''–E_0 curves obtained under stress would tend towards this 'initial' curve at sufficiently long ageing times.

The polarisation values obtained at high field levels did not show a tendency for saturation up to 3.5 kV mm^{-1}, as shown in fig. 3. These results indicate that the coercive field for this material was in the region of 3.0 kV mm^{-1}; in this case, it is expected that there would be no significant field-induced domain switching for field amplitudes below 2.0 kV mm^{-1}.

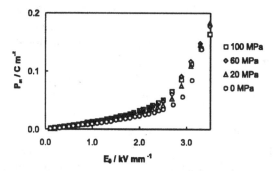

Figure 3. Maximum polarisation change P_m as a function of field amplitude after ageing for 20 hours under various levels of applied compressive stress.

3.3. AGEING OF HIGH FIELD DIELECTRIC COEFFICIENTS AFTER REMOVAL OF COMPRESSIVE STRESS

The ε_r' and ε_r'' values obtained after removal of the stress showed a qualitatively similar ageing behaviour to that observed on application of the stress, as can be seen by comparison of figures 1 and 4. The extent of 90 ° domain switching on removal of the stress was clearly larger as the stress level increased, leading to stronger initial increases in ε_r' and ε_r''.

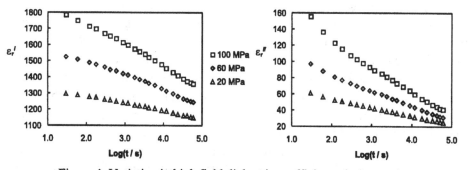

Figure 4. Variation in high field dielectric coefficients during ageing after removal of applied compressive stress.

The resulting reductions in ε_r' and ε_r'' during ageing were then more pronounced for higher stress levels. The ε_r'-t curves obtained for the different stress levels were again well-separated, but there was some indication that they might be tending towards a common ultimate value.

3.4. A MODEL FOR THE STRESS-INDUCED AGEING AND DEAGEING MECHANISMS

It was established by previous authors that the very low loss in well-aged acceptor-doped (hard) ferroelectric ceramics is due to a strong domain wall stabilisation effect, caused by the time-dependent reorientation of acceptor ion-oxygen vacancy defect associates (e.g. $A_{Ti}''-V_O^{\bullet\bullet}$) towards the local domain polarisation [8]. This results in a gradual increase in the domain wall force constant and thereby the domain wall contributions to the dielectric, elastic and piezoelectric coefficients are reduced. Therefore, it is apparent that the 90 ° domain switching induced by the applied compressive stress has the effect of partially negating the domain wall stabilisation process. For the regions within which the orientation of polarisation is switched by 90 °, the domain wall force constant is much reduced, yielding the observed increases in ε_r' and ε_r''. Immediately after the 90 ° domain switching has occurred, under the influence of the applied stress, the non-favourably oriented defect associates start to re-orient towards the new local domain polarisation and so a new ageing process is initiated. During ageing under load, the dielectric coefficients reduce in response to the gradually increasing degree of domain stabilisation.

When the load is released after a certain ageing period, most of the switched domains are able revert back immediately to their original configurations corresponding to the zero stress state [5], and so again a new ageing process is initiated as the defect associates tend to re-orient back towards their initial directions. The proposed ageing and deageing mechanisms induced by the applied compressive stress are illustrated schematically in fig. 5. Here, the model is based on the basic laminar 90° domain stack which has been observed many times in tetragonal perovskite ferroelectric ceramics.

In order to gain a more complete understanding of the observed phenomena, it is instructive to consider the different possible contributions to the dielectric properties of these hard PZT ceramics. At low fields, ε_r' is the sum of an intrinsic ionic contribution ε_i' and a contribution from small-scale reversible domain wall vibrations ε_{dv}', while the loss ε_r'' is due solely to the lossy domain wall vibration contribution ε_{dv}''. For field levels greater than the 'threshold field' E_t, additional field-dependent contributions arise as a result of a larger-scale domain wall translation mechanism, which can be denoted as ε_{dt}' and ε_{dt}''. We should expect that each of the domain-related contributions will reduce during ageing, whereas ε_i' should remain essentially constant at a fixed stress.

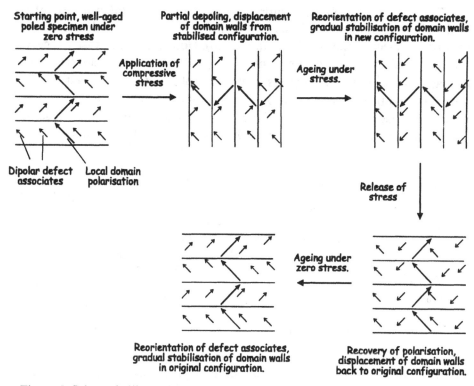

Figure 5. Schematic illustration of ageing and deageing mechanisms caused by 90 °
ferroelectric domain switching.

Previous authors have demonstrated that it is possible to separate the different contributions to ε_r' by plotting the dielectric results obtained during ageing on the complex permittivity plane [9]. Ideally, a single lossy polarisation mechanism should yield a linear relationship, which can be extrapolated to zero loss to give ε_i'. In the present case, a set of nonlinear curves was obtained as a function of ageing time, as shown in fig. 6(a). This can be interpreted as being due to the superposition of 2 lossy polarisation mechanisms (i.e. domain wall vibration and domain wall translation), both of which are affected by domain wall stabilisation during ageing. Each of the curves obtained was fitted with a quadratic function, yielding intrinsic ionic permittivity values of 897, 1032 and 1134 for compressive loads of 20, 60 and 100 MPa respectively. Therefore, it seems reasonable to deduce that the intrinsic ionic permittivity increases significantly under the applied compressive stress, as was concluded by previous authors [2,5]. This increase in ε_i' as a function of the applied stress may itself be due to a combination of 2 factors i.e. the effect of compressive stress applied along the tetragonal c-axis and the influence of partial 90 ° domain switching (since $\varepsilon_a > \varepsilon_c$).

156

The ε_r'-ε_r'' curves obtained during ageing at zero stress (after release of the load) showed similar evidence of contributions from 2 domain-related polarisation mechanisms, as shown in fig. 6(b). However, in this case the curves showed a tendency towards a common 'zero loss' value of around 900, which indicates that ε_i' in the 'zero stress' state was not affected significantly by the previously applied stress. Further measurements of ε_r' and ε_r'' in the low field region are necessary to verify these conclusions, since in that case the extrapolation to zero loss should be much more precise and reproducible.

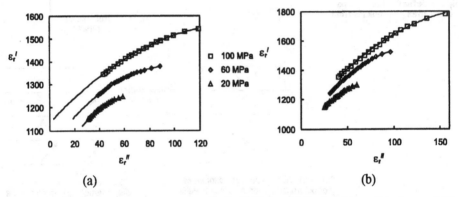

(a)   (b)

Figure 6. ε_r'-ε_r'' plots obtained as a function of ageing time after (a) application and (b) release of various compressive stresses ($E_0 = 1$ kV mm^{-1})

A complex permittivity plot was also constructed from the field-dependent dielectric data obtained at a fixed ageing time of 20 hours, as shown in fig. 7. In this case, the linear ε_r'-ε_r'' relationship can be attributed to the single lossy polarisation mechanism of domain wall translation. The domain wall vibration mechanism is assumed to be independent of field amplitude, and so does not distort the linear plot.

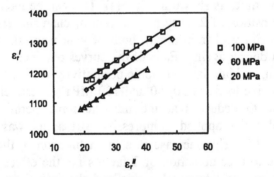

Figure 7. ε_r'-ε_r'' plots obtained as a function of field amplitude for various applied stresses (ageing time = 20 hours).

4. Conclusions

The initial increases in $\varepsilon_r^{'}$, $\varepsilon_r^{''}$ and tan δ on application of a compressive stress were due largely to a stress-induced deageing effect caused by ferroelastic 90 ° domain switching. Measurements of the high field dielectric properties during subsequent ageing under stress indicated that two distinct domain related polarisation mechanisms were active, both of which were affected strongly by time-dependent domain stabilisation processes; these mechanisms were identified as small scale domain wall vibration and large scale domain wall translation. The results also indicated that the intrinsic ionic permittivity of these tetragonal PZT ceramics increased significantly under stress, in agreement with previous studies.

Acknowledgements

This work was carried out as part of Technology Group 01 (Magnetics, Acoustics & Hydrodynamics) of the MoD Corporate Research Programme. The technical assistance of Mr I G Easdon, Mr G Ingham and Mr G Pickles of the Materials Science Centre is also gratefully acknowledged.

References

1. Damjanovic, D. (1998) Ferroelectric, dielectric and piezoelectric properties of ferroelectric thin films and ceramics, *Rep. Prog. Phys.* **61**, 1267-1324.
2. Kreuger, H.H.A. (1967) Stress sensitivity of piezoelectric ceramics: Part 1. Sensitivity to compressive stress parallel to the polar axis, *J. Acoust. Soc. Am.* **42**, 636-645.
3. Butler, J.L., Rolt, K.D. and Tito, F.A. (1994) Piezoelectric ceramic mechanical and electrical stress study, *J. Acoust. Soc. Am.* **96**, 1914-17.
4. Damjanovic, D. and Demartin, M (1997) Contribution of the irreversible displacement of domain walls to the piezoelectric effect in barium titanate and lead zirconate titanate ceramics, *J. Phys.: Condensed Matter 9*, 4943-4953.
5. Guyomar, D., Audigier, D. and Eyraud, L. (in press), Characterisation of piezoceramic under uniaxial stress, *Proc. ISAF '98*.
6. Hall, D.A., Ben-Omran, M.M. and Stevenson, P.J. (1998) Field and temperature dependence of dielectric properties in $BaTiO_3$-based piezoceramics, *J. Phys.: Condensed Matter* **10**, 461-476.
7. Hall, D.A. and Stevenson, P.J. (in press) High field dielectric behaviour of ferroelectric ceramics, *Ferroelectrics*.
8. Robels, U. and Arlt, G. (1993) Domain wall clamping in ferroelectrics by orientation of defects, *J. Appl. Phys.* **73**, 3454-3460.
9. Dederichs, H. and Arlt, G. (1986) Ageing of Fe-doped PZT ceramics and the domain wall contribution to the dielectric constant, *Ferroelectrics* **68**, 281-292.

STRESS DEPENDENCE OF THE PIEZOELECTRIC, DIELECTRIC AND ELASTIC PROPERTIES OF PZT CERAMICS

S.W. MAHON, D. MOLONEY, F. LOWRIE, A.R. BOWLES
Transducer Materials Group, Mechanical Sciences Sector, DERA
Farnborough, Farnborough, Hampshire, GU14 0LX, United Kingdom

Abstract

This study describes the experimental development of techniques to determine the transverse and longitudinal properties of piezoelectric materials under static uniaxial stress at low drive levels (<1 V) based on the resonant (IEEE) analysis. This technique offers the significant advantage that it is the only viable route that affords the possibility of calculating the full elasto-piezo-dielectric tensor, and therefore allows the development of predictive models of materials performance in high power devices.

Mechanical stress and its direction of application with respect to the polarisation axis dramatically affect piezoelectric materials. In most transducers the uniaxial compressive stress is applied parallel to the polarisation and this study shows that under these conditions the piezoactivity is enhanced ($\sim$ 40% in 4 MPa stress) in the polarisation direction but decreases laterally (*ie.* a reduction in d_{31}). These changes are accompanied by similar anisotropic variations in the elastic moduli, with c_{33}^E and s_{11}^E decreasing by $\sim$ 25%. These results can be explained by a consideration of the microscopic domain processes that occur under uniaxial compressive stress where the formation of 90° domain walls is enhanced. This new data is important in understanding the variation in materials properties that occur under operational conditions, and this information is essential in both materials selection and the design of high power devices.

C. Galassi et al. (eds.), Piezoelectric Materials: Advances in Science, Technology and Applications, 159–168.

1. Introduction

Piezoceramics remain the preferred mechanism of electromechanical transduction in many applications including high power acoustic projectors and precision actuation by virtue of their low cost and high efficiency. Existing piezoceramic materials have only a moderate strain which means they must be operated at increasingly higher electrical drive fields to achieve greater output power. To prevent the ceramic being exposed to tensile stress the material must be driven under a large static pre-stress.

The electromechanical performance of a material is described numerically by three sets of matrix coefficients describing the piezoelectric, dielectric and elastic properties of the material. These are the parameters required for finite element modelling and each coefficient may be a function of frequency, field, stress and temperature. If the performance of a device is to be accurately predicted then all the matrix coefficients are needed over the full range of conditions that the materials may experience in operation.

The standard techniques for determining these parameters (IEEE standard 176 [1]) in piezoelectrics are only applicable at very low fields and at zero stress where non-linearities and irreversibilities can be neglected. However, at the high stress levels, whether electrical and mechanical, which are often required in practical devices, a variation in the electromechanical coefficients that characterise the performance of an electroceramic material occurs. Using quasi-static techniques, several authors [2, 3] have shown that associated with an increase in applied stress is an increase in both the dielectric and mechanical losses that impose the limitations on acoustic power and actuation efficiency of a device. It is essential, therefore, in transducer design and materials selection that the piezoelectric material is characterised under the conditions of stress that it will experience in operation.

This paper describes the experimental development of techniques to determine the transverse and longitudinal properties of materials under static uniaxial stress at low drive levels (~1 V). Previous authors [4, 5] have reported the variation observed in selected materials parameters under high stress, but this is the first attempt to extend the resonant analysis to these conditions. This technique offers the significant advantage that it is the only route that permits the calculation of the full elasto-piezo-dielectric tensor and therefore allows the development of predictive models of materials performance in high power devices.

2. Background Theory

In piezoelectric materials the electrical and elastic variables are coupled and are described by a linear equation of the form

$$D = dT + \varepsilon^T E$$
$$S = s^E T + dE$$

(1)

where D, T, E and S are the dielectric displacement, stress, electric field and strain respectively. The constants d, ε^T and s^E are the piezoelectric constant, dielectric permittivity (at constant stress) and the elastic compliance (at constant field) respectively. The constitutive equations can be written in four other forms depending on the thermodynamic potentials used to derive them and thus four related piezoelectric constants can be defined d, g, e, h. These four piezoelectric constants are all related but each represents a different aspect of the piezoelectric relationship and is useful for a different set of conditions. For example, d measures the strain in a free crystal for a given applied field, e the stress developed by a given field when the crystal is clamped, g the open-circuit voltage for a given stress, and h the open-circuit voltage for a given strain.

The constitutive relationship can be expressed in more convenient matrix form by using the symmetrical aspects of a poled ceramic and the degree of redundancy that exists in the tensor quantities, such that;

$$
\begin{pmatrix} S_1 \\ S_2 \\ S_3 \\ S_4 \\ S_5 \\ S_6 \\ D_1 \\ D_2 \\ D_3 \end{pmatrix} = \begin{pmatrix} s_{11}^E & s_{12}^E & s_{13}^E & 0 & 0 & 0 & 0 & 0 & d_{13} \\ s_{12}^E & s_{11}^E & s_{13}^E & 0 & 0 & 0 & 0 & 0 & d_{13} \\ s_{13}^E & s_{13}^E & s_{33}^E & 0 & 0 & 0 & 0 & 0 & d_{33} \\ 0 & 0 & 0 & s_{55}^E & 0 & 0 & 0 & d_{15} & 0 \\ 0 & 0 & 0 & 0 & s_{55}^E & 0 & d_{15} & 0 & 0 \\ 0 & 0 & 0 & 0 & 0 & s_{66}^E & 0 & 0 & 0 \\ 0 & 0 & 0 & 0 & d_{15} & 0 & \varepsilon_{11}^T & 0 & 0 \\ 0 & 0 & 0 & d_{15} & 0 & 0 & 0 & \varepsilon_{11}^T & 0 \\ d_{13} & d_{13} & d_{33} & 0 & 0 & 0 & 0 & 0 & \varepsilon_{33}^T \end{pmatrix} \begin{pmatrix} T_1 \\ T_2 \\ T_3 \\ T_4 \\ T_5 \\ T_6 \\ E_1 \\ E_2 \\ E_3 \end{pmatrix} . \quad (2)
$$

For low loss materials the IEEE Std 176-1987 outlines the admittance resonance techniques necessary for the determination of the above tensor which is sufficient to fully describe the behaviour of a piezoelectric ceramic. Using this method a small AC voltage excites a piezoelectric sample and the phase and magnitude of the current with respect to the excitation voltage is monitored and the AC impedance of the sample as a function of frequency is found. The parallel resonance frequency f_p is defined to be the frequency at which the maximum resistance occurs. The maximum value of the conductance spectrum represents the series resonance frequency f_s. By using the series and parallel frequencies and their harmonics for a range of vibrational modes and sample geometries the IEEE procedure outlines the calculation of the elasto-piezo-dielectric matrix shown in equation 2 for zero loss materials. However, it is well established that on the application of a uniaxial stress the losses in a piezoceramic increase. To accommodate these non-negligible losses the admittance resonance equations can be derived in terms of complex material coefficients, and these methods based on the work of Smits [6] and Sherrit [7] have been automated in the Piezoelectric Resonance Analysis Program (PRAP) which has recently been reviewed elsewhere [8].

The PRAP has been used to analyse the resonance spectra determined for both transverse and longitudinal properties of a 'hard' PZT material, PZ26 (density = 7700 kg.m^{-3}) (Ferroperm Ltd.) which is suitable for use in high power transducers. Figure 1 shows a schematic diagram of the resonator geometries investigated in this study.

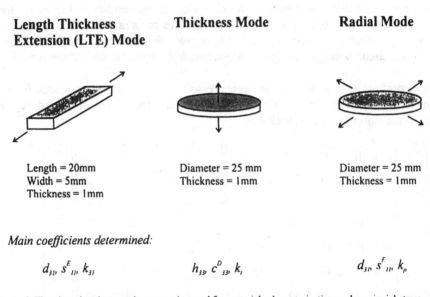

Length Thickness Extension (LTE) Mode	Thickness Mode	Radial Mode
Length = 20mm Width = 5mm Thickness = 1mm	Diameter = 25 mm Thickness = 1mm	Diameter = 25 mm Thickness = 1mm

Main coefficients determined:

d_{31}, s^E_{11}, k_{31} ⠀⠀⠀⠀ h_{33}, c^D_{33}, k_t ⠀⠀⠀⠀ d_{31}, s^F_{11}, k_p

Figure 1. The piezoelectric ceramic geometries used for materials characterisation under uniaxial stress. In each case the sample is poled normal to the electrode face.

An important parameter for a high power transducer material is the coupling coefficient (k) that expresses the efficiency of the electromechanical conversion process and determines the bandwidth over which a device may operate. This coefficient is readily calculated from the d coefficient, stiffness and permittivity, and is given as:

$$k^2 = \frac{d^2}{\varepsilon^T s^E}.$$ (3)

3. Development of Characterisation Techniques

To determine the resonant impedance spectra under high static stress a new experimental mounting fixture has been developed which is shown schematically in Figure 2. The admittance spectra were determined using a HP4192a impedance analyser. A load cell monitors the force exerted on the sample. The whole stainless steel assembly is small enough to fit inside a thermal chamber enabling materials properties to be determined as a function of temperature. To ensure that the impedance spectra are determined under the condition of static stress and without clamping or interference from the mounting system there are several features which must be addressed:

- The disc spring arrangement, which is mounted on guide shafts, is used to exert the static pre-stress. At resonance the vibrational amplitude of the piezoceramic will increase causing an increased compression of the springs, which in turn leads to an increase in applied force and thus pre-stress. However, this dynamic loading can be minimised by combining the springs in an alternating (series) arrangement such that their dynamic spring constant accommodates any dynamic stress generated during excitation. The spring arrangement used in this experiment reduced the dynamic stress to < 5% of the static loading.

- The sample must be acoustically decoupled from the stainless steel assembly to avoid exciting mechanical system resonances in the press that will interfere with the electromechanical impedance spectrum of the ceramic sample under test. This requires a decoupling medium with significantly different acoustic impedance to both ceramic and stainless steel, making syntactic foams suitable. Syntactic foams are closed cell foams that are manufactured by adding rigid spheres to a rigid epoxy resin matrix. They are designed to have low densities (e.g. 500 - 700 $kg.m^{-3}$) coupled with high stiffness. The two decoupling foams used in this study have a formulation based on a rigid epoxy resin system. This system is designed to have low viscosity and slow curing characteristics, facilitating the incorporation of large volume fractions (VF) of microspheres while retaining good rigidity. Two types of sphere were added to the resin in this investigation. Z1000 spheres are rigid ceramic spheres of fairly high density ($\sim$700 $kg.m^{-3}$) and GMB230 microspheres are a low density ceramic grade ($\sim$250 $kg.m^{-3}$). The properties of the decoupling media are summarised in Table 1. The 3480 foam containing the higher volume fraction of microspheres has been found to offer both better decoupling and compression resistance, and this composition was used for all the piezoelectric measurements presented in this study.

Table 1. Summary of the properties of the acoustic decoupling media

Database N°	Microsphere	VF Micro-sphere %	VF Air %	Density $kg.m^{-3}$	Velocity $m.s^{-1}$	ρc Rayls (x10⁶)
3485	unfilled	0	0	1120	-	-
3483	Z1000	39	22	680	2521	1.71
3480	GMB230	50	0	640	2455	1.57
PZT ceramic	-	-	-	7587	5188	39.36

- It is important to ensure that the sample experiences a uniaxial stress only and thus lateral clamping effects must be eliminated as these will introduce planar compressive

164

stresses and violate the assumption of a traction-free interface which is an explicit boundary condition for the derivation of the reduced matrix shown in equation 2. Lateral stresses arising from frictional forces at the sample-press interface act to superimpose an apparent hydrostatic pressure on the axial load. Thus, to achieve the desired condition of axial stress the coefficient of friction at this interface must be reduced, and this has been achieved by inserting a thin graphite disc between the sample and the aluminium foil electrode. By introducing the low friction graphite layer care must be taken to avoid corruption of the resonance spectrum by introducing a resistive component into the equivalent circuit. Therefore, before measuring impedance spectra with the HP4192a analyser both the open and short circuit corrections were performed to eliminate parasitic impedances due to the experimental mounting and graphite layer.

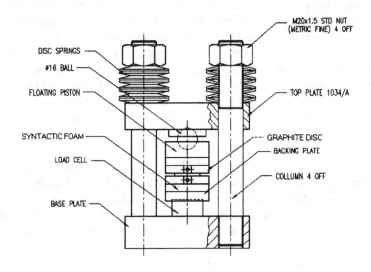

Figure 2. Experimental test fixture for resonance measurements under high static stress

4. Experimental Results

The dependence of the material parameters on moderate uniaxial compressive stress (< 5MPa) has been measured. The impedance spectra were determined at low field (<<1 V_{AC}) under uniaxial compressive loads using the experimental test fixture shown in figure 2 and analysed using PRAP. The investigated spectra included those from a disc oscillating in its thickness and radial modes and a long ceramic bar vibrating in its length thickness extension mode. These resonant modes allow the determination of both longitudinal and transverse components of the material constants. In all cases the uniaxial stress was parallel to the polarisation of the piezoceramic.

Figures 3a and 3b show the variation in the longitudinal piezoelectric (h_{33}) and elastic stiffness (c_{33}) constants determined from the electrical impedance spectra as a function of static stress. All parameters exhibit a strong pressure dependence. The piezoelectric constant (h_{33}) increases by 43% in only 4 MPa compressive stress and this is accompanied by a 20% decrease in the ceramic stiffness. These changes are complementary in that a more compliant material will offer larger strains and therefore increased piezoelectric coefficients.

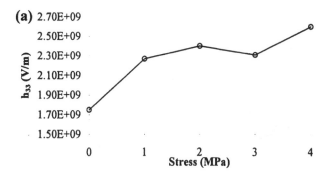

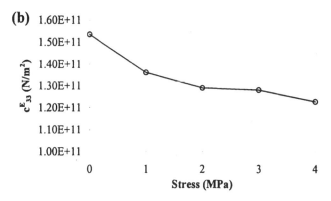

Figure 3. Variation in material properties of 'hard' PZT (PZ26) under uniaxial stress determined from the thickness resonance

Figure 4 (a-c) shows the variation in the lateral piezoelectric (d_{31}), elastic compliance (s_{11}) and lateral coupling (k_{13}) constants as a function of uniaxial static stress up to 4 MPa as determined from the LTE resonant mode. As in the longitudinal parameters the materials coefficients are strongly pressure dependent, particularly at low stress levels. The lateral piezoelectric constant (d_{31}) decreases by 38% while the elastic compliance (s_{11}) decreases by ~30%, i.e. the ceramic becomes stiffer laterally. The coupling coefficient (k_{13}) decreases by ~27%. Figure 4b compares the piezoelectric coefficient (d_{31}) determined from both the LTE and the radial modes, and they agree well and show the same monotonic decrease with applied load.

166

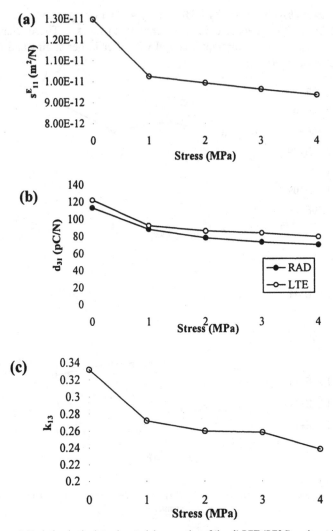

Figure 4. Variation in the lateral material properties of 'hard' PZT (PZ26) under uniaxial stress

5. Discussion

The results show that under moderate uniaxial stress applied parallel to the polarisation direction the piezoelectric activity increases longitudinally but decreases laterally. Similarly the ceramic becomes more compliant in the longitudinal direction but more stiff laterally. The efficiency of electromechanical conversion denoted by the coupling coefficients shows a significant increase parallel to the polarisation but decreases in the transverse direction. These results can be explained by a consideration of the microscopic domain processes that occur under uniaxial compressive stress.

Within the ceramic, each crystallite is composed of domains whose configuration reflects the underlying crystal symmetry. The domains orientate to minimise the total energy of the ceramic and in particular to reduce the electrostatic and domain wall energy contributions. In tetragonal ferroelectric materials there exist two types of domain boundary, 180° boundaries and 90° boundaries.

Applied stress fields cause 90° domain boundaries to be displaced and to allow growth of domains with the longer tetragonal axis most favourably oriented perpendicular to the applied stress. Compressive stress parallel to an elongated tetragonal axis in a crystallite will tend to cause growth of domains perpendicular to this axis by movement of 90° walls, with associated shrinkage in the compressive stress direction. The influence of mechanical stress on 90° domain walls is the main cause of the variations of parameters with mechanical stress fields in ferroelectric ceramics and the influence of stress on parameters can usually be predicted, at least qualitatively, from the expected effect on the 90° boundaries.

90° domain walls have large strains and stress fields associated with them and therefore lead to enhanced piezoactivity. The increased quantity of 90° domain walls formed under high stress having their a-axis in the polarisation direction leads to a large increase in the longitudinal piezoelectric constant (h_{33}). Under compressive stress few 90° domains exist with their a-axis perpendicular to the polarisation with a resultant decrease in the lateral piezoactivity (d_{31}). The mobility and stress field of 90° domain walls is well known to lead to lower stiffness values in ferromagnetic materials that have a closely analogous microstructure [9]. A similar phenomenon is observed in this study of ferroelectric ceramics in which an increased quantity of 90° domains perpendicular to the poling direction gives rise to a decrease in the longitudinal stiffness (c_{33}). The compressive uniaxial stress field does not promote the formation of 90° domains in the transverse direction and thus a reduced compliance would be expected laterally, as observed.

6. Conclusions

A system and procedure for the determination of the low field elasto-piezo-dielectric tensor for PZT ceramics under uniaxial stress is described. The results show that under uniaxial stress parallel to the polarisation the piezoelectric activity increases longitudinally but decreases laterally. Similarly the ceramic becomes more compliant in the longitudinal direction but more stiff laterally. These results can be explained by a consideration of the microscopic domain processes that occur under uniaxial compressive stress where the formation of 90° domain walls is enhanced. This new data is important in understanding the variation in materials properties that occur under operational conditions and is essential in both materials selection and the design of high power devices.

7. Acknowledgements

This work was carried out as part of Technology Group 01 (Magnetics, Acoustics & Hydrodynamics) of the MoD Corporate Research Programme. The authors would like to thank Dr R Lane for his valuable contribution to this research. The helpful advice of Prof. B. Mukherjee through TTCP is also acknowledged.

8. References

1. IEEE Standard on Piezoelectricity, IEEE Std-176 (1987). New York, IEEE/ANSI
2. Woollett R.S. & C.L. LeBlanc (1973) Ferroelectric nonlinearities in transducer ceramics, *IEEE Trans. on Sonics and Ultrasonics*, Vol. SU-20, No.1., 24-31
3. Sherrit S, R.B. Simpson, H.D. Wiederick & B.K. Mukherjee (1996) Stress and temperature dependence of the direct piezoelectric charge coefficient in lead zirconate titanate ceramics. *Proc.SPIE Far East and Pacific Rim Symposium on Smart Materials, Structures and MEMS, December 1996.*
4. Kreuger H.H.A. (1967) Stress sensitivity of piezoelectric ceramics: Part 1. Sensitivity to compressive stress parallel to the polar axis, *J. Acoust. Soc. Am.* 42(3), 636-645.
5. Kreuger H.H.A & D. Berlincourt (1961) Effects of high static stress on the piezoelectric properties of transducer materials *J. Acoust. Soc. Am.* 33 1339-1344.
6. Smits J.G. (1976) Iterative method for accurate determination of real and imaginary parts of the materials coefficients of piezoelectric ceramics. *IEEE Trans. on Sonics and Ultrasonics*, Vol. SU-23, 393-402.
7. Sherrit S., H.D. Wiederick & B.K. Mukherjee (1992) Non-iterative evaluation of the real and imaginary materials constants of piezoelectric resonators. *Ferroelectrics* 134, 111-119.
8. Kwok K.W., H.L.W. Chan & C.L. Choy (1997) Evaluation of the materials parameters of piezoelectric materials by various methods. *IEEE Trans Ultrasonics Ferroelectrics & Frequency Control* 44 (4) 733-742.
9. Bozorth R.M. (1951) *Ferromagnetism*, Van Nostrand, New York.

SWITCHING OF FERROELECTRIC NANOSTRUCTURES

C. HARNAGEA, M. ALEXE, A. PIGNOLET,
K. M. SATYALAKSHMI, D. HESSE, AND U. GÖSELE
Max-Planck-Institut für Mikrostrukturphysik
Weinberg 2, D-06120 Halle/Saale, Germany

ABSTRACT

Ferroelectric domain structures of epitaxial $SrBi_2Ta_2O_9$ (SBT), $BaBi_4Ti_4O_{15}$ (BBiT) and $Bi_4Ti_3O_{12}$ (BiT) thin films as well as PZT nanostructures have been imaged at the nanometer scale. The surface polarization state was monitored using scanning force microscopy (SFM) and lock-in techniques, by measuring small vibrations of the film surface induced by a small AC voltage applied between a bottom electrode underneath the film and the SFM tip. The local ferroelectric properties were characterized by acquiring local piezoelectric hysteresis loops. The hysteresis loop measurements were interrupted at significant points and the ferroelectric domain configuration was imaged to map the switching of polarization.

1. Introduction

Ferroelectric thin films are very attractive in view of their possible applications in nonvolatile memories, microelectromechanical devices and pyroelectric detectors. The decrease in size (down to tens of nanometers) of microelectronic devices requires an appropriate description of the basic local processes in ferroelectric thin films. Therefore, it is of both fundamental and practical interest to investigate whether ferroelectric structures with nanometer sizes still exhibiting ferroelectric, piezoelectric, and pyroelectric properties, and to study how these properties are affected by the size [1]. While the macroscopic dielectric, piezoelectric and ferroelectric characteristics of perovskite ferroelectric thin films (especially the PZT family and the Bi-oxide layered perovskite compounds) have intensively been investigated [2,3], only a few experimental data on their properties measured at the nanometer scale are available. Among several electric SFM techniques [4] the piezoresponse mode of SFM has proven to be the most suitable method to study and to *control* the ferroelectric domain structure at the nanometer scale. Systematic studies of domain dynamics, retention and fatigue effects have already been reported for thin films of PZT-based materials [5,6]. However, the first report concerning the observation of the ferroelectric domain structure of SBT thin films was only published last year [7].

The aim of this paper is to report first experiments evidencing the switching behavior of *individual* ferroelectric nanostructures of PZT, SBT, BiT and BBiT .

C. Galassi et al. (eds.), Piezoelectric Materials: Advances in Science, Technology and Applications, 169–177.
© 2000 *Kluwer Academic Publishers. Printed in the Netherlands.*

2. Experimental

2.1. SAMPLE PREPARATION

The ferroelectric SBT, BBiT and BiT thin films were grown on epitaxial LaNiO$_3$ (LNO) films on SrTiO$_3$(100) substrates by pulsed laser deposition. The SBT films were also deposited on LNO/CeO$_2$/YSZ/Si(100) 3''wafers. Details about the deposition, morphology and macroscopic ferroelectric properties can be found in previous papers [8,9]. The intermediate layers between the ferroelectric film and the substrate serves as a template for epitaxial growth. The 150 nm thick LNO layer also serves as a bottom electrode. The thicknesses of the ferroelectric films, as revealed by cross section TEM analysis, were 120 nm (SBT), 350 nm (BBiT) and 500 nm (BiT). The PZT regular structures with lateral sizes down to 100 nm were fabricated using electron beam direct writing as described elsewhere [10].

2.2. FERROELECTRIC DOMAIN IMAGING

The experimental set-up is basically the same as that reported by other authors [11,12]. In the experiments presented here a Dimension 5000 Scanning Probe Microscope (Digital Instruments) working in contact mode (repulsive force regime) was used. A conductive tip is scanned over the sample surface while maintaining a constant deflection of the cantilever (constant force mode). Simultaneously, a testing AC signal with a frequency $f = 14.2\,kHz$ and an amplitude $A = 2.8\,V$ is applied between the tip and the bottom electrode of the sample. The AC field-induced mechanical oscillations of the sample underneath the tip are transmitted to the tip and converted into an electrical signal by the optical detector of the SFM. The oscillations of the sample surface are extracted from the global deflection signal using a lock-in amplifier (EG&G Instruments, Model 7260). The image of the ferroelectric domain structure is obtained by simultaneously monitoring the topography of the sample and the first harmonic of the signal (further referred to as piezoresponse).

The voltages are applied to the sample using highly doped silicon cantilevers, with a resistivity of 0.02 Ωcm, a resonance frequency of 321 kHz and a spring constant of 40 N/m. The manufacturer (Nanosensors) guarantees a tip radius smaller than 15 nm.

2.3. NANOMETER SCALE PIEZOELECTRIC MEASUREMENTS

In order to apply bias voltages to the sample, as well as for piezoelectric hysteresis loop measurements, a DC source connected in series with the built-in AC source of the lock-in amplifier was used. Both the lock-in amplifier and the DC source are computer-controlled, allowing a good synchronization of the system during the measurements. Each point of the piezoelectric hysteresis loop was measured in the following manner [13,14]. First a DC pulse of a duration $T_{bias} = 100\,ms$ and of a corresponding voltage is applied. After a time interval of $T_{delay} = 2\,s$ from the suppression of the DC polarizing voltage the piezoresponse signal is recorded and stored. Using this procedure electrostatic interactions between tip/cantilever and the bottom electrode are avoided and only the *remanent* piezoelectric coefficient is measured as a function the voltage of the DC pulse *previously* applied to the sample. The hysteresis loops are acquired using an adequate software to simultaneously control the lock-in amplifier, the AC and DC source.

It is well known that in ferroelectric materials the piezoelectric constant first increases and then decreases with increasing mechanical stress applied to the sample. Thus an important parameter governing the piezoresponse mode imaging of ferroelectric domains is the contact force between the conductive tip and the surface of the sample. Zavala at al. [14] reported a 30% decrease of the piezoelectric constant of undoped PZT while increasing the contact force from 10 μN to 23 μN. Similarly Gruverman and Ykeda [7] reported that they could obtain piezoresponse images from samples with low piezoelectric constant (in their case SBT, $d_{33} \approx 20 \, pm/V$) only by using a very low contact force of 1 nN between tip and sample. Measurements on non-ferroelectric samples carried out by J. A. Christman et al. [15] showed the effect of the electrostatic interaction between tip/cantilever and sample/bottom electrode on to the first harmonic signal. The testing AC field produces a mechanical oscillation of the cantilever which is converted by the detector and therefore is also present in the global deflection signal. When the tip was used as the top electrode the amplitude of this oscillation was found to be very high (equivalent to a 6.3 pm/V piezoelectric coefficient) for soft cantilevers with a force constant < 0.3 N/m whereas it was much lower (under their noise level of 0.4 pm/V) for stiff cantilevers of a force constant 2.1 N/m.

Given the fact that a high spring constant of the cantilever is needed, the use of a high contact force between tip and sample is necessary. The experiments presented here were carried out using contact forces between 0.5 μN and 5 μN.

3. Experimental Results

3.1. DOMAIN STRUCTURE AND SWITCHING IN SBT GRAINS

The SBT film used in this study exhibits regions with mixed crystallographic orientation. These consist in a_t-oriented grains protruding 10 nm to 50 nm out of a c_t-oriented background in which they are embedded [9]. a_t and c_t are the unit cell parameters of the pseudotetragonal SBT unit cell. Fig. 1(a) and Fig. 1(b) show the topography and the as-formed domain structure of the sample, respectively. The phase offset of the lock-in amplifier was adjusted so that positive voltages applied to the bottom electrode produce a white contrast of the crystallites; therefore white regions of the image represent zones where the polarization is oriented upward. The flat c_t-oriented background displays a gray contrast in the corresponding piezoresponse image, indicating a very low piezoelectric activity (the signal variations, if any, are under the noise level). On the contrary, the a_t-oriented grains exhibit a strong (black or white) contrast evidencing the presence of a spontaneous polarization normal to the film surface. It should be mentioned that even small a_t-oriented grains (lateral size less than 200 nm) exhibit a stable ferroelectric domain structure.

A switching experiment carried out on a 400 nm x 250 nm grain is presented in the Figs. 1(c-g). After the simultaneous acquisition of images (c) and (d), the SFM tip was positioned at the point marked by the cross in (c) and a sequence of increasing

172

pulses was applied, each of them followed by the measurement of the remanent piezoelectric coefficient (part $\boxed{d}$ - $\boxed{e}$ of the hysteresis curve). The effect of this procedure can be seen in the image Fig. 1(e), acquired after the last voltage applied was $V_{max} = +12V$, equivalent with an electric field of 1MV/cm: the polarization of the grain was switched in the positive direction.

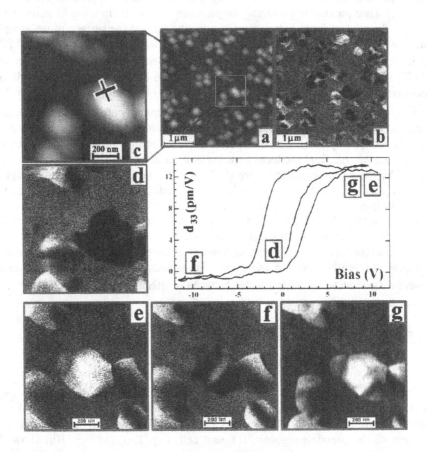

Figure 1. Hysteresis loop of a 400 nm x 250 nm *a*-oriented crystallite of SBT. The images show simultaneously acquired topography (a) and piezoresponse (b) of a 4 x 4 μm^2 area of the sample. Higher magnification (1 x 1 μm^2) images are presented showing the topography (c) and the domain structure (d) of the crystallite before performing the hysteresis measurement at the point marked by the cross in (c). The poling procedure was interrupted at $+V_{max}$ (e) and $-V_{max}$ (f) in order to image the corresponding domain structure of the grain. The piezoresponse image (g) was acquired after completing the hysteresis loop (the last voltage applied was $+V_{max}$).

Then the SFM tip was again fixed over the same point and successive decreasing pulses were applied (trace $\boxed{e}$ - $\boxed{f}$ of the loop). The image shown in Fig. 1(f) was recorded after the application of a pulse of $V_{min} = -12V$. Finally, the hysteresis curve was completed by applying increasing pulses from V_{min} to V_{max} with the tip fixed over the same point of the crystallite. It should be mentioned that in this experiment an offset of the signal equivalent to an effective piezoelectric constant $d_{off} = 5$ pm/V was always present, probably due to the constant interaction between cantilever and bottom electrode as discussed in Section 2.3. This offset is the reason for the shift of the hysteresis curve in the positive direction of the piezoelectric coefficient axis. If the offset is subtracted from the hysteresis data, the maximum value obtained for the piezoelectric coefficient (measured in remanent conditions) $d_{33}^{max} = 7 pm/V$ is comparable with the value of $d_{33}^{rem} = 9 pm/V$ reported by Kholkin et al. measured macroscopically for SBT thin films [16].

From Fig. 1(d) and Fig. 1(f) it can be noticed that the domain structure of the grain at the negative saturation is not the same as the initial state. While in the initial state all the grain has a negative polarization, after the poling procedures described above the grain contains a positive domain of about 50 nm x 250 nm. This might be a proof that the heterostructure has an initial imprint, i.e. the switching in the positive direction is favored.

3.2. DOMAIN STRUCTURE AND SWITCHING IN BBiT

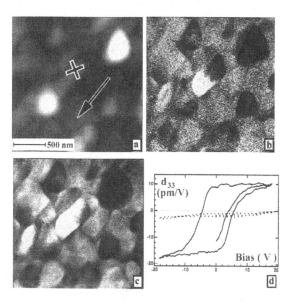

Figure 2. Switching of a ferroelectric domain in a 500 nm x 250 nm *a*-oriented crystallite of BBiT. The topographic (a) and piezoresponse (b) images are simultaneously acquired. The piezoresponse image taken after having performed the hysteresis loop measurement is shown in (c). Hysteresis loop acquired by positioning the SFM tip at the place marked by the cross in (a) is shown in (d, —) together with a hysteresis loop measurement (•) at the point marked by the arrow in (a).

In Fig. 2 an experiment performed on a BBiT/LNO/STO heterostructure is presented. First the images shown in Fig. 2(a) and Fig. 2(b) were simultaneously acquired. The central rectangular a_r-oriented grain is initially divided into two domains. Then the remanent hysteresis loop shown in Fig. 2(d) was taken with the SFM tip positioned at the point marked by the cross in (a). Since the last voltage applied was $+V_{max} = 20$ V, it can be seen that the entire grain became positive, resulting in the domain structure shown in Fig. 2(c), i.e. a ferroelectric domain as small as 300 nm could be switched in an a_r-oriented crystallite of BBiT. For comparison, a similar hysteresis loop measurement was performed over a point of the c_r-oriented background and plotted (•) on the same graph. It can be concluded that, as for SBT, no polarization normal to the surface could be detected in the c_r-oriented regions.

3.3.   SWITCHING BEHAVIOR OF A BiT CRYSTALLITE

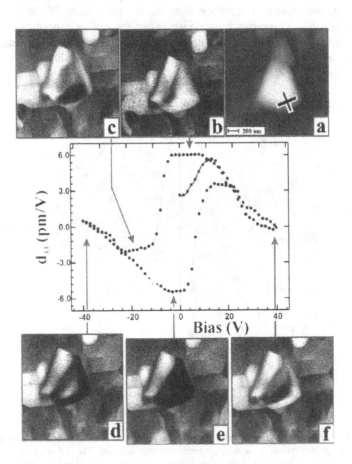

Figure 3. Piezoelectric hysteresis loop measurements performed on an a-oriented crystallite of BiT, at the point marked by the cross in the topographic image (a). The piezoresponse images (b - f) show the configuration of domains corresponding to the point (indicated by arrow) on the loop.

A switching experiment carried out on a crystallite of 300 nm height is shown in Fig. 3. The images 3(a) and 3(b) acquired simultaneously show the topography and the initial positive polarization state of the grain, respectively. The hysteresis loop measurement was interrupted in order to acquire the image of the domain structure and re-started from the same point at the place marked by the cross in (a), as in case of the SBT experiment. The shape of this curve indicates a strong depoling of the grain at high electric fields applied. Such behavior is often observed in macroscopic piezoelectric measurements [17,18] and it is usually explained by the decrease of permittivity at high electric fields, which is related to the piezoelectric constant by the linearized electrostriction expression: $d_{33} = 2Q\varepsilon_{33}P$. However, since the measurements presented here were done with zero bias applied, this explanation cannot be valid. Similar behavior of local depoling after poling with high electric fields was reported in PZT thin films [19]. Moreover, the poling in the opposite direction than the expected one and similar nanoscopic piezoelectric hysteresis loops (measured at zero bias) was very recently notified in thick $BaTiO_3$ single crystals [20]. Franke et. al. [19] explained the decrease of polarization at strong electric fields as a depolarization effect induced by the strong electrostatic force between the SFM tip and the bottom electrode, but this does not explain the poling in the opposite direction with respect to the applied field. Further experiments are needed to elucidate the origin of this behavior.

3.4. SWITCHING IN PZT POLYCRYSTALLINE STRUCTURES

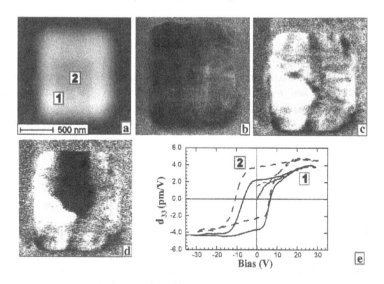

Figure 4. Switching experiments on a 1 μm x 1 μm x 100 nm PZT structure. Topographic (a) and piezoresponse (b) images simultaneously obtained before poling. (c): Piezoresponse image acquired after poling at the point marked ⬚1 (the last voltage applied was +30V). (d): Configuration of domains formed after poling at the point marked ⬚2 (last voltage applied was -30V). No change in the topography was observed. (e): Hysteresis loops measured at the regions marked ⬚1 (——) and ⬚2 (— —) in the topographic image (a).

1 μm x 1 μm x 100 nm PZT patterned structures were achieved on STO(100) substrates [18]. The behavior of one of these individual structures under an electric field can be

seen in Fig. 4. The topography (Fig. 4(a)) and the initial polarization state (Fig. 4(b)) were first acquired. The piezoelectric response is weak. However, an initial negative polarization can be noticed. Then the SFM tip was positioned across the region marked „ $\boxed{1}$ " in the topographic image and a hysteresis loop (Fig. 4(e)) was recorded. The last voltage applied was $V_{max} = +30V$. As result, a region of 500 nm x 350 nm located directly under the tip was fully positively poled and even extended to the upper edge, as can be seen in the image of Fig. 4(c).

The switching behavior was then probed in the region marked „ $\boxed{2}$ " in the topographic image by recording an other hysteresis loop (Fig. 4(e)). The last pulse applied at this region had an amplitude of $V_{min} = -30V$, so that it was left negatively poled. This is confirmed by the next piezoresponse image, Fig. 4(d) where it can be seen that the area poled in the center has grown up to the upper edge, the remaining negative domain having a size of 700 nm x 400 nm. By comparing the two loops acquired from points separated by 400 nm distance one can observe that they are slightly different: the hysteresis loop of the center shows a higher coercive field and has a better symmetry with respect to the d_{33}-axis than the other one. The low piezoelectric coefficient values computed of $d_{33}^{max} = 4$ pm/V can be explained by the very fine grains of the PZT and by the very high contact force used for probing (see Section 2.3). A thin water layer present on the sample surface may also reduce the effective voltage applied to the sample, giving rise to smaller values of the calculated piezoelectric coefficient than the real ones.

4. Conclusions

Imaging of the ferroelectric domain structure and switching of the polarization in SBT, BBiT, BiT thin films and in nanostructures of PZT have been achieved at the nanometer scale by using the piezoresponse mode of SFM. The sub-micron a-oriented crystallites exhibit a stable fine domain structure. The ferroelectric behavior depends on the probing point.

5. References

1. Scott, J. F., Alexe, M., Zakharov, N. D., Pignolet, A., Curran, C. and Hesse, D. (1998) Nano-phase SBT-family ferroelectric memories, *Integrated Ferroelectrics* **21**, 1-14.
2. Lappalainen, J., Frantti, J., and Lantto, V. (1997) Electrical and mechanical properties of ferroelectric thin films laser ablated from a $Pb_{0.97}Nd_{0.02}(Zr_{0.55}Ti_{0.45})O_3$ target, *J. Appl. Phys.* **82**, 3469-3477.
3. Subarao, E. C. (1962) A family of ferroelectric Bismuth compounds, *J. Phys. Chem. Solids* **23**, 665-676.
4. Auciello, O., Gruverman, A., Tokumoto, H., Prakash, S.A., Aggarwal, S., and Ramesh, R. (1998) Nanoscale scanning force imaging of polarization phenomena in ferroelectric thin films, *MRS Bulletin* **23**, 33-42.
5. Gruverman, A., Auciello, O., Tokumoto, H., (1996) Nanoscale investigation of fatigue effects in $Pb(Zr,Ti)O_3$ films, *Appl. Phys. Lett.* **69**, 3191-3193.
6. Gruverman, A., Tokumoto, H., Prakash, S.A., Aggarwal, S., Yang, B., Wuttig, M., Ramesh, R., Auciello, O., and Venkatesan, T (1997) Nanoscale imaging of domain dynamics and retention in ferroelectric thin films, *Appl. Phys. Lett.* **71**, 3492-3494.
7. Gruverman, A. and Ikeda, Y. (1998) Characterization and control of domain structure in $SrBi_2Ta_2O_9$ thin films by scanning force microscopy, *Jpn. J. Appl. Phys.* **37**, L939-L941.

8. Satyalakshmi, K. M., Alexe, M., Pignolet, A., Zakharov, N. D., Harnagea, C., Senz, S. and Hesse, D. (1999) $BaBi_4Ti_4O_{15}$ ferroelectric thin films grown by pulsed laser deposition, *Appl. Phys. Lett.* **74**, 603-605.

9. Pignolet, A., Satyalakshmi, K. M., Alexe, M., Zakharov, N. D., Harnagea, C., Senz, S., Hesse, D. and Gösele, U. (1999) Epitaxial Bismuth-layer structured perovskite thin films grown by pulsed laser deposition, *Integrated Ferroelectrics*, in press (Proceedings of the *11th International Symposium on Integrated Ferroelectrics*, ISIF 99).

10. Alexe, M., Harnagea, C., Hesse, D., and Gösele, U. (1999) Patterning and switching of nano-size ferroelectric memory cells, *Appl. Phys. Lett.*, **75**, 1793-1795

11. Franke, K., Besold, J., Haessler, W. and Seegebarth, C. (1994) Modification and detection of domains on ferroelectric PZT films by scanning force microscopy, *Surf. Sci. Lett.* **302**, L283-L288.

12. Gruverman, A., Auciello, O. and Tokumoto, H. (1998) Scanning force microscopy: application to nanoscale studies of ferroelectric domains, *Integrated Ferroelectrics* **19**, 49-83.

13. Hidaka, T. Maruyama, M. Saitoh, N. Mikoshiba, M. Shimizu, T. Shiosaki, L.A. Wills, R. Hiskes, Dicarolis, S.A., and Amano, J. (1996) Formation and observation of 50 nm polarized domains in $PbZr_{1-x}Ti_xO_3$ thin film using scanning probe microscope, *Appl. Phys. Lett.* **68**, 2358-2359.

14. Zavala, G., Fendler, J.H. and Trolier-McKinstry S. (1997) Characterization of ferroelectric lead zirconate titanate films by scanning force microscopy, *J. Appl. Phys.* **81** 7480-7491.

15. Christman, J.A., Woolcott, R.R. Jr., Kingon A.I. and Nemanich R.J. (1998) Piezoelectric measurements with atomic force microscopy, *Appl. Phys. Lett.* **73**, 3851-3853.

16. Kholkin, A.L., Brooks, K.G., and Setter, N. (1997) Electromechanical properties of $SrBi_2Ta_2O_9$ thin films, *Appl. Phys. Lett.* **71**, 2044-2046.

17. Nomura, S., Uchino, K. (1982) Electrostrictive effect in $Pb(Mg_{1/3}Nb_{2/3})O_3$-type materials, *Ferroelectrics* **41**, 117-132.

18. Kholkin, A.L., Calazda, M. L., Ramos, P., Mendiola, J., and Setter, N. (1996) Piezoelectric properties of Ca-modified $PbTiO_3$ thin films, *Appl. Phys. Lett.* **69**, 3602-3604.

19. Franke, K., Huelz, H., and Weihnacht, M. (1998) Stress-induced depolarization in PZT thin films, measured by means of electric force microscopy, *Surf. Sci.* **416**, 59-67.

20. Abplanalp, M., and Günter, P. (1999) Writing and dynamic observation of ferroelectric domains in barium-titanate by scanning force microscopy, presented at the 9th European Meeting on Ferroelectricity, Prague, July 12-16, 1999.

21. Alexe, M., Harnagea, C., Pignolet, A., Hesse, D., Gösele, U. (2000) Nano-size ferroelectric structures, this volume.

MEASUREMENT AND SIMULATION OF THE ELECTROMECHANICAL BEHAVIOR OF PIEZOELECTRIC STACK TRANSDUCERS

H. JANOCHA, B. CLEPHAS
Laboratory of Process Automation (LPA)
University of Saarland, D-66041 Saarbrücken, Germany

1. Characterization and application of piezoelectric stack transducers

Piezoceramics have been technically employed for a long time, for example as fast sensors in accelerometers. They are also used as actuators in resonator or ultrasonic applications. To a growing extent piezoelectric transducers are employed as fast and precise drives which are characterized by short reaction times on the order of milliseconds and high forces larger than a kilonewton at, however, small displacements in the range of microns [12]. To achieve a maximum displacement the transducers are driven with an electrical field of up to 2000 V/mm. Since driving voltages above 1000 V are difficult to handle the thickness of a single ceramic plate is limited to 0.5 mm.

Thus, several plates operating in d_{33}-mode are stacked together as shown in Figure 1.

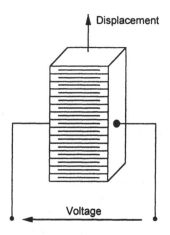

Figure 1. Schematic construction of a piezoceramic stack transducer

C. Galassi et al. (eds.), Piezoelectric Materials: Advances in Science, Technology and Applications, 179–190.

180

Classical stacks comprise layers of ceramic, metal electrodes and glue. So-called cofired stacks consist of metalloid ceramics which are sputtered together [4].

The ceramic layer thickness is often reduced to 0.1 mm which allows a driving voltage of only 200 V to achieve the maximum displacement. The basic considerations made in this paper, however, are also valid for high-voltage stacks.

A possible application field for such stack transducers is the active vibration control of truss structures in which one or more of the passive elements are substituted by active piezoelectric stacks. Piezo stacks are also used in precision positioning and there they are often combined with traditional spindle drives.

2. Piezoelectric stacks in large-signal operation

When applying an electric field in the range of $E = -200$ V/mm ... 200 V/mm, the behavior of a piezoceramic can be regarded as linear. In the large-signal range, however, with an electric field of up to 2000 V/mm, the strain-field characteristic is non-linear and hysteretic, as shown in Figure 2 [11].

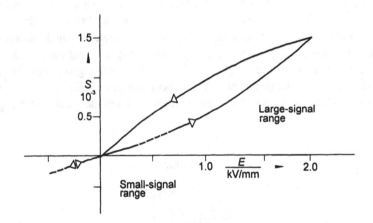

Figure 2. Strain-field characteristic of a piezoceramic

The non-linear behavior mainly depends on the fact that the original piezoelectric effect (the mechanical deformation of dielectric dipoles in an electric field) is superimposed by ferroelectric effects that also contribute to the strain. These ferroelectric effects are based on the orientation of dipoles in an electric field and show a non-linear and hysteretic behavior due to a delay in reorientation. Only an unidirectional electric field is allowed, otherwise the ceramic is irreversibly depolarized. The ferroelectric effects are also responsible for a creep causing an additional strain with an amount that is nearly proportional to the width of the hysteresis loop. Because of the long time-constant, however, this effect must only be considered in quasi-static operation [7].

For piezo stacks, mostly ceramics based upon PZT (lead zirconate titanate, with up to ten additional elements) are employed. These ceramics allow a strain in the range of 1.5 $\cdot 10^{-3}$ (cf. Figure 2) at a Curie temperature (this is the temperature above which the ceramic is thermally depolarized) of up to 350 °C.

Materials with additional lanthanum and artificially produced single crystals, both of which show higher strains, are still undergoing reasearch. In the following sections the electrical and mechanical properties of piezo stacks are discussed in more detail.

2.1 ELECTRICAL PROPERTIES

Due to their construction comprising layers of ceramic and metal electrodes, piezoelectric stacks electrically represent a capacitor. For ceramic plates of 0.5 mm thickness the capacitance is within the range of a nanofarad. Using plates of 0.1 mm thickness (with a driving voltage of 200 V) the capacitance for a stack of the same size is about 25 times higher and up to several microfarads. In sinusoidal operation the current is proportional to the frequency and also to the capacitance, and therefore low-voltage stacks afford a high current in dynamic operation. If the layer thickness was further reduced to achieve a lower driving voltage, the capacitance would reach even higher values. This is the reason why piezo stacks for a voltage below 200 V are technically less important. Moreover, such thin plates would be difficult to manufacture. Figure 3 shows the electrical field-dependent integral capacitance C (the capacitance the amplifier must be adapted to) of a piezo stack normalized by the capacitance in the small-signal range C_0.

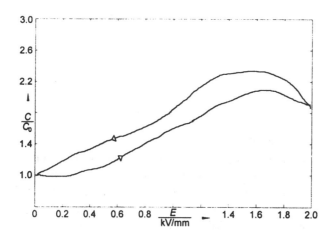

Figure 3. Capacitance of a piezoelectric stack depending on the applied electrical field

182

The capacitance changes with the applied field, its maximum value becomes more than twice as large as that in small-signal operation.

Electronic amplifiers can be designed according to two different operating principles: analogue and switching. Analogue amplifiers possess a very low harmonic distortion, but also high internal losses and no possibility to recover the field energy stored in the transducer. Thus it depends on the application which amplifier is best suited.

Still in a research state at the Laboratory of Process Automation is an amplifier which combines both principles. The output voltage is roughly set by a switching amplifier and is exactly controlled by an analogue output module. In this way the advantages of both concepts are combined.

2.2 MECHANICAL PROPERTIES

The strain-field characteristic is the most important one to describe the behavior of piezoelectric stacks in large signal operation. Figure 4 shows a typical characteristic of a glued stack and one of a cofired stack.

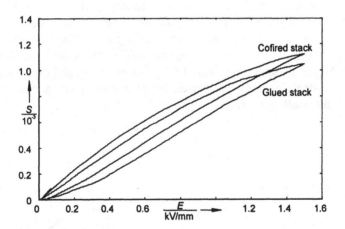

Figure 4. Strain-field characteristics of different stacks

There are no significant differences between the different stacks with regard to the maximum strain or the hysteresis, which has a remarkable width. On the one hand, this hysteresis limits the continuous operating frequency to a range below one kilohertz, otherwise the internal losses would cause a heating that destroys the ceramic thermally.

On the other hand, the hysteresis represents a problem in precision-positioning and must be compensated either in closed loop or by a suitable hysteresis modeling.

Figure 5 shows the strain-stress characteristic when a glued stack is mechanically stressed by an external force. Like any other solid, due to Hooke's law the stack is compressed and thus the strain S has negative values. The strain-stress characteristic shows a hysteresis which can partly be explained by mechanical polarization effects, partly it depends on the assembly of the stack. The inhomogeneous assembly with layers of ceramic, electrodes and glueing material is also responsible for the effect that the stiffness at low forces is smaller than that at high ones. The different layers cannot be manufactured to be completely parallel and thus they are pressed together at low stresses first. Consequently the stiffness is smaller at low forces than at high ones.

The stress-dependence of the stiffness is slightly smaller if regarding cofired stacks, otherwise its strain-stress characteristic is similar to that in Figure 5.

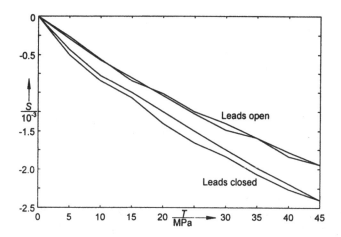

Figure 5. Strain-stress characteristic of a piezoelectric stack

Another aspect to be seen in Figure 5 is the effect that the stiffness possesses different values if the contacts are open or closed. With open contacts a mechanical stress causes a charge on the electrodes because of the direct piezoelectric effect. This charge itself induces an electrical field that due to the indirect piezoelectric effect causes a mechanical stress opposite to the external one. Thus altogether the stiffness is higher than in the case where the electrodes are connected and the electrical field is short-circuited [8].

In the same way as the electrical parameters influence the mechanical ones they themselves depend on the external mechanical load because of the direct piezoelectric effect which causes a charge on the electrodes if an external force is applied. The electronic amplifier must be able to accommodate this charge. Especially if the power transistors within the output stage are used as voltage supplies, compensating circuitry for this charge is necessary. Therefore the output stage of an amplifier for a piezo stack is often designed as a current supply controlled by the difference between the desired and measured output voltage at the stack.

184

The external mechanical force also has an influence on the capacitance of the transducer. Figure 6 represents the mean integral capacitance C normalized to its value at zero prestress C_Z. The capacitance was determined over a whole period of the input signal and therefore displays no hysteresis. The capacitance monotonically increases with the external mechanical load.

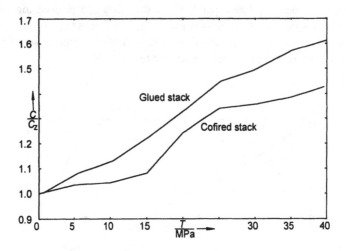

Figure 6. Capacitance as a function of mechanical stress

On the one hand, the influence of the mechanical force on the electrical parameters is a disturbing effect which must be compensated by the power amplifier. On the other hand, however, this ability of piezoelectric transducers can be utilized to get information about the mechanical system without an additional sensor, just by measuring the electrical quantities. Transducers which are simultaneously used for sensing and actuation are often called smart actuators. They have already been successfully employed, for example in precision positioning or active vibration control.

All realizations use the direct piezoelectric sensory effect; the characteristics in Figure 6 show another possible approach which could be evaluated for a smart actuator that is based on the dependence of the capacitance C on the mechanical load [3].

2.3 MODELING

To design systems with piezoelectric actuators a suitable modeling is necessary. There are many different models for piezoelectric transducers, of which a rather simple one is often used. It consists of an electrical circuit with a main capacitance C and a series resonating circuit of L_M, C_M, R_M representing the mechanical properties. This model and most others usually employed are based on a linear energy conversion which, however, cannot be assumed for piezoceramics in large-signal operation (cf. Figure 4).

We have developed an energy-based network-model for piezoelectric transducers as shown in Figure 7 [2].

The electrical properties of the transducer are modeled by the admittance G and the non-linear, hysteretic capacitance C which can be taken from a characteristic similar to that in Figure 6, which, however, must be extended to consider the dependence on the applied electrical field. Since such characteristics are normally not provided by the manufacturers, they have to be determined by the users themselves with the help of measurements.

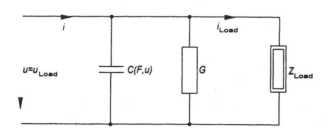

Figure 7. Energy-based network-model of a piezoelectric stack

The electromechanical energy conversion and the whole mechanical system are represented by the non-linear impedance Z_{Load}. If the differential equations of the mechanical system are known, the network can be calculated on the physical law that the mechanical power at the transducer is equal to the electrical one (internal losses are considered by the admittance G). If the system can be transformed into discrete-time, it can be solved iteratively taking into account all non-linearities.

In our experiments, the basic characteristics of several different stacks such as the strain-field characteristic or the variable capacitance were measured in advance and afterwards their mechanical and electrical behavior in large-signal range were calculated with the help of the model in Figure 7. The results showed a very good agreement with the measurements.

2.4 SIMULATION

Examinations concerning the aging of piezo stacks have proved that the mechanical stress inside the ceramic is a main reason for micro cracks. Depending on the electrical and mechanical conditions some of these cracks become larger and finally lead to a disruptive discharge and a destruction of the ceramic.

To calculate the internal stress within a piezo stack a Finite Element Method (FEM) is well-suited. Although many crystalline systems have not yet been physically evaluated, a simple model based on the basic piezoelectric equations yields valuable results.

Due to the layer construction of piezo stacks (cf. Figure 1) the metal electrodes cannot cover the whole ceramic plate and consequently the electrical field and the piezoelectric strain are zero at the edges. A high internal stress occurs in these regions which can be diminished if the stack is externally prestressed by a suitable mechanical force. This prestressing force can be determined by an FEM simulation. It is, for example, necessary for piezoceramics which have to perform billions of operating cycles in large-signal operation, such as in injection valves for combustion engines.

These ceramics are embedded into their application and prestressed before being polarized. The prestress is sustained during the lifetime of the component. If it is lost, the ceramic will be damaged [1].

3. Standardization for piezoceramics

If piezo stacks are to be employed in new applications, it is necessary to have commonly accepted, expressive and comparable characteristics describing their behavior in the large-signal range. Such characteristics are not only essential for sophisticated modeling (cf. section 2.3) but also very helpful for a fast estimation of the system behavior.

3.1 SMALL-SIGNAL RANGE

In 1958 the Institute of Radio Engineers (IRE), New York, published recommendations to define elastic, piezoelectric and dielectrical material parameters as well as an electromagnetic coupling factor for piezoelectric transducers. In 1961 the IRE also proposed measurement methods for these parameters, and in 1976 they edited the 'Guide to Dynamic Measurements of Piezoelectric Ceramics with High Electromechanical Coupling' as an IEC-standard. The measurement instructions, valid for quartz resonators, were extended to ceramics to include the impedance of the ceramic, the electromechanical equivalent circuit, the relation between coupling factor and resonant frequencies. Electronic circuits to measure these values as well as elastic parameters are also defined [6].

According to the IRE recommendations, the parameters are determined at the electromechanical resonant frequency at a driving voltage of 1 V. Using piezoceramics as actuators, however, these are driven up to their maximum voltage (typically 200 V), and the geometry of the stacks differs from the specification of IRE. The recommendations do not consider these operating conditions; they refer to the small-signal range of piezoceramics and only include a hint that the linear connection between electrical and mechanical parameters does not apply in high electrical fields.

Up to now these recommendations have been the only ones available as guidelines for manufacturers to characterize their actuators.

3.2 LARGE-SIGNAL RANGE

In many application fields for piezo stacks the largest achievable displacement of the transducers is required. Prediction of their behavior in this large-signal range is not possible if only the small-signal parameters are known. Using these parameters it can only be estimated whether a stack fits the requirements of the application. The necessary electrical power cannot be calculated correctly. Beyond this, there are many different materials in various geometries, which are difficult to choose from if there are no comparable characteristics.

Thus, development engineers have to rely on self-made measurements in the large-signal range and on their experience to design actuator systems and suitable electronic drives. An exact calculation is not possible when only knowing the small-signal parameters of the piezo stack.

Due to the lack of commonly accepted guidelines concerning the large-signal behavior of piezoceramics there are many different methods to characterize them.

One of these measurement methods uses the piezoelectric transducer as the frequency determining capacitance in a resonant circuit. Non-linear parameters can be determined from the frequency spectrum if the driving field is up to 100 V/mm [5].

Another measurement is based on tempered ceramics which are inserted into a load cell. A defined change in temperature causes a charge and an electrical field which itself leads to an additional pressure because of the piezoelectric effect. The amount of the electrical field is determined from the dependence on the thermal power.

Other measurement methods characterize the dynamic behavior of piezoelectric actuators that are mechanically embedded into a mass-spring system. Here ceramics of different crystal structure showed that chemical inhomogenities and defects are the main origin for a different operating behavior.

Measurements concerning the dielectric and mechanical properties of low-voltage stacks with driving voltages up to several 100 V as a function of the mechanical load are also known. The data are compressed to loss-hyperbolas similar to those used for power semiconductors [10].

3.3 PROPOSALS

Summing up, there is an urgent need for a standardization of piezoceramics. Table 1 shows some basic aspects which must be considered.

An entire discussion of all aspects mentioned in Table 1 cannot be provided in this short article, and therefore only a few annotations are made.

Characteristic values	Influencing values	Methods
Electrical parameters (capacitance, admittance)	Electrical operating conditions (power amplified)	Measurement equipment
Mechanical parameters (geometry, assembly stiffness)	Mechanical system (load)	Measurement conditions
Electromechanical parameters	Temperature	Software
Hysteresis, creep	Operating conditions	Simulation, modeling
Power consumption, efficiency		Accuracy
Quality		Scaling

Table 1. Important aspects for a standardization of piezoceramics

3.3.1 Characteristic values
Fundamental for standardization is a definition of the main characteristics of piezo stacks. These surely include the strain-field characteristic and strain-stress characteristic in the large-signal range which, however, both show a hysteresis whose form and width due to aging effects depend on the history of the transducer as well as on the measurement environment. Thus, the operating conditions must be precisely defined. It becomes obvious that already these comparatively simple basic characteristics require a remarkable effort.

The capacitance of a piezo stack is the determining parameter for designing the power amplifier. Therefore there must be relevant guidelines to determine the large-signal capacitance of the transducers. Moreover, for many applications the power consumption and the efficiency not only of the electromechanical coupling but of the whole actuator are important factors.

Questions concerning the quality of produced stacks should also be taken into consideration. For this purpose, it must be clarified how the quality of a piezoelectric ceramic can be defined (maybe by a reference stack) and what tolerance must be maintained within a manufactured series. To guarantee a defined quality simple measurement methods for output control must be provided.

3.3.2 Influencing values
As discussed earlier in this paper the characteristics for piezo stacks depend on many influencing factors. Therefore for all characteristics the specification has to state the electrical and mechanical operating conditions under which they were measured. To cover a large range of influencing factors a huge number of characteristics would have to be provided which would be impossible to handle. So suitable ways for simplification must be found.

An influencing factor that is very difficult to capture is aging. There are only few publications about the reasons for aging and its influence on the behavior of the transducers, such as irreversible changes of the parameters. Therefore the parameters influencing aging must be evaluated with the aim to provide limits for the permissible electrical, mechanical and thermal conditions for stacks in order to achieve their specified life-time.

3.3.3 *Methods*

Since the characteristics for piezo stacks depend on several influencing factors the measurement equipment also has a considerable effect on the results. Thus, first of all, construction guidelines for the measurement equipment need to be proposed. The equipment must be suitable to determine the dielectrical and dynamic large-signal behavior of the transducers as well as the mechanical properties.

In addition computer programs to automate the measurement and evaluate the data must be developed. These programs should also include subroutines to illustrate the measured values, for example the power consumption in a similar way as the loss-hyperbolas of power transistors.

The capacitance of piezo stacks is often determined by a Sawyer-Tower circuit which needs a very high auxiliary capacitance. So it is the task for the software to calculate the capacitance from the measured voltage and the current in the transducer.

Quantitative values to characterize the accuracy of the measurement results must be defined. This is difficult because the behavior of the transducers is non-linear and time-varying. Therefore all characteristics must contain a statistical divergence as well as a description of the time-dependence.

A last aspect to be mentioned concerns guidelines for scaling. It will be necessary to define reference geometries which cannot cover all possible forms of assembly. Due to the non-linear behavior of piezoceramics it cannot be expected that there is a simple relation between such reference geometries and the stacks used for a special application. Thus, an important task is to find scaling rules.

4. Conclusion and outlook

The aim of this contribution has been to present some characteristics for piezoelectric stacks which are relevant for applications. We also wanted to show that the behavior of the transducers is strongly dependent on the mechanical and electrical operating conditions.

In contrast to this fact there are no commonly defined and widely accepted characteristics describing piezoceramics in large-signal operation. Therefore users have to rely on their own measurements and experiences if they have to choose a suitable ceramic for their application.

From our point of view it is essential for the wider use of piezoelectric transducers that manufacturers, users and research institutions cooperate in defining characteristics for the large-signal operation. Only then will the excellent abilities of these materials be fully utilized.

5. References

1. CeramTec (1995) *Datenblätter Monolithische Vielschichtaktoren.* Lauf, Germany.
2. Clephas, B. and Janocha, H. (1998) Mathematical methods for calculation and simulation of adaptive systems, *Proceedings Smart Mechanical Systems - Adaptronics,* Warrendale, USA.
3. Clephas, B. and Janocha, H. (1998) Simultaneous sensing and actuation of a magnetostrictive transducer, *Proceedings SPIE Smart Structures and Materials* (Ed. Marc E. Regelbrugge), San Diego, USA, 174-184.
4. Hellebrand, H. and al. (1994) Large piezoelectrical monolithic multilayer actuators, *Proceedings Actuator 94* (Ed. H. Borgmann, K. Lenz; Publ. AXON Technologie Consult GmbH), Bremen, Germany, 119-122.
5. Hennig, E. and Pertsch, P. (1996) New methods for measuring mechanical stiffness of piezo actuators, *Proceedings Actuator 96* (Ed. H.Borgmann; Publ. AXON Technologie Consult GmbH), Bremen, Germany, 249-252.
6. IRE and IEE Standards on Piezoelectrical (1958-1961 and 1965-1991) *Standard definition and methods of measurement for piezoelectrical vibrators.*
7. Janocha, H. (Ed., 1992) *Aktoren,* Springer, Berlin, Germany.
8. Janocha, H. (Ed.,1999) *Adaptronics and Smart Structures,* Springer, Berlin, Germany.
9. Janocha, H. and al. (1997) Messung und Simulation des elektromechanischen Verhaltens von piezoelektrischen Stapelwandlern unter realen Betriebsbedingungen, *Proceedings DGM Symposium Langzeitverhalten von Funktionskeramiken* (Ed. P. Otschik; Ed. Werkstoff-Informationsgesellschaft), Dresden, Deutschland, 65-72.
10. Pertsch, P. and al. (1996) Properties of piezoelectrical multilayer under uniaxial compressive stress, *Proceedings Actuator 96* (Ed. H.Borgmann; Publ. AXON Technologie Consult GmbH), Bremen, Germany, 217-220.
11. Queensgate Instruments (1997) *Nano positioning book,* Berkshire, UK.
12. Uchino, K. (1997) *Piezoelectric actuators and ultrasonic motors.* Kluwer, Boston.
13. Zickgraf, B. (1996) *Ermüdungsverhalten von Multilayer-Aktoren aus Piezokeramik.* VDI Fortschrittsberichte 191, Düsseldorf, Germany.

0.9PMN-0.1PT CERAMICS :
FROM ELECTROMECHANICAL PROPERTIES TO APPLICATIONS VIA NANOSTRUCTURAL CHARACTERIZATION

M. LEJEUNE, S. KURUTCHARRY, E. LATTARD,
M. OUDJEDI, P. ABELARD
SPCTS – UMR 6638, E.N.S.C.I
47 à 73 Avenue Albert Thomas, 87065 Limoges Cedex - FRANCE

Abstract

The electromechanical responses of the 0.9PMN-0.1PT ceramics are studied under different types of electric field (static or dynamic) and mechanical stress. Relationships between the sensitivity of 0.9PMN-0.1PT nanostructure to these different excitation parameters, and the associated electromechanical responses are reported. The potential of 0.9PMN-0.1PT ceramics for active damping applications is investigated.

1. Introduction

Based lead magnesium niobate materials have been extensively studied for the last ten years to develop electrostrictive actuators for commercial applications requiring large field induced-strains associated with low hysteresis (1). $(1-x)PbMg_{1/3}Nb_{2/3}O_3 - x\ PbTiO_3$ compositions with $x = 0, 0.05, 0.1$ are well-known as perovskite-structure relaxor materials. In contrast to a "normal" ferroelectric which shows a sharp ferroelectric-paraelectric phase transition at a well defined temperature (called the Curie temperature), they exhibit a diffuse phase transition (DPT) appearing in the weak field permittivity versus temperature curve (2). The evolution of the 0.9PMN-0.1PT-0.12MgO nanostructure with decreasing temperature which is responsible for this relaxor behavior has been studied in previous investigations through the frequency dependence of the dielectric permittivity(3-5). Low frequency dielectric characterization has revealed that the Curie-Weiss law is no longer observed below a temperature Td around 200°C, which is well above the temperature of the maximum permittivity (Tm ~ 40°C). This means that a local polarization appears at this temperature corresponding to the nucleation of polar clusters (6-9). Moreover, a dielectric relaxation observed for 0.9PMN-0.1PT-0.12MgO over a large frequency range (100Hz-1GHz), corresponds to a multi-Debye process with broadening of the relaxation time distribution as the temperature decreases. This distribution can again be correlated to a nucleation and growth mechanism of polar clusters below Td. Different workers have commonly suggested that this dielectric behavior could result from chemical heterogeneities(10,11). This hypothesis has been confirmed using EDX and EELS

191

C. Galassi et al. (eds.), Piezoelectric Materials: Advances in Science, Technology and Applications, 191–202.
© 2000 *Kluwer Academic Publishers. Printed in the Netherlands.*

spectroscopic analysis techniques, which have revealed a large distribution of chemical composition around the nominal one on a nanometric scale(3). Therefore the nucleation and growth mechanism of polar clusters with decreasing temperature can be attributed to the successive transitions of the different local compositions. However, due to its strong chemical heterogeneity, the material remains primarily paraelectric down to very low temperature.

The purpose of this paper is (i) to study the electromechanical response of 0.9PMN-0.1PT ceramics under different types of stimuli, from which their main interests will be highlighted and (ii) to carry out investigations on the potential of 0.9PMN-0.1PT ceramics for active damping applications. Relationships between the sensitivity of 0.9PMN-0.1PT nanostructure to the different electric or mechanical excitation parameters, and the associated electromechanical responses will be reported.

2. 0.9PMN-0.1PT ceramics behavior under static mechanical or electric excitations

2.1. PURE MECHANICAL STRAIN

The effect of mechanical strain has been evaluated by ultrasound techniques. Since the mechanical strain is defined by the following relation : $\Delta x_m = s^{E=0}.X = s_0.X$ where s_0 is the elastic compliance of the material for zero applied electric field, and X is the uniaxial compressive stress, the mechanical strain can be deduced from ultrasound propagation measurements of the Young's modulus, Y_0 $(1/s_0)(12)$.

0.9PMN-0.1PT ceramic Young's modulus measurements were performed as a function of the compressive uniaxial stress (-5 to -100MPa) for different temperatures from 8°C to 45°C and for zero electric field applied. No significant dependence of Y_0 with stress and temperature was found so that the associated mechanical strains (Δx_m) corresponds to a single quasi-linear function of the stress (figure 1).

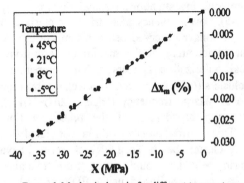

This shows that the evolution of the 0.9PMN-0.1PT ceramic nanostructure with decreasing temperature or increasing stress without electric field applied is not significant enough to lead to a modification of the elastic compliance of the material.

Figure 1. Mechanical strain for different temperatures

2.2. FIELD-INDUCED STRAIN

The quasi-static (20mHz) field induced longitudinal strains were measured for given uniaxial stress X and temperature, i.e. $\Delta x^X_T(E)$, using a linear variable differential transformer (LVDT) (resolution 20nm).

2.2.1. *Dependence on the temperature*
The 0.9PMN-0.1PT field-induced longitudinal strains reported in figure 2 at different temperatures for zero uniaxial stress i.e. $\Delta x^0_T(E)$ show that the maximum longitudinal strain increases with decreasing temperature. Furthermore, saturation occurs at lower electric field as the temperature decreases, and the hysteresis increases correlatively.

Before the application of an electric field, the material corresponds to a distribution of nanopolar domains distributed in a paraelectric matrix. Previous investigations (3) have shown that the application of an electric field promotes the transition of the paraelectric matrix, such that the polar volume rapidly increases below 20°C when a 0.5kV/mm electric field is applied.

Consequently, the faster saturation of the field-induced strain as the temperature decreases can be associated to a significant correlated evolution of the nanostructure. The latter can be described as the reorientation of the spontaneous polarization vectors of the initial polar domains to become parallel with E, and to the nucleation and growth of well-oriented polar domains (i.e. to the displacement of the interface between polar and paraelectric phases). These different mechanisms induce a dissipation of energy due to defects in the material which explains a larger hysteresis of the field-induced strains observed at low temperatures. This phenomena is consistent with a quasi-macroscopic ferroelectric field-induced transition of the material below 20°C (13-15)

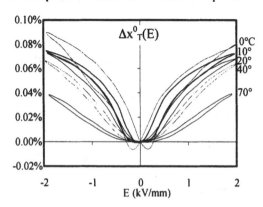

Figure 2. Field-induced strain at different temperatures for zero compressive stress applied, $\Delta x^0_T(E)$.

2.2.2. *Dependence on the compressive uniaxial stress*
These measurements were carried out at room temperature for an uniaxial compressive stress ranging from -5 to -100MPa. Figure 3 shows that the field-induced strain, $[\Delta x^X_{20°C}(E)]$, notably decreases when a stress is applied. Correlatively, the hysteresis and the saturation phenomena are reduced, which means that the field-induced evolution of the nanostructure previously mentioned is inhibited by the application of an uniaxial stress.

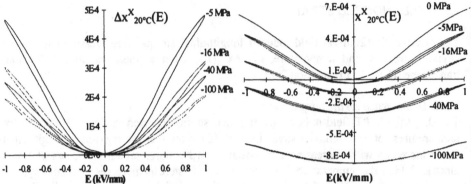

Figure 3. Field-induced strain at room temperature for different stresses, $\Delta x^X_{20°C}(E)$.

Figure 4. Total strain, $x^X_{20°C}$ (E), induced by simultaneous application of electric field and uniaxial stress

2.3. APPARENT ELASTIC COMPLIANCE

2.3.1. *Total strain $x^X_T(E)$, induced by simultaneous application of electric field (E) and stress(X)*

The total strain is the sum of (i)the mechanical strain, Δx_m i.e. the strain of the material as a function of the compressive stress without electric field applied and (ii)of the electrical strain, measured once a given uniaxial stress X is applied, i.e. $\Delta x^X(E)$. In this way, the total strain $x^X_T(E)$ induced by simultaneous application of electric field (E) and uniaxial stress (X) has been calculated for 0.9PMN-0.1PT at room temperature. This is reported in figure 4.

More explicit curves, $x^E_T(X)$, have been deduced from the $x^X_T(E)$ ones, by reporting the average value of the longitudinal strains corresponding to different magnitudes of the electric field as a function of the stress (figure 5). This method, which neglects the hysteresis of the field-induced strain curves (see figure 4) becomes invalid for temperatures below 20°C, according to the results mentioned previously in figure 2. Consequently, another approach must be taken to appropriately describe the field-induced strain characteristics of 0.9PMN-0.1PT ceramics with large hysteresis. This will be further presented in this paper.

For temperatures above 20°C, it appears (figure 5) that 0.9PMN-0.1PT ceramics exhibit a complex macroscopic non-linear electro-mechanical behavior, which is more pronounced if the electric field is high or the temperature is low. The experimental $x^E_T(X)$ curves can be well fitted as polynomial functions of E and X :

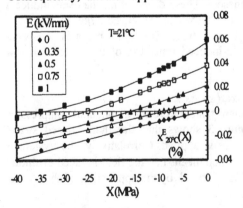

Figure 5. Total strain, $x^E_{20°C}$ (X), induced by simultaneous application of electric field and uniaxial stress.

$$x_T(X,E) = \sum_i \left(\sum_j M_{ij}{}^T * E^j \right) * X^i$$

with i=0...2   and j=0...3

for X and E ranging respectively from 0 to -40MPa and from 0 to 0.5kV/mm, with M_{ij} function of the temperature.

which reveals a strong dependence of the apparent elastic compliance upon the electric field and stress, according to the range of temperature.

2.3.2. Adaptive behavior of 0.9PMN-0.1PT ceramics

The apparent elastic compliance s(1/Pa) is defined by the following relation :

$$s_T^E(X) = \left. \frac{\partial \, x_T(E,X)}{\partial \, X} \right|_E$$

Therefore, the apparent elastic compliance can be deduced from the functions $x_T(E,X)$ previously determined at different temperatures and which corresponds to polynomial functions of E and X such as

$$s_T(X,E) = \sum_i \left(\sum_j A_{ij}{}^T * E^j \right) * X \quad (II)$$

with i = 0...1 and j = 0...3

for X and E ranging respectively from 0 to -40MPa and from 0 to 0.5kV/mm.

In figure 6, the curves $s_{20°C}(X,E)$ reveal that :

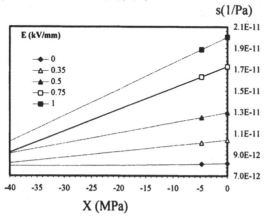

(i)the elastic compliance of the material increases with the amplitude of the electric field for low stresses,
(ii)the elastic compliance of the material decreases with the magnitude of the stress for high electric field.

Moreover, the dependence of the elastic compliance s upon the electric field and the stress is enhanced as the temperature decreases. Similar results have been previously observed by

Figure 6. Dependence of the elastic compliance s $_{20°C}$ with the amplitudes of the electric field and stress

other investigators : in particular, Nakajima et al (16) and Brown et al (17) have found that the apparent modulus decreases with the magnitude of the applied DC electric field.

In fact, the variations of the 0.9PMN-0.1PT-0.12MgO elastic compliance with the electric field, temperature and stress can be correlated to the evolution of its nanostructure with these different parameters. As previously shown, a high electric field induces the nucleation and growth of polar clusters distributed in a paraelectric matrix. This mechanism is more pronounced when the temperature or the uniaxial compressive stress is low. Therefore, as the development of the polar volume is associated to an increase of the elastic compliance, one can conclude that nanopolar clusters are responsible of this phenomena and consequently that the polar phase exhibits a higher elastic compliance than the paraelectric matrix. This result is in good agreement with Vielhand (18) who reported that the application of an electric field should soften the material because it would stabilize more polarized regions.

Consequently, according to these conclusions, the apparent elastic compliance of the material should increase with the amplitude of the electric field up to a maximum value corresponding to the state of maximum transition of the paraelectric matrix for a given temperature and stress. In fact, Brown et al (17) have found a minimum in the apparent modulus for an electric field as a function of the composition in the [(1-x)(0.9PMN-0.1PT)-xBaTiO$_3$)] system. They correlate this minimum to a saturation of the field-induced polarization, which is in agreement with the nucleation and growth phenomena of well oriented polar domains as proposed in this paper. In our case, we don't observe a saturation of the elastic compliance of the material in the range of the electric field we have studied (E<0.5kV/mm). However, we can estimate the amplitude of the electric field corresponding to its saturation from the expression of the apparent elastic compliance as a polynomial function of E and X (Equation II). We find in this way, that the elastic compliance at room temperature for no stress applied exhibits a maximum for a value of the electric field equal to 0.709kV/mm.

As previously mentioned (point ii), the application of an increasing stress stiffens the material once it is polarized under a 0.5kV/mm electric field (figure 6). This means, as found previously, that the development of the polar volume induced by the electric field can be balanced by the application of a stress. Therefore, it appears that 0.9PMN-0.1PT ceramics exhibit some similarities with the behavior of PZT ceramics under stress. Previous studies (19-20) have shown that a compressive stress parallel to the polarization direction induces a decrease on the remanent polarization, which can be partially or even completely eliminated. These polarization changes correspond to domain switching and are responsible for non linear deformation. Conversely, stresses applied perpendicular to the poling direction stabilizes the polarization and even favors a field-induced transition by compression perpendicular to the polarization field direction. As far as 0.9PMN-0.1PT ceramics are concerned, they are initially unpoled but the application of a compressive stress parallel to the field direction inhibits the field-induced transition from disoriented nanopolar clusters to well oriented macropolar regions. Consequently, their non linear deformation with increasing stress can be attributed to this mechanism, which is very similar to the effect of PZT depolarization under parallel compressive stress. By analogy, it could be interesting to study for 0.9PMN-0.1PT ceramics, the effect of a compressive stress perpendicular to the field direction on the field induced transition mechanism.

Finally, we can also note that these different results are close to the analysis of 0.9PMN-0.1PT electromechanical behavior by Schmidt (21-22). In particular, Schmidt considered that the ferroelectric transformation upon increasing field or decreasing temperature in PMN is similar to the martensitic transformation in alloys. Creation of plate-like martensitic regions is observed within the parent phase at a temperature denoted by M_s with increase of the martensitic volume fraction upon further cooling by formation of new plate-like regions. Correlatively, martensitic transformation leads to a decrease on the stiffness.

2.4. MAIN FEATURES OF 0.9PMN-0.1PT CERAMICS ELECTROMECHANICAL BEHAVIOR - CONSEQUENCES

As shown previously, around 20°C 0.9PMN-0.1PT ceramics exhibit high field-induced strains (10^{-3}) associated to low hysteresis (only 4% compared to 12% for PZT ceramics) at low frequency. Moreover, as 0.9PMN-0.1PT ceramics are unpoled, very weak long term creep is observed (only 3% compared to 15% for PZT ceramics). Consequently, these ceramics are excellent candidates for precise submicron positioning.

They are also able to generate high forces (around 50MPa upon the application of a 2kV/mm electric field). Consequently associated to high displacements, they can supply high mechanical energy densities so that miniaturized actuators can be developed.

Nevertheless, the main interest of 0.9PMN-0.1PT ceramics is an adaptive electromechanical behavior. This material is able to adjust its response to a stress (i.e. its strain) by a variation of the electric field and consequently is promising for active damping applications.

However, in any case, due to the strong dependence of the electromechanical behavior with temperature, it will be necessary to stabilize the temperature of the devices. In addition, because of the large magnitude of the applied electric fields, multilayer structures will be used to limit the applied voltage (between 20 to 200V).

3. 0.9PMN-0.1PT ceramics behavior under dynamic electric solicitations

This part of the study is the first step to be carried out to evaluate 0.9PMN-0.1PT ceramics for active vibration control applications. Active vibration control using ceramics consists in canceling the vibration of a structure thanks to a feedback electric field so that the positioning of the system remains unchanged. The ceramic component has a double function of

(i) pressure sensor, capable of detecting vibrations,

(ii) actuator, used in a feedback loop, to compensate the strain of the structure by electric field variation.

Therefore, it is necessary to determine

(i) the dielectric response as a pressure sensor [i.e. the sensitivity of the polarization with the stress according to the applied electric field, $P(X,E)$], to extract the vibration $X(t)$ through the measured polarization variation $P(t)$ of the ceramic component.

(ii) the electromechanical behavior as actuator [i.e. the variation of the strain with applied electric field for different compressive stresses, $x(X,E)$] to calculate the electric feedback $E(t)$, allowing recovery of the initial positioning x_0 of the system (i.e. $x[X(t),E(t)]$= constant).

198

Polarization and strain measurements were performed by using, respectively, a Sawyer-Tower circuit and a laser interferometer system (with resolution of 10nm), at room temperature for different compressive stresses ranging from 0 to -30MPa, by applying a sinusoidal signal of variable amplitude (from 0.1 to 0.5 kV/mm) and frequency (200mHz to 1kHz) in addition to a 0.6kV/mm DC electric field. The value of the polarization electric field E_0 (0.6kV/mm) was deduced from strain characteristics already carried out under quasi-static electric field (20mHz) with no applied load. This corresponds to the highest sensitivity of the material, that is to say, to the maximum value of the apparent piezoelectric coefficient ($\delta x/\delta E$).

Moreover, the mechanical resonance of the measuring device, defined by the relation :

$$f_r = \frac{1}{2\pi}\sqrt{\frac{k}{m}}$$, with k and m, corresponding respectively to the stiffness of

the measuring device, and to the moving mass during experiments, was found to be equal to 2.8kHz. Therefore, as this value is outside of the frequency range of study, i.e. 200mHz-1kHz, the frequency dependence of the polarization and strain characteristics observed further, will be attributed to intrinsic effects to 0.9PMN-0.1PT material.

3.1. DEPENDENCE OF THE 0.9PMN-0.1PT POLARIZATION AND STRAIN ON THE ELECTRIC AND MECHANICAL STIMULATION CONDITIONS

3.1.1 *Influence of the frequency of the sinusoidal electric field (f=ω/2π)*

As shown in figure 7, increase of the frequency from 20mHz to 1kHz for a given amplitude of the sinusoidal signal (0.5kV/mm) and of the compressive stress (~ -1MPa) leads to enhancement of the hysteresis (surface of the curves).

The polarization and longitudinal strain maximum remains stable respectively up to 50Hz and 250Hz, then decreases gradually in the case of polarization and rapidly as the strain is concerned.

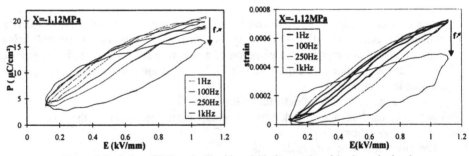

Figure 7. Evolution of P(E) and x(E) with variable frequencies of the dynamic signal

3.1.2 *Influence of the amplitude dE of the sinusoidal electric field*

Figure 8 reveals that the more the amplitude dE increases, the more the characteristics P(E) and x(E) saturate. Correlatively, the hysteresis increases. This evolution is observed whatever is the frequency of the sinusoidal signal and the prestress.

3.1.3. *Influence of the compressive static stress X*

For a given electric field (f=200mHz, dE=0.5kV/mm), the field-induced polarization and strain are plotted for an uniaxial compressive stress ranging from -1MPa to -32MPa. The maximum value of polarization is weakly reduced with increasing compressive stress (Figure 9) whereas in the same conditions, the peak longitudinal strain decreases more strongly.

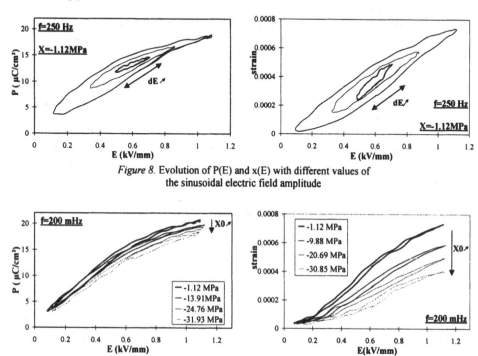

Figure 8. Evolution of P(E) and x(E) with different values of the sinusoidal electric field amplitude

Figure 9. Evolution of P(E) and x(E) with compressive stresses

3.2. SENSITIVITY OF THE 0.9PMN-0.1PT FIELD-INDUCED TRANSITION MECHANISMS TO THE ELECTRIC AND MECHANICAL SOLICITATION CONDITIONS

Before application of a sinusoidal electric field dE and a compressive stress X, the nanostructure of 0.9PMN-0.1PT corresponds to nanopolar clusters inside a paraelectric matrix. As shown in previous studies (3), the size distribution of the polar clusters at room temperature is only fixed in this case by the magnitude of the initial DC electric field.(i.e. 0.6kV/mm in our case). Moreover, this size distribution has been correlated to a multi-Debye dielectric relaxation mechanism observed through weak-field permittivity measurements over a large frequency range. The contribution at the lowest frequency (around 100Hz) corresponds to the largest polar domains, because of their higher stability. Consequently, the gradual decrease of the maximum polarisation from a frequency of the sinusoidal electric field dE equal to 50Hz associated to the increase of the hysteresis (figure 7) results from the progressive relaxation of the displacement of the interface between the nanopolar domains and the paraelectric matrix.

Furthermore, for a given frequency of the sinusoidal electric field and compressive stress, the displacement of the interface between the nanopolar domains and the paraelectric matrix is enhanced by the increase of the amplitude dE of the sinusoidal electric field and the consequent polarisation and strain saturation. The corresponding hysteresis is also enhanced.

We have shown (§ 2.2.2) that the application of a compressive prestress inhibits the development of the polar volume induced by a DC electric field. Consequently, this leads to the decrease of the initial polarisation and strain in the material at $E=E_0=$ 0.6kV/mm, just before the application of the sinusoidal electric field. Added to this modification of the initial nanostructure (at $E=E_0=0.6$kV/mm), the increase of the prestress limits the displacement of the interface between the nanopolar domains and the paraelectric matrix induced by a given dE sinusoidal electric field. Hence, one can observe correlatively a decrease of the polarisation and strain maximum (figure 9).

From these results, it appears that the dependence of the polarisation and strain (maximum value and hysteresis) on the sinusoidal electric field (amplitude and frequency) and prestress can be understood through (i)the modification of the initial polar volume according to the applied prestress, (ii)the sensitivity of this nanostructure to the application of a sinusoidal electric field according to its amplitude and frequency for variable prestresses.

3.3. DETERMINATION OF 0.9PMN-0.1PT BEHAVIOR LAWS AS SENSOR AND ACTUATOR

3.3.1. *Behavior as sensor*

The treatment of the experimental data defines the polarization as a function of the electric field and mechanical stress in terms of an hyperbolic tangent formulation i.e.:

$$P(X,E)=P_s.(f,X)tanh[k(f).(E_0+dE.sin(\omega t+\varphi(f)))]$$

This fitting was obtained successively

(i) by adding a phase delay φ to the previous hyperbolic formulation proposed by other workers for similar compositions (23-24), in order to take into account the hysteresis observed as the frequency increases.

(ii) by taking k and φ parameters independent of the stress, to easily extract the stress from the polarization law, with no hypothesis on the Ps dependence.

(iii)

Under these conditions,

The k factor was fitted by the relationship : $k(f) = \dfrac{\alpha}{\beta + \gamma.f}$

and the phase delay φ can be decomposed in terms of a multi-Debye function, i.e. :

$$\varphi(f) = \sum_{i=1}^{3} \alpha_i.\varphi_i(f)$$

$$\text{with } \varphi_i(f) = \varphi_\infty - \frac{\varphi_\infty}{1+(2\pi.f)^2.\tau_i^2} \qquad (\varphi_\infty = -\pi)$$

with α_i (weight coefficient) corresponding to the contribution of each Debye term ϕ_i, characterized by a relaxation time τ_i.

Finally, by analogy with other works (25), longitudinal field-induced strain was defined by the relation

$$x\,(E) = Q.P^2, \text{ such that}$$

$$x = \frac{1}{k^2}.(2.r.X + m).\tanh^2(k.E)$$

and the following fit of P_s was carried out successfully : $P_s = \frac{1}{k(f)}.B(X,f)$

Thus, the equation of the sensor function can be written in the following form :

$$P(E,\,X) = \frac{1}{k(f)} \cdot \sqrt{\frac{a(f).X + b(f)}{c.X + d}} \cdot \tanh\left[k(f) \cdot \left(E_0 + dE\sin\left(2\pi.f.t + \varphi(f)\right)\right)\right]$$

By this means, the stress X(t) corresponding to the vibration can be deduced from P(t) measurements through this last equation using numerical methods.

3.3.2. *Behavior as actuator*
The field-induced strain is defined by x (E)= Q.P² i.e.

$$x^x(E) = Q.\frac{1}{k^2(f)}.B^2(X,f).\tanh^2\left[k(f).\left(E_0 + dE.\sin\left(\omega t + \varphi(f)\right)\right)\right]$$

The Q factor is fitted well by a second order function over the whole frequency range such that :

$$Q_{33}(f,X) = \frac{T(X)}{1 - 6\cdot10^{-4}.f + 5\cdot10^{-7}.f^2}, \text{ with } T(X) = 1.65\cdot10^{-2}\cdot\exp\left[1.5\cdot10^2.X\right]$$

Therefore, after evaluation of the total strain corresponding to the following expression

$$x(X,E) = s^{E=0}.X + x^x(E) = s_0.X + x^x(E)$$

the electric feedback $\delta E(t)$ allowing a constant positioning of the system can be calculated through this last equation, i.e. x[X(t),E(t)] = constant.
It remains to validate the attenuation of a low frequency vibration by using 0.9PMN-0.1PT ceramic as a pressure sensor and actuator according to these expressions.

4. Acknowledgments

This study has been mainly supported by the Defense Ministry (DGA)

5. References

1. Uchino, K. (1986) Electrostrictive Actuators : Materials and Applications, *Am. Ceram. Soc. Bull.* **65**[4], 647-652
2. Smolenskii, G. A. (1958) On the mechanism of polarization in solid solutions of PNN-PMN, *J. Tech. Phys. USSR* **28**, 7
3. Lattard, E., Lejeune, M., Guinebretière, R., Imhoff, D., Abelard, P. (1997) Study of 0.9PMN-0.1PT dielectric behaviour in relation to the nanostructure, *J. Phys. III.* **7**, 1173-1196
4. Lu, Z.G. and Calvarin, G. (1995) Frequency dependence of the complex dielectric permittivity of ferroelectric relaxors, *Physical Review B* **51**[5], 2694-2702
5. Cross, L.E. (1987) Relaxor Ferroelectrics, *Ferroelectrics* **76**, 241-267
6. Burns, G. and Dacol, F.H. (1990) Ferroelectrics with a glassy polarization phase, *Ferroelectrics* **104**, 25-35
7. De Mathan, N., Husson, E., Calvarin, G., Gavarri, J.R., Hewat, A.W. and Morell, A. (1991) A structural model for the relaxor Pb($Mg_{1/3}Nb_{2/3}$)O_3 at 5 K, *J. Phys.: Condens. Matter* **3**, 8159-8171
8. Bonneau, P., Garnier, P., Calvarin, G., Husson, E., Gavarri, J.R., Hewat, A.W. and Morell, A. (1991) X-ray and neutron diffraction studies of the diffuse phase transition in Pb($Mg_{1/3}Nb_{2/3}$)O_3 ceramics, *J. Solid State Chemistry* **91**, 350-361
9. Isupov, V.A. (1956) On the question of causes of formation of a Curie range of temperatures in certain ferroelectric solid solutions, *Soviet Physics, Technicals Physics* **1**, 1846-1849
10. Smolenskii, G.A. (1970) Physical phenomena in ferroelectrics with diffused phase transition, *J. Phys. Soc. Jpn* **28**, 26-37
11. Chabin, M., Malki, M., Husson, E. et Morrel, A. (1994) Etudes diélectriques de la transition ferroélectrique induite par application d'un champ électrique dans les céramiques Pb($Mg_{1/3}Nb_{2/3}$)O_3 (PMN), *J. Phys. III.* **4**, 1151-1163
12. Sommer, R., Yushin, N.K. and Van Der Klink, J.J. (1992) Dielectric susceptibility of PMN under DC bias, *Ferroelectrics* **127**, 235-240
13. Butcher, S.J. and Daglish, M. (1989) A field-induced phase transition in PMN ceramics, *Ferroelectrics Letters* **10**, 117-124
14. Nakajima, Y., Hayashi, T., Hayashi, I. and Uchino, K. (1985), Electrostrictive properties of a PMN stacked actuator, *Jpn. J. Appl. Phys.* **24**[2], 235-238
15. Brown, S.A., Hom, C.L., Massuda, M., Prodey, J.D., Bridger, K., Shankar, N., and Winzer, S.R. (1996) Electromechanical testing and modeling of a PMN-PT-BT Relaxor Ferroelectric, *J. Am. Ceram. Soc.* **79**[9], 2271-2282
16. Viehland, D., Jang, S.J., Cross, L.E., and Wuttig, M. (1991) Anelastic relaxation and internal strain in lead magnesium niobate relaxors, *Philosophical Magazine A* **64**[4], 835-849
17. Arndt, H., Schmidt, G., and Vogel, N. (1984) Influence of uniaxial pressure on the properties of PLZT ceramics, *Ferroelectrics* **61**, 9-18
18. Cao, H. and Evans, A.G. (1994) Electric-field-induced fatigue crack growth in piezoelectrics, *J. Am. Ceram. Soc.* **77**[7], 1783-1786
19. Schmidt, G. (1990) Diffuse ferroelectric phase transitions in cubically stabilised perovskites, *Phase Transitions* **20**, 126-162
20. Schmidt, G., Arndt, H., Borchhardt, G., Cieminski, J.V., Petzsche, T., Borman, K., Sternberg, A., Zirnite, A. and Isupov, V.A., (1981) Induced phase transitions in ferroelectrics with diffuse phase transition, *Phys. Stat. Sol. (a)* **63**, 501-510
21. Zhang, X.D. and Rodgers, C.A. (1993) A macroscopic phenomenological formulation for coupled electromechanical effects in piezoelectricity, *J. of Intell. Mater. Syst. and Struct.* **4**, 307-316
22. Hom, C.L. and Shankar, N. (1994) A fully coupled constitutive model for electrostrictive ceramic materials, *J. of Intell. Mater. Syst. and Struct.* **5**, 795-801
23. Fripp, M. (1995) Distributed structural actuation and control with electrostrictors, S.M. Thesis, Massachusetts Institute of Technology, Departement of Aeronautics and Astronautics, Cambridge, Massachusetts (USA)

SOLID-STATE ^{25}Mg, ^{45}Sc and ^{93}Nb MAS NMR STUDIES: LOCAL B-SITE CHEMICAL ENVIRONMENTS AND ORDERING IN PB(MG$_{1/3}$NB$_{2/3}$)O$_3$ (PMN) AND RELATED RELAXOR FERROELECTRICS

JOHN J. FITZGERALD,[1] JIONG HUANG[1] AND HERMAN LOCK[2]
[1]*Department of Chemistry and Biochemistry, South Dakota State University, Brookings, SD, 57007*
[2]*Department of Chemistry, Colorado State University, Fort Collins, CO 80523.*

Abstract

Solid-state ^{25}Mg, ^{45}Sc and ^{93}Nb NMR spectroscopic studies of the local chemical environments and relative degree of B-site ordering in lead-based perovskite relaxor ferroelectrics are reported. ^{93}Nb MAS NMR spectra of PMN and PZN provide conclusive evidence for significant B-site disorder, while ^{93}Nb and ^{45}Sc MAS NMR indicate B-site ordering in PSN. ^{93}Nb MAS NMR spectra of PMN show resonances with different C_Q values assigned to different types of Nb(V) B-sites of pseudo-cubic, axial and rhombic symmetry. For PSN, ^{93}Nb and ^{45}Sc MAS NMR spectra support the presence of predominantly a single Nb(V) and Sc(III) site with intermediate C_Q values assigned to axially distorted B-sites. ^{25}Mg MAS NMR of PMN also show evidence for at least two different Mg(II) sites of local cubic and axial symmetry due to partial B-site disorder in PMN. ^{25}Mg and ^{93}Nb NMR spectra of La^{3+}-substituted PMN (at 25 mol%) show that A-site substitution by La(III) leads to increased B-site ordering in the Mg-rich and Nb-rich nanodomains of PMN. By contrast, the ^{25}Mg and ^{93}Nb NMR spectra of BMN, a Ba^{2+} A-site analog of PMN, demonstrate that BMN has a more highly ordered Mg/Nb B-site distribution than PMN. These multinuclear NMR approaches demonstrate the enormous potential of solid-state NMR spectroscopy to provide both qualitative and quantitative information about the chemical environments, cation ordering and motional behavior of the B-site cations in a diverse range of perovskite relaxor ferroelectrics.

1. Scientific background – Relaxor ferroelectrics

The dielectric properties of the class of complex lead perovskite materials with the general formula Pb(B'$_{1/3}$B"$_{2/3}$)O$_3$, where B' = Zn^{2+}, Mg^{2+}, Ni^{2+} or Cd^{2+}; B" = Nb^{5+}, and PbB'$_{1/2}$B"$_{1/2}$O$_3$, where B' = Sc^{3+} or In^{3+}; and B" = Nb^{5+} or Ta^{5+}), are dependent on both the composition and short-range and intermediate-range chemical ordering of the multiple B-sites [1-4]. Perovskites with partial or complete B-site ordering with short coherency (2-50

C. Galassi et al. (eds.), Piezoelectric Materials: Advances in Science, Technology and Applications, 203–218.

nm), are relaxor ferroelectrics that exhibit broad, diffuse phase transitions and associated diffuse Curie temperature regions [2, 5-11].

Lead magnesium niobate, $Pb(Mg_{1/3}Nb_{2/3})O_3$, (PMN) and lead zinc niobate, $Pb(Zn_{1/3}Nb_{2/3})O_3$, (PZN), are non-stoichiometric B'B" lead perovskites[1, 5, 9]. PMN undergoes a diffuse phase transition with a Curie temperature (T_C) from -7 to -15°C [12]. PMN ceramics show short, coherent long-range order on the nanoscale (40-100 A), with both Mg^{2+}-rich and Nb^{5+}-rich nanodomain regions [2, 10, 11, 13-24]. Neutron diffraction studies [25] have shown average ion shifts from the ideal cubic lattice are Δ_{Pb} = 0.32A; $\Delta_{Nb,Mg}$ = 0.09A; Δ_O = 0.16A. Addition of lead titanate (PT) to PMN increases T_C by 5°C/mol%, with compositions greater than the morphotropic phase boundary (MPB) (34 mol% PT) undergoing pseudo-cubic (Pm3m) to tetragonal (R3m) change [12-14]. Lead zinc niobate, $Pb(Zn_{1/3}Nb_{2/3})O_3$ (PZN), also has a complex multiple B-site perovskite lattice where Zn^{2+} and Nb^{5+} ions are postulated to be disordered causing local compositional variations and fluctuations of the transition temperature (140°C) [25-27]. PZN is cubic above T_C (Pm3m) and has rhombohedral (R3m) symmetry below T_C. PZN forms (1-x)[Pb(Zn$_{1/3}$Nb$_{2/3}$)O$_3$]/x[PbTiO$_3$] solid solution to x = 0.1, where a phase transition between rhombohedral and tetragonal occurs at 25°C [17, 18, 28]. While 100% PZN is rhombohedral, 0.9PZN/0.1PT is proposed to be a compositional mixture of rhombohedral and tetragonal phases, with single-crystal 0.93PZN/0.07PT materials exhibiting unusually high electrostrictive properties [29].

Stoichiometric lead-based perovskites $Pb(B'_{1/2}B''_{1/2})O_3$, where B' = Sc^{3+} or In^{3+}, and B" = Nb^{5+} or Ta^{5+}, are also relaxor ferroelectrics with moderate to complete B-site ordering based on XRD and HRTEM studies [30-40]. Both $Pb(Sc_{1/2}Nb_{1/2})O_3$ (PSN), and $Pb(In_{1/2}Nb_{1/2})O_3$ (PIN), have diffuse phase transitions with T_C near 90°C,[29] and PSN can exist with ordered B-sites, depending on annealing conditions and crystalline forms [30,31]. Neutron diffraction shows that PSN is cubic (Pm3m) in the paraelectric state (400 K), and is rhombohedral (R3m) in the ferroelectric state (200 K) [32]. Comparative HRTEM studies [35] of PSN and the tantalum analog show that PSN has significant, but less {111}-type ordering than $Pb(Sc_{1/2}Ta_{1/2})O_3$, with small domains of {110}-type ordering. Calculated absolute displacements in the ferroelectric state along the {111} axis are Δ_{Pb} = 0.094A, $\Delta_{Sc/Nb}$ = -0.148 A and Δ_O = -0.195 A [32].

In this report, solid-state magic-angle spinning (MAS) ^{93}Nb MAS NMR spectra of single crystal PMN and PZN, and polycrystalline PSN and PIN are described. In addition, ^{25}Mg and ^{93}Nb NMR spectra of polycrystalline PMN, La^{3+}-substituted PMN and BMN, and ^{45}Sc MAS NMR spectra of PSN are also reported. These solid-state NMR studies demonstrate the utility and applicability of ^{25}Mg, ^{45}Sc and ^{93}Nb NMR to obtain detailed atomic-level information about the chemical environments and cation ordering of the local Nb(V), Sc(III) and Mg(II) B-sites in these perovskite systems. These unique approaches thus provide direct means to qualitatively and quantitatively probe the local B-site cation chemical environments, short-range and intermediate-range ordering, composition-and temperature-dependent structural and electrical properties of these materials.

2. Experimental

2.1 REAGENTS AND MATERIALS

PbO (Alfa, 99.9995%), Nb_2O_5 (Alfa, 99.9985%), Sc_2O_3 (Aldrich, 99.999%), In_2O_3 (Aldrich, 99.995%), $BaCO_3$ (Fisher, 99%) and $(MgCO_3)_4Mg(OH)_2 \cdot H_2O$, Aldrich, 99%) were used as purchased. MgO (Aldrich, 99+%) was fired at 950°C for 6 hrs prior to use.

2.2 SYNTHESIS OF RELAXOR FERROELECTRIC SYSTEMS.

2.2.1 Single Crystal PMN and PZN.
Single crystal PMN and PZN were synthesized by the lead oxide flux method, [29] and donated by Professor T. Shrout and Dr. S. E. Park, Pennsylvania State University.

2.2.2 Synthesis of PMN, La^{3+}-substituted PMN and BMN Powders.
PMN powders were synthesized by a modified columbite method, [41] where MgO in the form of $(MgCO_3)_4Mg(OH)_2 \cdot H_2O$ was used in a 2% excess. The $(MgCO_3)_4Mg(OH)_2 \cdot H_2O$ powder was slurried with Nb_2O_5 in ethanol, the mixture ground for 2 hrs, oven dried (140 °C) overnight and fired at 1000 °C for 6 hrs. The precursor $MgNb_2O_6$ was then mixed with PbO in ethanol for 1.5 hrs based on the PMN stoichiometry. A two-stage calcination process as summarized by Gupta and Kulkari [42] included: firing at 800 °C for 2 hrs, reslurrying in EtOH for 1 hr, oven drying (140 °C) overnight, and firing at 900 °C for 2 hrs. BMN was prepared by a similar procedure where the columbite precursor was mixed with $BaCO_3$ in ethanol for 1 hr, oven dried at 140°C, and fired at 1100°C for 2 hrs. La^{3+}-substituted PMN was prepared by the modified columbite method [41] by slurry mixing magnesium niobate (prepared from Nb_2O_5 and $(MgCO_3)_4Mg(OH)_2 \cdot H_2O$ by slurry mixing and firing at 1100 °C for 2 hrs) with La_2O_3 in ethanol, oven drying (140 °C) overnight, and firing at 1000 °C for 6 hrs.

2.2.3 Synthesis of $Pb(Sc_{1/2}Nb_{1/2})O_3$ (PSN), and $Pb(In_{1/2}Nb_{1/2})O_3$ (PIN).
These materials were prepared by the columbite method using Sc_2O_3/Nb_2O_5 and In_2O_3/Nb_2O_5 powder mixtures based on the formula of the columbite precursors, by mixing in an ethanol slurry for 1 hr, oven drying at 140°C, and firing for 4 hrs at 1350°C for $ScNbO_4$ and $InNbO_4$. The columbite-type precursors were then mixed with PbO and fired at 800°C for 2 hrs and 1000°C at 2 hrs for PIN; and at 800°C for 2 hrs, followed by 900°C firing at 2 hrs for PSN.

2.3 CHARACTERIZATION METHODS.

2.3.1 Power X-ray Diffraction Analysis.
Powder X-ray diffraction analysis were performed at the Engineering and Mining Experiment Station, South Dakota School of Mines and Technology, Rapid City, SD, using a Phillips X-ray diffractometer at CuK_α (40 kV and 20 mA). The XRD patterns agree with the literature, as discussed elsewhere [43].

2.3.2 Solid-State ^{93}Nb NMR Measurements.
All solid-state MAS NMR spectra were obtained using a Chemagnetics Infinity-600 NMR

TABLE 1. Nuclides of significance for solid state NMR studies of the structure and dynamics studies of Pb($Mg_{1/3}Nb_{1/3}$)O_3 (PMN) and related relaxor ferroelectrics

A-Site Cations	B-Site Cations	Anion Sites
^{207}Pb (I = 1/2) Abundance = 22.60% v_0 = 125.52 MHz at 14.1T Dominated by CSA	^{93}Nb (I = 9/2) Abundance = 100% v_0 = 146.64 MHz at 14.1T Large Quad. + Moderate CSA Q (moment) = -0.22 Barns	^{17}O (I = 5/2) Abundance = 0.037% v_0 = 81.36 MHz at 14.1T Small Quadrupole Q (moment) = -0.026 Barns
^{137}Ba (I = 3/2) Abundance = 11.32% v_0 = 66.72 MHz at 14.1T Moderate Quadrupole Q (moment) = 0.28 Barns	^{25}Mg (I = 5/2) Abundance = 10.1% v_0 = 36.72 MHz at 14.1T Mod. to Large Quadrupole Q (moment) = 0.20 Barns	
	^{45}Sc (I = 7/2) Abundance = 100% v_0 = 145.98] at 14.1T Mod. to Large Quadrupole Q (moment) = -0.22 Barns	

spectrometer equipped with two "home-built" probes with 3.2 mm and 4.0 mm sample spinning systems. The ^{25}Mg, ^{45}Sc and ^{93}Nb NMR spectra were measured at 36.73 MHz, 145.48 MHz and 146.75 MHz, with 6 kHz, 16.7 kHz and 1.67 kHz sample spinning speed, and using a 4.2 mm, 3.2mm and 3.2 mm sample spinning system, respectively. Single-pulse ^{25}Mg, ^{45}Sc and ^{93}Nb MAS NMR spectra were typically obtained at 1.50 µsec, 0.50 µsec and 0.50 µsec pulse lengths, 1.500 sec, 0.200 sec and 0.200 sec delay times, with a 50 kHz, 2000 kHz and 2000 kHz spectral width (20.0 µsec, 0.5 µsec and 0.5 µsec dwell), respectively. Spectra were Fourier-transformed with seven linear projection points, and exponential line broadening of 200 Hz for ^{45}Sc and ^{93}Nb spectra. A line broadening of 60 Hz was used for the ^{25}Mg NMR spectra. The solid-state ^{93}Nb MAS NMR spectra were referenced by sample substitution to a solution of $NbCl_5$ in "wet" acetonitrile, assigned a chemical shift value of 0.00 ppm; the ^{25}Mg and ^{45}Sc NMR were referenced to aqueous 3M $MgSO_4$ and 1M $Sc(ClO_4)_3$, with the single sharp signals being assigned a chemical shift of 0.0 ppm.

3. Results and discussion

3.1 ^{93}Nb NMR OF NON-STOICHIOMETRIC PMN AND PZN SINGLE CRYSTALS.

^{93}Nb is a 100% natural abundant quadrupolar nuclide with a nuclear spin (I) value I = 9/2, and a large electric quadrupole moment of -0.22 barns (Table I). The ^{93}Nb NMR spectra

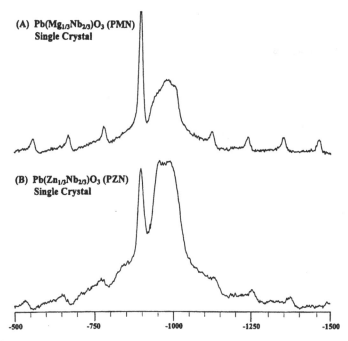

Figure 1. ^{93}Nb MAS NMR spectra of powdered samples of single crystal (A) lead magnesium niobate, Pb(Mg$_{1/3}$Nb$_{2/3}$)O$_3$, (PMN), and (B) lead zinc niobate, Pb(Zn$_{1/3}$Nb$_{2/3}$)O$_3$, (PZN), measured at 14.1 Tesla.

are affected by chemical shift and quadrupolar interactions and are generally characterized by the isotropic or observed chemical shift (δ_{iso} or δ_{obs}), the quadrupolar coupling constant (C_Q) and asymmetry parameter (η_Q). Solid-state ^{93}Nb NMR studies of niobium oxides have been reported for the ferroelectric alkali niobates of lithium, sodium and potassium using principally static, wide line NMR [44-47]. Recently, both ^{93}Nb static and MAS NMR spectra for a wide range of niobium oxides, alkali, alkaline earth, trivalent and lead niobates have shown that ^{93}Nb NMR spectroscopy is highly sensitive to the local Nb(V) electronic environment [48, 49] The observed ^{93}Nb frequency shifts vary from -900 to -1250 ppm and are assigned to local pseudo-octahedral NbO$_6$ environments. The ^{93}Nb quadrupolar coupling constants range from 12 to 22 MHz for corner-shared NbO$_6$ octahedra in alkali and lead niobates to greater than 50 MHz for corner/edge-shared octahedra in MgNb$_2$O$_6$ and ZnNb$_2$O$_6$. The chemical shifts are also somewhat sensitive to next-nearest neighbor B-site ions, e.g., Nb(ONb)$_6$ versus Nb(OMg)$_6$.

The ^{93}Nb MAS NMR spectra for single crystal PMN and PZN perovskites measured at 14.1 Tesla are shown in Figure 1, and have been extensively described [50]. Based on measured ^{93}Nb chemical shifts and quadrupolar coupling constants determined for model Nb(V)-oxygen systems, the two ^{93}Nb MAS resonances in PMN and PZN were assigned to a range of different Nb(V) B-sites of local cubic (for the sharp ^{93}Nb peak at -900 ppm), and local Nb(V) B-sites of axial and rhombic symmetry (associated with broad overlapping peaks centered near -980 ppm). The Mg/Nb (and Zn/Nb) B-sites in PMN (and PZN), are partially ordered and not randomly occupied. The local cubic Nb(V) sites

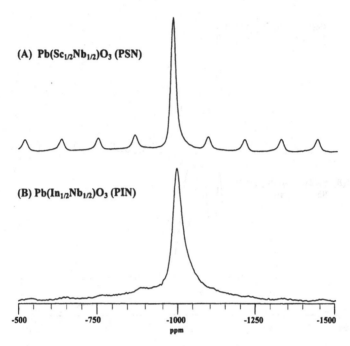

Figure 2. ^{93}Nb MAS NMR spectra of polycrystalline (A) lead scandium niobate, Pb(Sc$_{1/2}$Nb$_{1/2}$)O$_3$ (PSN), and (B) lead indium niobate, Pb(In$_{1/2}$Nb$_{1/2}$)O$_3$ (PIN) measured at 14.1 Tesla.

in PMN are assigned to next-nearest neighbor (nnn) Nb(OMg)$_6$ configurations in the Mg-rich regions, and the axial and rhombic symmetry Nb(V) B-sites of nnn Nb(ONb)$_{6-x}$(OMg)$_x$ configurations in Nb-rich regions of PMN [50,51]. Recent extensive two-dimensional ^{93}Nb nutation spectroscopy for PMN, [51,52] by contrast, show three ^{93}Nb resonances due to the different types of Nb(V) B-sites in PMN, a symmetric, near-cubic Nb(V) B-site (C$_Q$ < 0.8 MHz) assigned to nnn Nb(OMg)$_6$ in the Mg-rich regions of PMN, and a range of distorted Nb(V) sites (C$_Q$ ≈ 17 MHz) similar to the Nb(V) B-sites in the alkali niobates that are assigned to axial nnn Nb(ONb)$_{6-x}$(OMg)$_x$ configurations, and more highly distorted Nb(V) B-sites (C$_Q$ > 62 MHz) due to Nb(ONb)$_{6-x}$(OMg)$_x$ configurations of rhombic symmetry. The latter two types of Nb(V) B-sites are proposed to exist in the Nb-rich regions of PMN [41, 50-57].

3.2 ^{93}Nb MAS NMR SPECTRA OF STOICHIOMETRIC PSN AND PIN POWDERS.

The ^{93}Nb MAS NMR spectra acquired at 14.1 Tesla for PSN and PIN powders are shown in Figure 2. For PSN, a single, narrow, unsymmetrical ^{93}Nb resonance is observed at -983 ppm (FWHH ≈ 2,900 Hz), with an extensive array of spinning sidebands. A single broader ^{93}Nb peak is observed for PIN than for PSN centered at -1000 ppm (FWHH ≈ 6,700 Hz), with lower intensity spinning sidebands. The static ^{93}Nb NMR spectra of PSN and PIN (not shown) do not have well-defined singularities, and both materials show a single ^{93}Nb peak at -980 ppm (FWHH ≈ 9,500 Hz) and -1000 ppm (FWHH ≈ 17,000 Hz), respectively. The decreased linewidth at MAS conditions for PSN and PIN indicates that ^{93}Nb quadrupolar interactions are the major line broadening mechanism, while the broader linewidth seen in the ^{93}Nb MAS NMR spectra of PIN indicates that additional inhomogeneous broadening is also present due to chemical shift dispersion effects as a result of In/Nb disorder [34-36]. The ^{93}Nb NMR spectra for PSN and PIN show that these materials contain principally a single type of Nb(V) site, with PSN having a lower degree of dispersion due to both local Sc/Nb B-site ordering and possibly higher crystallinity [30-33] In addition, the ^{93}Nb MAS and static NMR spectra of PSN show an underlying broader ^{93}Nb resonance, thus demonstrating the need for more detailed solid-state NMR investigations of the B-site ordering behavior of PSN perovskite as a function of synthesis and annealing conditions. Similarily for PIN, the ^{93}Nb MAS NMR spectra of the synthesized PIN and PIN annealed at 1200 °C for 20 hours show linewidths of 6,700 Hz and 5,200 Hz, consistent with a reduction in inhomogeneous line broadening as a result of increased B-site ordering and/or increased sample crystallinity. In contrast to the multiple Nb(V) B-site perovskites such as PMN and PZN that exist with principally Mg/Nb B-site disorder, PSN powders have significant Sc/Nb B-site ordering, while PIN prepared at intermediate reaction temperatures likely exists with principally In/Nb B-site disorder [34-36]. The local Nb(V) B-sites in PSN are of nnn Nb(OSc)$_6$ configurations, while PIN would have a wide range of nnn Nb(ONb)$_{6-x}$(OIn)$_x$ configurations [32, 35]. Based on recent neutron diffraction studies [32] of single crystal PSN, it is proposed that the nnn Nb(OSc)$_6$ B-site configurations originate from Sc/Nb ordering down the {111} axis, while minor ordering of the {110}-type would result in nnn Nb(ONb)$_4$(OSc)$_2$ B-site configurations.

3.3 ^{93}Nb AND ^{25}Mg MAS NMR SPECTRA OF PMN, BMN AND La^{3+}-PMN (PMN25La).

^{93}Nb and ^{25}Mg MAS NMR spectra of polycrystalline PMN, its Ba(II) analog, Ba(Mg$_{1/3}$Nb$_{2/3}$)O$_3$ (BMN), and a La^{3+}-substituted PMN were measured at 14.1 Tesla as shown in Figures 3 and 4, respectively. The ^{93}Nb MAS NMR spectrum of polycrystalline PMN shown in Figure 3 is nearly identical to single crystal PMN (Figure 1) showing two resonances, a sharp ^{93}Nb signal at -902 ppm, and a broader resonance centered at -980 ppm as previously assigned to different types of Nb(V) B-sites in PMN. By contrast, the ^{93}Nb MAS NMR spectra of BMN (Figure 3) shows an intense, narrow resonance at -925 ppm (3,900 Hz), with a broader underlying resonance centered near -972 ppm. Previous high-resolution TEM studies of BMN indicate that this dielectric material has more highly ordered Mg/Nb B-sites than PMN [13-18, 56-58, 64]. The most intense ^{93}Nb resonance is assigned to higher symmetry Nb(V) B-sites, while the latter is assigned to lower symmetry Nb(V) sites with a range of different nnn Nb(ONb)$_{6-x}$(OMg)$_x$ configurations. While insufficient information is presently available to make definitive assignments, the higher symmetry Nb(V) B-sites are likely of axial symmetry, in contrast to the lower content, more distorted Nb(V) B-sites of rhombic symmetry. Possible Nb(V) B-site configurations that correspond to the bulk stoichiometry of BMN include the nnn cis- and trans- Nb(ONb)$_2$(OMg)$_4$ B-site configurations. The ^{93}Nb MAS NMR spectrum of BMN thus

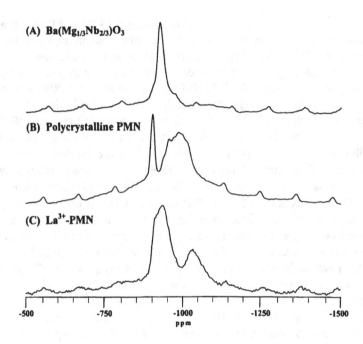

(A) $Ba(Mg_{1/3}Nb_{2/3})O_3$

(B) Polycrystalline PMN

(C) La^{3+}-PMN

| -500 | -750 | -1000 | -1250 | -1500 |

ppm

Figure 3. ^{93}Nb MAS NMR spectra of polycrystalline materials: (A) PMN, (B) barium magnesium niobate, $Ba(Mg_{1/3}Nb_{2/3})O_3$ (BMN), and (C) La^{3+}-substituted PMN at 25 mol% La_2O_3 (PMN25La) measured at 14.1 Tesla.

establishes that this mixed B'B" system is more highly ordered than PMN (or PZN) based on comparison of their corresponding ^{93}Nb MAS NMR spectra (see Figure 1).

The ^{25}Mg MAS NMR spectra of PMN, BMN and La^{3+}-PMN measured at 14.1 Tesla are given in Figure 4. ^{25}Mg is also a quadrupolar nuclide ($I = 5/2$) with a moderate quadrupole moment, a low Lamor frequency and a lower natural abundance (Table I) that leads to significantly lower NMR sensitivity than for ^{93}Nb and ^{45}Sc (Table I). Relatively few ^{25}Mg MAS NMR studies have been reported, although studies of selected magnesium oxide systems, [59] the Mg/Si ordering in magnesium orthosilicate, [60] various Mg-containing minerals [61] and $MgO/MgCO_3$ mixtures [62] have been reported. The ^{25}Mg NMR measurements of MgO systems show that the chemical shift range for ^{25}Mg NMR is relatively narrow, with six-coordinate MgO_6 sites observed in the chemical shift range 0-30 ppm and four-coordinate MgO_4 sites from 50 to 60 ppm [59]. The quadrupolar parameters C_Q and η_Q have been determined for few Mg-O systems, thereby precluding a detailed assessment of the influence of local electronic structure on the magnitude and orientation of the ^{25}Mg quadrupolar interactions.

The ^{25}Mg MAS NMR spectrum of PMN as shown in Figure 4 (center) has two major resonances at 25.8 ppm and 1.2 ppm. The sharp, low intensity ^{25}Mg peak at 25.8 ppm has a frequency shift similar to cubic MgO_6 sites in MgO of the NaCl lattice ($\delta_{iso} = 26.0$ ppm) and is tentatively assigned to near-cubic or highly symmetry Mg(II) B-sites of

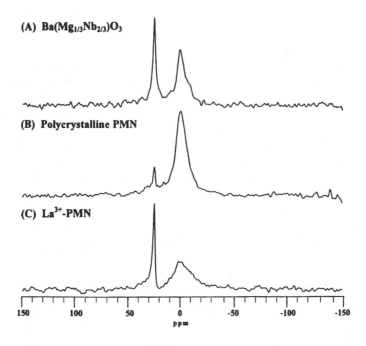

(A) Ba(Mg$_{1/3}$Nb$_{2/3}$)O$_3$

(B) Polycrystalline PMN

(C) La^{3+}-PMN

| 150 | 100 | 50 | 0 | -50 | -100 | -150 |

ppm

Figure 4. ^{25}Mg MAS NMR spectra of polycrystalline materials: (A) PMN, (B) barium magnesium niobate, Ba(Mg$_{1/3}$Nb$_{2/3}$)O$_3$ (BMN), and (C) La^{3+}-substituted PMN at 25 mol% La$_2$O$_3$ (PMN25La) measured at 14.1 Tesla.

nnn Mg(ONb)$_6$ configurations in the Mg-rich regions of PMN. The Mg(II) B-sites corresponding to this ^{25}Mg resonance are analogous to the Nb(V) B-sites with nnn Nb(OMg)$_6$ configurations attributed to the ^{93}Nb peak at -902 ppm. The broader ^{25}Mg resonance at 1.2 ppm with poorly-defined singularities is assigned to distorted six-coordinate Mg(II) B-sites with nnn Mg(OMg)$_{6-x}$(ONb)$_x$ configurations in the Nb-rich nanodomain regions of PMN. As noted from ^{93}Nb nutation results for PMN, [51, 52] where a third, broad resonance is observed due to highly distorted Nb(V) B-sites of rhombic symmetry, the ^{25}Mg NMR spectrum of PMN also shows a broad underlying ^{25}Mg resonance distinct from the two readily observable ^{25}Mg peaks at 25.8 and 1.2 ppm. The ^{25}Mg MAS NMR spectrum of PMN, therefore, provides collaborative evidence for Mg/Nb B-site disorder in PMN. Limited 1:1 Mg/Nb B-site ordering occurs as evidenced by the Mg(II) B-sites of nnn Mg(ONb)$_6$ configurations (ca. 10% intensity) that correspond to the sharp ^{25}Mg signal at 25.8 ppm.

The ^{25}Mg MAS NMR spectrum of BMN as shown in Figure 4 (top) may be contrasted with the ^{25}Mg spectrum of PMN. The ^{25}Mg MAS NMR spectrum of BMN shows increased intensity for the ^{25}Mg resonance at 25.6 ppm and a less intense, broader ^{25}Mg peak at 1.8 ppm. Assuming similar assignments for these ^{25}Mg resonances as noted for PMN, the ^{25}Mg MAS NMR spectrum of BMN indicates that BMN has a higher abundance of Mg(II) B-sites of nnn Mg(ONb)$_6$ configurations associated with 1:1 Mg/Nb ordering in the Mg-rich regions of BMN. Extensive ordering of Mg/Nb in BMN has been previously reported; [56-58] however, the ^{25}Mg and ^{93}Nb NMR results for the BMN studied here do not support complete B-site ordering for BMN. These results suggest that the degree of Mg/Nb B-site ordering in BMN is substantially influenced by synthesis and reaction conditions, including the annealing process and temperatures used [53, 54, 56-58, 63, 64].

The ^{25}Mg and ^{93}Nb MAS NMR spectra of PMN modified by substitution of 25 mol% La(III) for Pb(II) in the A-sites of PMN are shown in Figures 3 and 4 (bottom), respectively. Substitution of La(III) ion into PMN causes significant changes in the ^{93}Nb NMR spectra, including increased intensity for the sharp ^{93}Nb peak at -910 ppm, the appearance of a distinct resonance at -936 ppm, the apparent disappearance of the broad peak centered at -980 ppm, and the appearance of a new Gaussian-like resonance at -1030 ppm. The Mg/Nb B-site configurations in PMN are thus dramatically altered following La(III) ion A-site substitution, resulting in concomitant atomic-level B-site modifications of PMN. The appearance of the ^{93}Nb NMR peak at -1030 ppm, in particular, suggests that the Nb-rich regions in PMN contain an increased amount of Nb(V) B-sites with nnn Nb(ONb)$_6$ B-site configurations, that are analogous to those Nb(V) B-sites in the alkali niobates [48]. Recent ^{93}Nb NMR of alkali niobates show ^{93}Nb resonances in the -1017 to -1102 ppm chemical shift region [48]. The increased Mg/Nb B-site ordering is also accompanied by additional increase in the number of Nb(V) B-sites with nnn Nb(OMg)$_6$ B-site configurations (-910 ppm signal), decreased content of Nb-rich regions containing Nb(V) B-sites of nnn Nb(ONb)$_{6-x}$(OMg)$_x$ configurations (associated with signal at -980 ppm), and the formation of appreciable Nb(V) B-sites of nnn Nb(ONb)$_{6-x}$(OMg)$_x$ configurations in the Nb-rich regions of PMN (-936 ppm resonance). The complex redistribution process for Mg/Nb cations in the B-sites of PMN following La(III) substitution in the A-sites is consistent with the increased Mg/Nb B-site ordering as reported for La^{3+} ion-substituted PMN [63,64]. Most importantly, these results indicate that the Mg/Nb B-site redistribution process is exceedingly complex based on the ^{93}Nb MAS NMR results reported here.

The ^{25}Mg MAS NMR spectrum of La^{3+}-substituted PMN as given in Figure 4 (bottom) also show very complex changes in the Mg/Nb B-site distribution. The ^{25}Mg NMR peak at 25.7 ppm for the La^{3+}-substituted PMN increases substantially compared to the corresponding ^{25}Mg peak intensity at 25.8 ppm for PMN as a result of increased amount of Mg(II) B-sites with nnn Mg(ONb)$_6$ configurations in the Mg-rich nanodomain regions in PMN. The redistribution of Mg/Nb cations in the B-sites of PMN is also accompanied by a significant decrease in the ^{25}Mg NMR signal at 2.9 ppm that is associated with Mg(II) B-sites with nnn Mg(OMg)$_{6-x}$(ONb)$_x$ configurations found in the Nb-rich nanodomain regions of PMN. Significant lineshape and frequency shift changes for the ^{25}Mg resonance near -0.1 ppm are also observed, suggesting further more complex alterations not currently understood. These ^{25}Mg MAS NMR results suggest additional directions for more extensive ^{25}Mg NMR studies, including investigations to provide correlations between the ^{25}Mg NMR and structural parameters for Mg(II)-oxygen chemical environments in model magnesium-oxygen perovskite systems [43].

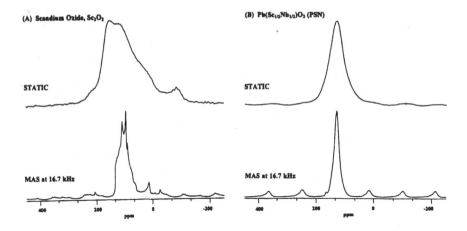

Figure 5. ^{45}Sc static and MAS NMR spectra of polycrystalline (A) scandium oxide, Sc_2O_3 and (B) lead scandium niobate, $Pb(Sc_{1/2}Nb_{1/2})O_3$ (PSN) measured at 14.1 Tesla.

3.4 ^{45}Sc MAS NMR SPECTRA OF POLYCRYSTALLINE ScNbO$_4$ AND PSN.

The ^{45}Sc static and MAS NMR spectra of polycrystalline ScNbO$_4$ and PSN measured at 14.1 Tesla are shown in Figure 5. ^{45}Sc is also a 100% natural abundant quadrupolar nuclide ($I = 7/2$) with a high NMR receptivity (Table I). Comparison of the ^{45}Sc static and MAS NMR spectra for Sc_2O_3 shows that MAS leads to significant line narrowing for the broad quadrupolar powder pattern observed due to the central transition ($1/2 \leftrightarrow -1/2$) resonance of ^{45}Sc that spans from -100 to 235 ppm (FWHH = 21,000 Hz). A second, lower intensity ^{45}Sc resonance is also seen in the ^{45}Sc static and MAS NMR spectra near -84 ppm due to another ^{45}Sc resonance. The broader ^{45}Sc signal for Sc_2O_3 shows a typical doublet (94 and 104 ppm) characteristic of a quadrupolar powder pattern due to incomplete removal of the second-order quadrupolar line broadening at MAS conditions, consistent with an asymmetry parameter of $\eta_Q \approx 0.5$. The second, lower intensity ^{45}Sc resonance also shows a doublet (15 and -24 ppm) powder pattern due to incomplete averaging of the second-order quadrupolar interactions. The two ^{45}Sc resonances seen in the ^{45}Sc static and MAS NMR spectra of Sc_2O_3 (Figure 5A, lower spectrum) are assigned to two different six-coordinate Sc(III)-oxygen sites (in a 3:1 ratio), consistent with the two distorted ScO_6 sites reported from crystal structure results for the $C-M_2O_3$ lattice of Sc_2O_3.[65]

The ^{45}Sc static and MAS NMR spectra of $Pb(Sc_{1/2}Nb_{1/2})O_3$ (PSN) are shown in Figure 5B. While the ^{45}Sc static NMR spectrum (upper) shows a single, broad ^{45}Sc resonance centered at 131 ppm (FWHH = 8,500 Hz), the ^{45}Sc MAS NMR spectrum (lower) of this intense peak is further narrowed. A minor sharper ^{45}Sc NMR signal is also seen at 167 ppm. The principal ^{45}Sc resonance is assigned to Sc(III) B-sites of nnn Sc(ONb)$_6$ configurations due to 1:1 Sc/Nb ordering down the {110} axis in PSN, [30, 32] while the minor ^{45}Sc signal is assigned to an unknown impurity. The ^{45}Sc MAS NMR spectrum obtained using a Hahn-echo sequence (not shown) also shows this additional broad underlying ^{45}Sc resonance due to highly distorted symmetry Sc(III) B-sites of possible rhombic symmetry. These Sc(III) B-sites are tentatively assigned to lower symmetry Sc(III) B-sites of nnn Sc(ONb)$_4$(OSc)$_2$ configurations corresponding to Sc/Nb B-site ordering down the {111} axis as reported for single crystal PSN [32]. Previous ^{45}Sc static NMR measurements of single crystal PSN [33] have shown a broad and sharp ^{45}Sc NMR signal for PSN, although the latter is likely substantially narrowed by Sc(III) B-site motion as also reported for Nb(V) B-sites associated with the broad ^{93}Nb NMR peak observed for PMN [33, 66]. The sharp and broad ^{45}Sc NMR resonances observed for PSN (Figure 5B) are consistent with significant Sc/Nb B-site ordering in PSN, as also observed from the ^{93}Nb NMR results for PSN as previously discussed.

4. Conclusions

The applicability of solid-state ^{93}Nb, ^{45}Sc and ^{25}Mg NMR spectroscopy to explore the local Nb(V), Sc(III) and Mg(II) B-site oxygen chemical environments and the degree of B'B" ordering in both stoichiometric and non-stoichiometric Pb-based B'B" perovskites is discussed. Non-stoichiometric PMN and PZN relaxor ferroelectric systems are shown using ^{93}Nb MAS NMR spectroscopy to exist as primarily disordered B'B" perovskites, with ca. 10% of 1:1 Mg/Nb and Zn/Nb B-site ordering. The ^{93}Nb NMR resonances for PMN have been assigned to near-cubic Nb(OMg)$_6$ B-site configurations in Mg-rich regions, and distorted Nb(ONb)$_{6-x}$(OMg)$_x$ B-site configurations of axial and rhombic symmetry in Nb-rich regions of PMN. By contrast, the stoichiometric relaxor ferroelectrics PSN and PIN show primarily a single ^{93}Nb resonance assigned to Nb(V) B-sites of Nb(OSc)$_6$ and Nb(ONb)$_{6-x}$(OIn)$_x$ configurations due to Sc/Nb ordering and In/Nb B-site disorder, respectively. The ^{45}Sc MAS NMR spectrum for PSN is also consistent with the presence of Sc(III) B-sites of nnn Sc(ONb)$_6$ B-sites due to 1:1 Sc/Nb ordering down the {110} axis, and possible nnn Sc(ONb)$_4$(OSc)$_2$ site configurations as a result of ordering along the {111} axis.

The ^{25}Mg MAS NMR for PMN also indicates the presence of 1:1 Mg/Nb B-site ordering with cubic Mg(II) B-sites of nnn Mg(ONb)$_6$ B-site configurations in the Mg-rich regions and Mg(II) B-sites of nnn Nb(ONb)$_{6-x}$(OMg)$_x$ configurations with axial and/or rhombic symmetry in the Nb-rich regions. By contrast, the ^{93}Nb and ^{25}Mg NMR of BMN shows significant, but incomplete Mg/Nb B-site ordering in contrast to its Pb(II) analog, lead magnesium niobate. The ^{93}Nb and ^{25}Mg MAS NMR spectrum of La^{3+}-substituted (25 mol% for Pb(II)) PMN is unique compared to the related spectra of PMN, demonstrating that A-site substitution of Pb(II) by La(III) ion leads to increased Mg/Nb B-site ordering. A complex Mg/Nb B-site cation redistribution process occurs within both the Mg-rich and Nb-rich nanodomains; however, both qualitative and quantitative NMR descriptions of the distributions of the Mg(II)/Nb(V) B-sites of different Nb(ONb)$_{6-x}$(OMg)$_x$ configurations are needed to completely understand this process.

The solid-state ^{93}Nb, ^{45}Sc and ^{25}Mg NMR studies reported here illustrate the extensive applicability of these spectroscopic approaches to explore the Nb(V), Sc(III) and Mg(II) B-site oxygen environments and degree of B'B" cation ordering in relaxor ferroelectrics. In addition, these results demonstrate the utility of solid-state NMR to quantify the degree of B'B" cation ordering in relaxor ferroelectrics as affected by both A- and B-site substitutions and doping. The use of these solid-state NMR approaches should also enable further detailed study of how synthesis and annealing conditions alter the B'B" -site cation distributions in PMN, PZN and related relaxor ferroelectrics of the perovskite structure.

5. Aknowledgements

This work was supported by a Department of Defense University Research Initiative Support Program (URISP) grant administered by the Office of Naval Research, Grant 0014-96-1-0781. The authors acknowledge helpful discussions with Professor G.E. Maciel and the use of facilities at the Colorado State University NMR Center, Fort Collins, Colorado.

6. References

1. Jona, F.; Shirane, G. (1993) *Ferroelectric Crystals*, Dover, New York.
2. Cross, L. E., (1994) Relaxor Ferroelectrics: An Overview, *Ferroelectrics*, **151**, 305.
3. Galasso, F. S., (1963), *Structure, Properties and Preparation of Perovskite-Type Compounds*, Pergamon Press, Oxford pp. 1-121.
4. Galasso, F.; Pyle, J., (1963) Ordering in Compounds of the A(B'0.33Ta0.67)O3 Type *Inorg. Chem.*, **2**(3), 482-484.
5. Bursill, L. A.; Peng, J., (1991) Ferroelectric and Ferroelastic Domain Structures in Piezoelectric Ceramics, *NIST Spec. Publ.* **804**, , 67-76.
6. Dougherty, T. P.; Wiederrecht, G. P.; Nelson, K. A.; Garrett, M. H.; Jensen, H. P.; Warder, C. (1992) Femtosecond Resolution of Soft Mode Dynamics in Structural Phase Transitions *Science*, , 258, 770-774, and references cited therein.
7. Glazer, A. M.; Roleder, K.; Dec, J. (1993), Structure and Disorder in Single-Crystal Lead Zirconate"*Acta Cryst.*, **B49**, 846-852.
8. Viehland, D. (1995) Transmission Electron Microscopy Study of High-Zr-Content Lead Zirconate Titanate *Phys. Rev. B*, **52**, 778-791.
9. Cross, L. E. (1987), Relaxor Ferroelectrics *Ferroelectrics*, **76**, 241-267.
10. Shrout, T.R.; Halliyal, A. (1987) Preparation of Lead-Based Ferroelectric Relaxors for Capacitors", *Am. Ceram. Soc. Bull.*, **66**(4), 704-711..
11. Uchino, K. (1986) Electrostrictive Actuators: Materials and Applications *Am. Ceram. Soc. Bull.*, **65**(4) 647-652.
12. Choi, S. W.; Shrout, T. R.; Jang, S. J.; Bhalla, A. S (1989). Dielectric and Pyroelectric Properties in the Pb(Mg1/3Nb2/3)O3-PbTiO3 System *Mater. Lett.*, **8**(67), 29-38.
13. Bursill, L.A.; Qian, H.; Peng, J. L.; Fan, X. D. (1995) Observation and Analysis of Nanodomain Textures in the Dielectric Relaxor Lead Magnesium Niobate *Physica B*, **216**, 1-23.
14. Ye, Z. G. (1996) Relaxor Ferroelectric Pb(Mg1/3Nb2/3)O3: Properties and Present Understanding *Ferroelectrics*, **184**, 193-208.
15. Depro, L. E.; Sangaletti, L. (1997) Structural Models for Lead Magnesium Niobate *Solid State Comm.*, **107**, 615-620.
16. Viehland, D.; Jang, S.; Cross, L.E. (1991) The Dielectric Relaxation of Lead Magnesium Niobate, *Phil. Mag. B.*, **64**(3), 335-344.

216

17. Chen, J.; Chan, H. M.; Harmer, M. P. (1989), Ordering, Structure and Dielectric Properties of Undoped and La/Na Doped $Pb(Mg_{1/3}Nb_{2/3})O_3$ *J. Am. Ceram. Soc.* **72**(4), 593-598.

18. Hilton, A.D.; Barber, D.J.; Randall, C.A.; Shrout, T.R. (1990) On Short Range Ordering in the Perovskite Lead Magnesium Niobate *J. Mater. Sci.*, **25**, 3462-3466.

19. Smolenskii, G. A.; Agronovskaya, A. I. (1959) Dielectric Polarization of a Number of Complex Compounds *Sov. Phys. Sol. State*, , 1(10), 1429-1437.

20. Newnham, R. E. (1991) Tunable Transducers: Nonlinear Phenomena in Electroceramics, *Chemistry of Electronic Materials*, NIST Publication **804**, 39-52.

21. Yan, M.F.; Ling, H.C.; Rhodes, W.W. (1989) Preparation and Properties of $PbO-MgO-Nb_2O_5$ Ceramics Near the $Pb(Mg_{1/3}Nb_{2/3})O_3$ Composition, *J. Mater.Res.*, **4**(4), 930-944.

22. de Mathan, N.; Husson, E.; Calvarin, G.; Gavarri, J. R.; Hewat, A. W.; Morell, A. (1991) Structural Model for the Relaxor $Pb(Mg_{1/3}Nb_{2/3})O_3$ at 5K *J. Phys. Condens. Mat.*, **3**, 8159-8171.

23. Caranoni, C.; Thuries, F.; Leroux, C.; Nihoul, G. (1994) Superstructures of Lead Niobium Oxides: Diffraction Experiments and Modelling *Phil. Mag. A*, **69**(3), 633-643.

24. Bonneau, P.; Garnier, P.; Calvarin, G.; Husson, E.; Gavarri, J. R.; Hewat, A. W.; Morell, A. (1991) X-ray and Neutron Diffraction Studies of the Diffuse Phase Transition in $Pb(Mg_{1/3}Nb_{2/3})O_3$ Ceramics *J. Solid State Chem.*, **91**, 350-361.

25. de Mathan, N.; Husson, E.; Calvarin, G.; Morell, A. (1991) Structural Study of a Poled $Pb(Mg1/3Nb2/3)O3$ Ceramic at Low Temperature *Mat. Res. Bull.*, **26**, 1167-1172.

26. Mulvihill, M.L.; Cross, L.E.; Uchino, K. (1995) Low Temperature Observation of Ferroelectric Domains in Lead Zinc Niobate *J. Am. Ceram. Soc.*, **78** (12), 3345-3351.

27. Sada, T.; Umezawa, C.; Uchino, K. (1989) Temperature Dependence of Electrostriction in Lead Zinc Niobate *Jpn. J. Appl. Phys.*, **28**(1), 46-49.

28. Takenaka, T.; Bhalla, A.S.; Cross, L.E. (1989) Dielectric, Piezoelectric and Pyroelectric Properties of Lead Zirconate-Lead ZincNiobate *J. Am. Ceram. Soc.*, **72**(6), 1016. -1023.

29. Park, S.E.; Shrout, T.R. (1996) Characteristics of Relaxor-BasedPiezoelectric Single Crystals for Ultrasonic Transducers *IEEE Ultrasonics Symp.*, 935-942.

30. Stenger, C.G.F.; Burggraaf, A.J. (1980) Order-Disorder Reactions in the Ferroelectric Perovskite $Pb(Sc_{1/2}Nb_{1/2})O_3$ and $Pb(Sc_{1/2}Ta_{/2})O_3$*Phys. Status Solidi A*, **61**, 275 285.

31. Stenger, C.G.F.; Scholten, F.L.; Burggraaf, A.J., (1979) Ordering and Diffuse Transitions in $Pb(Sc_{0.5}Ta_{0.5})O_3$ Ceramics *Solid State Comm.*, **32**, 989-992.

32. Knight, K.S.; Baba-Kishi, K.Z. (1995) Crystal Structure Refinements of Disordered $Pb(Sc1/2Nb1/2)O3$ in the Paraelectric and Ferroelectric States *Ferroelectrics*, **173**, 341-349.

33. Glinchuk, M.D.; Bykov, I.P.; Laguta, V.V.; Nokhrin, S.N. (1997) NMR Investigations of Mixed Relaxors xPMN(1-x)PSN *Ferroelectrics*, **199**(1-4), 173-185.

34. Yasuda, N.; Shibuya, S., (1990) Ferroelectricity in Disorder Lead Indium Niobate *Ferroelectrics*, **106**, 281-286.

35. Park, S.S.; Choo, W.K., (1991) Dielectric Properties of Lead Indium Niobate Ceramics *Ferroelectrics*, **118**, 117-122.

36. Yamashita, Y.; Shimanuki, S. (1996) Synthesis of Lead Scandium Niobate-Lead Titanate Pseudo Binary System Single Crystal *Mater. Res. Bull.*, **31**(7), 887-895.

37. Caranoni, C.; Lampin, P.; Siny, I.; Zheng, J.G.; Li, Q.; Kang, Z.C.; Boulesteix, C. (1992) Comparative Study of the Ordering of B-site Cations in Pb2ScTaO6 and Pb2ScNbO6 Perovskites *Phys. Stat. Sol.*, **A130**, 25-37.

38. Chakare, L.; Dambekalne, M.; Bormanis, K.; Sternberg, A.; Brante, I. (1997) Modified Pb(Sc1/2Nb2/3)O3 Ferroelectric Ceramics *Key. Eng. Mag.*, **132-136**, 1068-1071.

39. Tennery, V.J.; Hang, K.W.; Novak,R.E. (1968) Ferroelectric and Structural Properties of $(Pb(Sc_{1/2}Nb_{1/2})O_3)_{1-x}Ti_xO_3$ *J. Am. Ceram. Soc.*, **51**(2), 671-674.

40. Kodama, U.; Osada, M.; Kumon, O.; Nishimoto, T. (1969), Piezoelectric Properties and Phase Transition of $Pb(In_{1/2}Ta_{1/2})O_3-PbTiO_3$ Solid-Solution Ceramics *Am. Ceram. Soc. Bull.* **48**(12), 1122-1124.

41. Huang, J., Chasteen, N.D., Fitzgerald, J.J. (1998) X-band EPR Studies of Lead Magnesium Niobate (PMN) and PMN/PT Powders at 10 and 85 K *Chem. Mater.*, **10**(12), 3848-3855.

42. Gupta, S. M.; Kulkarni, A. R. (1994) Synthesis and Dielectric Properties of Lead Magnesium Niobate—A Review *Mater. Chem. Phys.*, **39**, 98-109.

43. Huang, J.; Fitzgerald, J.J., (1999) Synthesis and 93Nb MAS NMR and 93Nb Nutation Studies of Lead Magnesium Niobate Systems Prepared by Different Synthetic Methods Chem. Mater. 1999, submitted.

44. Peterson, G.E.; Carnevale, A. (1972) 93Nb NMR Linewidths in Nonstoichiometric Lithium Niobate *J. Chem. Phys.*, **56** (10), 4848-4851

45. Yatsenko, A.V.; Ivanova, E.N. (1995) Nuclear Magnetic Resonance Investigation of Nonstoichiometric Lithium Niobate Single Crystal,*Phys. Solid State*, , **37** (8) 1237-1240.

46 Ivanova,E.N.; Yatseako, A.V.; Sergeev, N.A. (1995) Nuclear Magnetic Resonance of 93Nb in LiNbO3: Effect of Structural Distortions on Lineshapes *Solid-State NMR*, **4**, 381-385.

47. Wolf, F.; Kline, D.; Story, H.S. (1970) ^{93}Nb and ^{23}Na NMR of Polycrystalline Sodium Niobate, *J. Chem. Phys.*, **53** (9), 3538-3543.

48. Shore, J. S.; Zhao, P.; Prasad, S.; Huang, J.; Fitzgerald, J.J. (1999) Niobium-93 NMR Spectroscopic Study of Inorganic Niobates, in preparation.

49. Fitzgerald, J. J.; Huang, J.; Zhao, P.; Prasad, S.; Shore, J. S. (1999) Synthesis and Structural Studies of Lead Niobates by Solid-State Nb-93 and Pb-207 NMR, *Inorg. Chem.*, submitted.

50. Fitzgerald, J. J.; Huang, J.; Shore, J. S. (1999) Solid-State 93Nb MASNMR Studies of Single Crystal Pb(Mg1/3Nb2/3)O3 (PMN) andPb(Zn1/3Nb2/3)O3 (PZN) Relaxor Ferroelectrics and Related Lead Titanate Solid-Solutions *Ferroelectrics*, **233**, 3-4, 187-210.

51. Fitzgerald, J.J.; Huang, J. Prasad, S.; Zhao, P.; Shore, J. (1999) Solid-State Static and MAS 93Nb NMR and 93Nb Nutation Spectroscopy of PMN and PMN/PT Powders of 5 to 50 Mol% PT *J. Am.Chem. Soc.*, in press.

52. Prasad, S.; Zhao, P.; Huang, J.; Fitzgerald, J.J.; Shore, J. (1999) Pure-Phase Two-Dimensional Niobium-93 Nutation Spectroscopy of Lead Metaniobate and the Piezoelectric Lead Magnesium Niobate *J Solid State NMR*, **14**, 231-235.

53. Rosenfeld, H.D.; Egami, T. (1994) A Model of Short and Intermediate-Range Atomic Structure in the Relaxor Ferroelectric PMN, *Ferroelectrics*, **158**, 351-356.

54. Egami, T.; Dmowski, W.; Teslic, S.; Davies, P.K.; Chen, I.W.; Chen, H. (1998) Nature of Atomic Ordering and Mechanism of Relaxor FerroelectricPhenomena in PMN *Ferroelectrics*, **231**, 206-217.

55. Boulesteix, C.; Varnier, F.;LIberia, A.; Husson, E. (1994) Numerical Determination of Local Ordering of Pb(Mg 1/3 Nb 2/3)O3 (PMN) From High Resolution Electron Microscopy Imaging *J. Solid State Chem.*, **108**, 141-147.

56. Yan, Y.; Pennycook, S.; Xu, Z.; Viehland, D. (1998) Z-Contrast Imaging of Lead Magnesium Niobate and Barium Magnesium Niobate *Appl. Phys. Lett.*, **72**, 3145-3147.

57. Viehland, D.; Kim, N.; Xu, Z.; D.A. Payne, D.A. (1995) Structural Studiesof Ordering in the (Pb1-xBax)(Mg1/3Nb2/3)O3 Crystalline Solid-Solution Series *J. Am. Ceram. Soc.*, **78**, 2481-2489.

58. Davies, P.; Tong, J.; Negas, T. (1997) Effect of Ordering-Induced Domain Boundaries on Low Loss Ba(Zn1/3Ta2/3)O3-BaZrO3 Perovskite Microwave Dielectrics *J. Am. Ceram. Soc.*, **80**, 1727-1740.

59. Dupree, R.; Smith, M.E. (1988) Solid-State Magnesium-25 NMR Spectroscopy *J. Chem Soc. Chem. Comm.*, 1483-1485.

60. Stebbins, J.F. (1996) Magnesium Site Exchange in Forsterite: A Direct Measurement by High Temperature 25Mg NMR Spectroscopy *Am. Mineral.*, **9**, 1315-1320.

61. MacKenzie, K.J.D.; Meinhold, R.H. (1994) 25Mg Nuclear Magnetic Resonance Spectroscopy of Minerals and Related Inorganics: A Survey Study *Am. Mineral.*, **79**, 250-260.

62. MacKenzie, K.J.D.; Meinhold, R.H.(1993) A Mg-25 MAS NMR Study of the Thermal Decomposition of Magnesium Carbonate *J. Mat. Sci. Lett.*, **12**, 1669-1698.

63. de. Mathan, N.; Husson, E.; Gaucher, P.; Morell, A. (1990) Modification of the B-site Order of Ceramics by thermal Annealing or by La-Doping *Mater. Res. Bull.*, **25**, 427-434.

64. Chen, J.; Chan, H.M.; Harmer, M.P. (1989) Ordering, Structure and Dielectric Properties of Undoped and Lanthanum/Sodium-Doped Lead Magnesium Niobate, Pb(Mg1/3Nb2/3)O3 *J. Am. Ceram. Soc.*, **72**(4), 593.

218

65.   Wells, A.F. (1984) *Structural Inorganic Chemistry*, 5th Ed., Clarendon Press, Oxford, pp. 545-546.
66.   Glinchuk, M.D.; Bykov, I.P.; Laguta, V.V. (1993) Dynamics of Niobium Ionsin PMN Diffused Phase Transition Region and Its NMR Investigation *Ferroelectrics*, , 143, 39-47.

PIEZOELECTRIC PROPERTIES OF PROTON-EXCHANGED OPTICAL WAVEGUIDES

D.ČIPLYS, R.RIMEIKA
Vilnius University, Laboratory of Physical Acoustics
Faculty of Physics, Saulėtekio 9, 2040 Vilnius, Lithuania
daumantas.ciplys@ff.vu.lt

Abstract

The proton exchange in lithium niobate has become a very popular technique for fabrication of high-quality optical waveguides, which find numerous applications in integrated optics and acoustooptics. Many acoustooptic devices are based on the interaction between surface acoustic waves and guided optical waves. The surface acoustic wave propagation conditions strongly depend on crystal surface properties, which are considerably affected by the proton exchange and post-exchange annealing procedures. The present paper deals with the piezoelectric properties of protonated structures, which differ significantly from those of initial crystals. The results of direct measurements of the electromechanical coupling coefficient using two methods are presented. The first method is based on the comparison of acoustic velocities on a free and metallised surface of the crystal, and the second is based on *in situ* measurements of acoustic attenuation during evaporation of a thin metal film on the acoustic propagation path. The degradation of piezoelectric properties in a surface layer due to the proton exchange is demonstrated, and they could not be restored by the post-exchange annealing.

1. Introduction

Proton-exchanged (PE) lithium niobate integrated optical waveguides have attracted considerable attention because of their relatively simple fabrication technique and the high refractive index changes obtained [1]. Since the interaction of surface acoustic waves (SAWs) with guided optical modes offers important applications in the field of communications and data processing, the investigations of SAW propagation in PE LiNbO$_3$ crystals deserve great interest. The SAW propagation in pure lithium niobate is well studied since this material is one of the most widely employed in acoustic applications [2]. The proton exchange is a process in which the lithium ions of the crystal lattice are partially substituted by hydrogen ions (protons) supplied from an ambient medium. As a result, a layer of H$_x$ Li$_{1-x}$NbO$_3$ is formed at the sample surface with crystal structure and physical properties differing from those of LiNbO$_3$. The

219

C. Galassi et al. (eds.), Piezoelectric Materials: Advances in Science, Technology and Applications, 219–230.
© 2000 *Kluwer Academic Publishers. Printed in the Netherlands.*

extraordinary refractive index of this layer is enhanced with respect to that of pure LiNbO$_3$ in the bulk of substrate what makes the propagation of guided optical waves possible. In order to reduce the number of defects introduced by the exchange process and to improve the temporal stability of the guides formed, the post-exchange annealing [3] is often performed, typically, at several hundreds centigrade degrees, mostly in air, for duration from several minutes to several hours. The distribution of protons in the crystal may be substantially affected by the latter procedure.

Due to formation of a surface layer with crystal structure and material properties differing from those of the substrate, the SAW propagation characteristics are affected by the proton exchange. The surface acoustic wave velocity perturbation by the proton exchange was first observed in [4] and later studied in details in [5-12]. Among various crystal cuts, the 128° rot. Y-cut LiNbO$_3$ substrates are widely used for fabrication of SAW devices. The SAW propagation in PE 128° rotated Y-cut LiNbO$_3$ has been studied in [11,12]. The interaction of surface acoustic waves and guided optical modes in PE samples of this cut has been first demonstrated in [13].

As the factor of primary importance for many SAW devices is the electromechanical coupling coefficient, it is important to know its behavior in lithium niobate subjected to the proton exchange. For YZ configuration, it has been shown [14] that the exchange reduces significantly the K^2 implying that the H$_x$Li$_{1-x}$NbO$_3$ layer possess no piezoelectric effect. This assumption was used in [9] when evaluating the elastic properties of the protonated layer. Concerning the other cuts, some contradictory results have been obtained. For the Z-cut, a noncomplete reduction of piezoelectric activity in the layer has been reported [8] , though one might be confused that authors reported the value of K^2 for pure (non-exchanged) crystal which was twice as large as the commonly reported values. Similarly, for 128° rot. Y-cut substrate, the piezoelectric effect in a protonated layer has been reported to be reduced but not completely vanished [11]. In [14], the change in the electromechanical coupling coefficient has been evaluated from the increase in acoustic transmission losses between two interdigital transducers, and in [8,11] the K^2 values were determined by measuring the radiation impedance of an interdigital transducer at resonance frequency. Here we present another approach for measuring the electromechanical coupling coefficient in proton-exchanged lithium niobate: first, by measuring the SAW velocities on a free and metallised surface [15]; secondly, by measuring the SAW attenuation during evaporation of a metal film [16,17].

2.   Definition of electromechanical coupling coefficient

The electromechanical coupling coefficient for surface acoustic waves is defined as (see e.g. [18])

$$\frac{K^2}{2} = \frac{V_f - V_m}{V_f} \tag{1}$$

where V_f is the SAW velocity on a free surface of the piezoelectric, and V_m is the SAW velocity on a surface covered with an infinitely thin perfectly conducting layer. The idea for such a definition is that the conductivity layer shortens the tangential component of the electric field created by the wave due to the piezoelectric effect reducing the contribution of the latter. The parameter defined in this manner is of great importance as it serves as a quantitative characteristic of piezoelectric materials with respect to surface acoustic waves. The values of K^2 in lithium niobate have been calculated for various crystal cuts and SAW propagation directions [19]. For orientations along the crystallographic axes, as well for the practically important rotated cut, the results of calculations are shown in Table 1.

TABLE 1. Electromechanical coupling coefficients for surface acoustic waves in LiNbO₃ for various cuts and propagation directions

Crystal cut	SAW propagation direction	Free surface velocity, m/s	Metallized surface velocity, m/s	Electromechanical coupling coefficient, 10^{-2}
X	Y	3748	3681	3.58
X	Z	3483	3396	5.00
Y	X	3769	3740	1.54
Y	Z	3488	3404	4.82
Z	X	3798	3788	0.53
Z	Y	3903	3859	2.25
128° rot Y	X	3994	3888	5.31

Such a definition of the electromechanical coupling coefficient may be employed for characterization not only of simple semi-space substrates but also of more complex layered structures. It is the purpose of present paper to study the electromechanical coupling coefficient in lithium niobate substrates subjected to the proton exchange procedure, which changes considerably the properties of a crystal layer near the surface.

3. Proton exchange process

In our experiments, we performed the proton exchange in pure benzoic acid. The temperatures and times of the treatment are listed in Table 2. Some of the samples were annealed after the exchange. In order to evaluate the characteristics of the layers formed, guided optical waves have been excited and their effective refractive indices measured. From these data, the refractive index profiles (depth dependencies) in proton-exchanged samples have been reconstructed. Here we present only some results of optical characterization of the layers, not going into details of the methods which may be found elsewhere. The examples of refractive index profiles for as-exchanged and post-exchange annealed samples are shown in Fig. 1. The profiles in as-exchanged samples exhibit a step-like shape, and one is able to use in further considerations the model of simple layer-on-substrate structure with layer thickness directly determined from the profile. These thicknesses are also given in Table 2. The refractive index profile experiences a considerable change due to the post-exchange annealing. The step is transformed to the graded distribution function as the initial amount of protons penetrates, due to diffusion, into deeper regions. The single layer approximation is not applicable in this case, however, one may introduce an effective thickness in order to enable a comparison with step-like profile data. An effective thickness of a graded layer can be defined as a thickness of the step which has the same area as the graded function and the height equal to the function value at the surface.

TABLE 2. a)   Times of proton exchange and layer thicknesses for different lithium niobate samples. Proton exchange temperature 230°C for all samples.

Sample	1	2	3	4	5
Exchange time, hours	1	5.5	6	10	25
Layer thickness, µm	1.2	2.5	3.1	3.9	6.2

b)   Temperatures of consecutive steps of post-exchange annealing for sample N 5. Annealing time 1 h for each step.

Annealing step	1	2	3	4
Temperature, °C	300	360	410	450

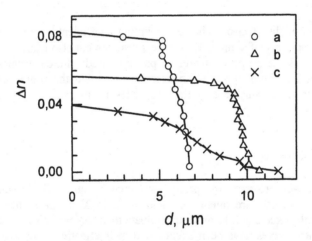

Figure 1. Refractive index profiles measured in lithium niobate sample after consecutive treatment procedures: (a) proton exchange at 230°C for 25 h, (b) two annealings at 300 and 360°C, and (c) third annealing at 410°C, each for 1 h.

4.   Measurements of electromechanical coupling coefficient

Evaluation of electromechanical coupling coefficient K^2 in proton-exchanged lithium niobate from acoustooptic measurements of SAW velocities is described in Chapter 4.1, and from SAW attenuation measurements during metal film evaporation, in Chapter 4.2.

4.1.   ACOUSTOOPTIC MEASUREMENTS OF SAW VELOCITY ON FREE AND METALLISED SURFACE

The surface acoustic wave velocities on free and covered with a thin metal film surfaces of PE samples have been measured by acoustooptic diffraction technique. The scheme of measurements is sketched in Fig.2.

The sample is placed on the rotary table of the precision goniometer. Light from the

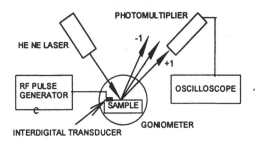

Figure 2. Schematic diagram of acoustooptic measurements of SAW velocity .

He-Ne laser with wavelength $\lambda = 632.8$ nm falls on the sample surface at angle θ_0. When reflecting from the surface at the same angle, the light is diffracted by the SAW into +1 and -1 orders at angles θ_{+1} and θ_{-1} , respectively. The photomultiplier mounted on the alidade of the goniometer serves as angle-sensitive light receiver. In the experiment, it is convenient to measure the angle 2Δ between the +1 and -1 diffraction maxima. From this angle, the velocity of SAW is evaluated using the expression

$$V = \frac{f\lambda}{\left(1 - \frac{\sin^2 \theta_o}{\cos^2 \Delta}\right)^{1/2} \sin \Delta} \tag{2}$$

which follows from the phase matching condition for longitudinal components of wavevectors of interacting optical and acoustic waves. The accuracy of velocity measurements increased with growing SAW frequency and was estimated to be of order 10^{-3} at frequencies above 100 MHz. At first, the reference measurements in the pure LiNbO$_3$ substrate have been performed. The results are shown in Fig.3. For a free surface, we obtained $V_f = 3982$ m/s. This value is slightly lower than the calculated one given in Table 1. Deposition of the metal layer leads to two effects. First, shorting of piezoelectric fields of the acoustic wave causes the frequency indepent drop in the velocity value. Secondly, the surface is mechanically loaded and dispersion in the form of linear velocity dependence on frequency appears.   The contributions from both effects can be separated by extrapolation of the $V(f)$ line to zero frequency where mechanical loading vanishes. The difference between values of the velocity on free surface and the extrapolated zero-frequency velocity on metallized surface directly yields the electromechanical coupling coefficient according to Eq.(1). We obtained $V_m(0) = 3877$ m/s and, consequently, $K^2 = 0.053$ which is in a good agreement with the value given in Table 1. The contribution of mechanical loading depends on the film thickness as well as on the metal mass density. This is illustrated in Fig.3 by different slopes of $V(f)$ dependencies corresponding to different thicknesses of relatively heavy copper ($\rho = 8.93$ g/cm^3) layers. By depositing much lighter aluminum ($\rho = 2.7$ g/cm^3) film, one is able to make the contribution of mechanical loading negligible. This is important for measurements in proton-exchanged samples where the method of extrapolation to zero can not be applied. It should be noted that all metal films used in acoustooptic velocity measurements were sufficiently thick for the acoustoelectronic attenuation, which is discussed below, to be negligible.

The SAW velocity dependencies on frequency were then measured on free and metallised surfaces of PE samples. As one can see from Fig.3, the strong acoustic dispersion is now observed even on a free surface. It arises because the parameters

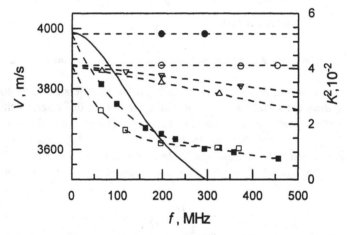

Figure 3. Surface acoustic wave velocities against frequency on differently treated surfaces of lithium niobate. Experimental values are for free (solid dots) and covered with ~0.1 µm thick Al film (open dots) surface of non-exchanged (circles) and proton-exchanged (squares) samples, as well for surface of non-exchanged sample covered with Cu films of two different thicknesses (triangles). Dashed lines are linear or polynomial approximations of experimental frequency dependences. Electromechanical coupling coefficient dependence on frequency is shown by solid line. Sample was exchanged at 230°C for 5.5 h.

responsible for the acoustic velocity, namely, elastic stiffness tensors, mass densities, and piezoelectric tensors, in the layer formed by the proton exchange are different from those in the rest of the substrate. The most important contribution arises from the differences in elastic stiffness coefficients [9]. The velocity behavior can be understood in terms of the SAW energy distribution between the protonated layer and the substrate. In the low frequency limit , the SAW velocity tends to that in pure LiNbO$_3$ as the amount of SAW energy confined in the layer becomes negligible in comparison with that in the substrate. In the high frequency limit, the SAW velocity tends to saturate at the value of the SAW velocity in $H_xLi_{1-x}NbO_3$ semi-space as the acoustic energy becomes confined within the layer.

A metal film deposited on the surface of a proton-exchanged sample gives an additional contribution to the SAW velocity by introducing, in general, the electrical and mechanical perturbations like in the above considered case of a non-exchanged crystal. As it was shown above, the contribution of mechanical loading can be neglected when the thin Al film is deposited. To ensure the same film thickness, the metal was simultaneously evaporated on the surface of both non-exchanged and exchanged samples in our experiment. In this case, the only mechanism responsible for the velocity change due to metallisation is the short-circuiting of piezoelectric fields, and the electromechanical coupling coefficient of PE sample can be directly found from the difference of velocities at any frequency. The experimental data were processed in the following way. The velocity values measured on free and metallised surfaces have been fitted to polynomial functions of frequency the difference of which was then substituted into Eq. (1). The resulting K^2 dependence on frequency is shown in Fig.3. It is plotted also in Fig.6 as a function of acoustic wavenumber-protonated layer thickness product kd. The observed reduction of electromechanical coupling coefficient with growing kd is consistent with the assumption that the piezoelectric fields are created only in the substrate and decay in the layer. At $kd>1$, the shorting layer appears to be far enough from the piezoelectric substrate not to affect the SAW propagation. In contrast, at $kd<<1$, the influence of $H_xLi_{1-x}NbO_3$ layer becomes substantially reduced, and the K^2 value approaches that of non-exchanged $LiNbO_3$.

4.2. MEASUREMENTS OF SAW ATTENUATION DURING METAL FILM EVAPORATION

4.2.1. *As-exchanged samples*

The method chosen is based on *in situ* measurements of the SAW attenuation during the evaporation of a metal film on the substrate surface [20]. This technique enables, in addition, to evaluate the dielectric permittivity of the substrate "seen" by the acoustic wave [21]. The scheme of experiment is shown in Fig. 4. The samples prepared for measurements had middle parts subjected to the proton exchange. The interdigital transducers for SAW excitation and detection were deposited in the nonexchanged regions, and the strip electrodes for film resistance measurements were deposited at the sides of exchanged area. A sample was then covered with a mask having a rectangular window in the middle through which copper was thermally evaporated at a rate less than 1 nm/s onto the surface area between the strip electrodes. The rf pulses of duration about 1 μs were fed into the input transducer to excite the SAW and the variation in the pulse amplitude U_{out} at the output transducer was monitored during evaporation. The attenuation of SAW in dB per length unity is

$$A = \frac{1}{W} 20 \lg \frac{U_{in}}{U_{out}} , \qquad (3)$$

where U_{in} is the pulse amplitude at the output transducer in the absence of the metal film, W is the length of metallised region in the SAW propagation direction. The resistance of the film during evaporation was measured as the ratio of dc current I and voltage U, and the sheet resistivity (resistance of an arbitary square of the surface area) of the film was determined as

$$R_s = \frac{U}{I}\frac{W}{L} ,$$ (4)

where L is the distance between dc electrodes .

The dependences of SAW attenuation upon the sheet resistivity measured for several samples exchanged for different times, as well for the non-exchanged sample, are presented in Fig. 5. They are fitted to the curves calculated according to the theory of interaction between a SAW propagating on a piezoelectric substrate and free charge

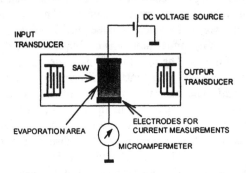

Figure 4. Schematic diagram of SAW attenuation vs evaporated film resistance measurements.

carriers in an adjacent thin conductive film [22]. The attenuation arising due to the creation of alternating currents in the film by piezoelectric fields of the SAW can be expressed as

$$A = 4.34 K_{eff}^2 \frac{\omega}{V} \frac{\omega_C/\omega}{1+\left(\omega_C/\omega\right)^2} ,$$ (5)

where $\omega = 2\pi f$ is the angular frequency and V is the velocity of the SAW, and the dielectric relaxation frequency is defined as

$$\omega_c = \frac{\omega}{\varepsilon_0 \varepsilon_{eff} VR_s} ,$$ (6)

where ε_0 is the dielectric permittivity of free space. The subscript "eff" emphasizes that electromechanical coupling coefficient K^2_{eff} and the relative dielectric permittivity ε_{eff} are effective parameters of the structure "seen" by the surface acoustic wave and depend both on LiNbO$_3$ substrate and H$_x$Li$_{1-x}$NbO$_3$ layer properties. It may be omitted below. The maximum attenuation occurs when $\omega_c = \omega$ or at sheet resistivity

$$R_S = \frac{1}{\varepsilon_0 \varepsilon_{eff} V} .$$ (7)

The maximum attenuation value is

$$A_{max} = 2.17 K_{eff}^2 \frac{2\pi f}{V}. \tag{8}$$

The effective electromechanical coefficient is evaluated from the experimentally measured value of attenuation maximum using Eq. (8), and the effective dielectric permittivity is found from the position of the maximum in R_S scale using Eq. (7). For non-exchanged lithium niobate we obtain $\varepsilon_{eff}=54\varepsilon_0$ and $K^2=0.053$ what is in good

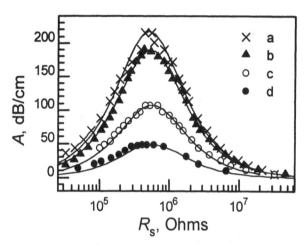

Figure 5. Change in SAW attenuation against sheet resistivity of Cu film during its evaporation. Dots are measured values in lithium niobate samples: non-exchanged (a), and proton-exchanged at 230°C for 1 (b), 10 (c), 25 (d) hours. Solid curves are calculated from Eq. (5) with best-fit values of K^2.

agreement with data commonly used in the literature. No shift of the attenuation maximum in the R_S scale is observed for proton exchanged samples implying that the dielectric permittivity of the $H_xLi_{1-x}NbO_3$ layer remains close to that of the $LiNbO_3$ substrate. The attenuation value at its maximum considerably decreases with increasing the thickness of protonated layer or the SAW frequency. In Fig. 6, all the measured values of K^2 are plotted as a function of the normalized parameter kd, where $k=\omega/V$ is the acoustic wavenumber. The fact that all the experimental points lay on the same curve serves as evidence for validity of the single homogenous layer model used. With increasing kd, the electromechanical coupling coefficient tends to zero implying that piezoelectric fields created by the surface acoustic wave decay significantly within the $H_xLi_{1-x}NbO_3$ layer. This observation confirms the assumption that $H_xLi_{1-x}NbO_3$ is nonpiezoelectric. If the piezoelectric effect in the layer would exist (though reduced), the dependence would be saturated at non-zero K^2_{eff} value with growing kd. It should be noted that the maximum kd value attained in [11] was only 0.7 where K^2 is really about 0.03, whereas for its reduction by the factor of order from the initial value, kd values twice as large are necessary.

4.2.2. *Influence of annealing on electromechanical coupling coefficient*

The post-exchange annealing procedure, which is frequently used in the PE optical waveguide technology has a strong influence on the proton distribution in the crystal

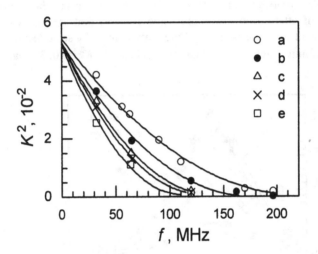

Figure 6. Influence of annealing on electromechanical coupling coefficient in proton-exchanged lithium niobate. Sample was proton-exchanged at 230°C for 25 h (a) and annealed for 1 h at 300 (b), 360 (c), 410 (d), 450 (e) °C.

and the refractive index profile of the waveguide [3,23]. It may cause the changes in the lattice structure $H_xLi_{1-x}NbO_3$ layer as different phases of the latter may exist depending on the relative amount of protons x [24,27]. The question now arises what is the influence of post-exchange annealing on the electromechanical coupling coefficient in the structure. The sample, which had been proton-exchanged at 230 °C for 25 h was subjected to consecutive annealing in air at temperatures 300, 360, 410, and 450°C, each of 1h duration. The comparatively long time of the initial exchange was chosen in order to obtain the multi-mode optical waveguide whose profile could be reliably determined. As usually, the evolution from the step-like profile to the graded one due to the redistribution of protons with consecutive steps of annealing is observed (Fig.1). It should be noted that no guided mode propagation was observed after the fourth annealing step, possibly due to the formation of thin $LiNb_3O_8$ layer at the very surface of the crystal as suggested in [23].

The dependence of electromechanical coupling coefficient upon acoustic frequency are shown in Fig.6. As the shape of the layer profile experiences significant transformations at various stages of treatment, the slight monotonous drop of the electromechanical coupling coefficient at given frequency is observed. This suggests that the layer formed at the surface by the proton exchange remains non-piezoelectric. It should be noted that partial restoring of the electromechanical coupling coefficient in Y-cut PE samples has been reported in [14].

5. Conclusions

The electromechanical coupling coefficient K^2 in proton-exchanged 128° rotated Y-cut lithium niobate has been experimentally studied. Two methods have been used. First, the surface acoustic wave velocities on the free surface of a sample and on the surface coated with thin Al film have been measured by the acoustooptic diffraction method. Secondly, the variation in SAW attenuation has been measured during evaporation of Cu film on the wave propagation path. As-exchanged as well post-exchange annealed samples have been studied. Due to the proton exchange, the electromechanical coupling coefficient is reduced as compared to that in a non-exchanged sample. The decrease in K^2 value depends on the acoustic frequency and protonated layer thickness: the greater is the kd product, the weaker is the coupling coefficient. The latter tends to zero at sufficiently high frequencies or thick layers, implying that the $H_xLi_{1-x}NbO_3$ layer possess no piezoelectric activity. The post-exchange annealing leads to a slight further decrease of electromechanical coupling coefficient indicating that the piezoelectric properties are not restored.

6. References

1. Jackel, J.L., Rice, C.E., and Veselka, J.J. (1982) Proton exchange for high-index waveguides in LiNbO₃, *Appl. Phys. Lett.* **41**, 607–608.
2. Shimizu, Y., (1993) Current status of piezoelectric substrate and propagation characteristics for SAW devices, *Jpn. J. Appl. Phys.* **32**, 2183–2187.
3. De Micheli, M., Botineau, J., Neveu, S., Sibillot, P., Ostrowsky, D.B., and Papuchon, M. (1983) Independent control of index and profiles in proton-exchanged lithium niobate guides, *Optics Letters* **8**, 114–115.
4. Hinkov, V. and Ise, E. (1985) Surface acoustic velocity perturbation in LiNbO₃ by proton exchange, *J. Phys. D: Appl. Phys.* **18**, L31–L34.
5. Hinkov, V., Barth, M., and Dransfeld, K. (1985) Acoustic properties of proton exchanged LiNbO₃ investigated by Brillouin scattering, *Appl. Phys. A*, **38**, 269–273.
6. Burnett, P.J., Briggs G.A., Al-Shukri S.M., Duffy J.F., and De La Rue, R.M. (1986) Acoustic properties of proton-exchanged LiNbO₃ studied using the acoustic microscopy $V(z)$ technique, *J. Appl. Phys.* **60**, 2517–2522.
7. Hinkov, V. (1987) Proton exchanged waveguides for surface acoustic waves in LiNbO₃, *J. Appl. Phys.* **62**, 3573–3578.
8. Chen, Y.C., Cheng, C.C. (1996) Proton-exchanged Z-cut LiNbO₃ waveguides for surface acoustic waves, *IEEE Trans. Ultrason. Ferroelectr. Freq. Control* **43**, 417–421.
9. Biebl, E.M. and Russer P. (1992) Elastic properties of proton exchanged lithium niobate, *IEEE Trans. Ultrason. Ferroelectr. Freq. Control* **39**, 330–334.
10. Čiplys, D., Rimeika, R., Korkishko, Yu.V., and Fedorov, V.A. (1998) Velocities of surface acoustic waves in proton exchanged lithium niobate with different $H_xLi_{1-x}NbO_3$ phases, *Ultragarsas(Ultrasound)* **29**, 24–28.
11. Kakio, S., Matsuoka, J., and Nakagawa, Y. (1993) Surface acoustic wave properties on proton-exchanged 128°-rotated Y-cut LiNbO₃, *Jpn. J. Appl. Phys. Part 1* **32**, 2359–2361.
12. Paškauskas, J., Rimeika, R., and Čiplys, D. (1995) Velocity and attenuation of surface acoustic waves in proton-exchanged 128° rotated Y-cut LiNbO₃, *J. Phys. D: Appl. Phys.* **28**, 1419–1423.
13. Saiga, N., Ichioka, Y. (1987) Acousto-optic interaction in proton-exchange 128° rotated Y-cut LiNbO₃ optical waveguides, *J. Appl. Phys.* **61**, 1230–1233.
14. Hickernell, F.S., Ruehle, K.D., Joseph, S.J., Reese, G.M., and Weller, J.F. (1985) The surface acoustic wave properties of proton exchanged YZ lithium niobate, *Proc. IEEE Ultrason. Symp.*, 237–240.
15. Čiplys, D. and Rimeika, R. (1998) Velocity of surface acoustic waves in metallised PE LiNbO₃, *Electron. Lett.*, **34**, 1707–1709.

230

16. Čiplys, D. and Rimeika, R. (1998) Electromechanical coupling coefficient for surface acoustic waves in proton-exchanged 128°-rotated Y-cut lithium niobate, *Appl. Phys.Lett.* **73**, 2417–2419.

17. Rimeika, R. and Čiplys, D. (1998) Influence of annealing on electromechanical coupling coefficient in proton exchanged 128°-rotated Y-X LiNbO$_3$, *Phys. Stat. Sol. (a)* **168**, R5–R6.

18. Auld, B.A. (1973) *Acoustic fields and waves in solids*, John Wiley and Sons, New York.

19. Slobodnik, A.J.Jr., Conway, E.D., and Delmonico, R.T. (1973) *Microwave Acoustics Handbook* **1A**, Air Force Cambridge Research Laboratories, Hanscom.

20. Sereika, A.P., Garska, E.P., Milkeviciene, Z.A., and Jucys, A.J. (1974) Electronic attenuation of surface acoustic wave in piezoelectric-metal film structure, *Solid State Physics (in Russian)* **16**, 2415–2417.

21. Kotelianskii, I.M., Krikunov, A.I., Medved, A.V, and Miskinis R.A. (1981) Measurement of effective electromechanical coupling constant and effective dielectric permittivity of piezoelectric- thin dielectric layer structures, *Microelectronics(in Russian)* **10**, 543–545.

22. Ingebrigtsen, K.A. (1970) Linear and non-linear attenuation of acoustic surface waves in a piezoelectric coated with a semiconducting film, *J. Appl. Phys.* **41**, 454–459.

23. Ganshin, V.A. and Korkishko, Yu.N. (1991) H:LiNbO$_3$ waveguides: effects of annealing, *Optics Comm.* **86**, 523–530.

24. Rice, C.E. (1986) The structure and properties of Li$_{1-x}$H$_x$NbO$_3$, *J. Solid State Chem.* **64**, 188–191.

25. Korkishko, Yu. V., Fedorov, V.A., and Kostrickii, S.M. (1998) Optical and x-ray characterization of H$_x$Li$_{1-x}$NbO$_3$ phases in proton-exchanged LiNbO$_3$ optical waveguides, *J. Appl. Phys.* **84**, 2411–2419.

A NOVEL TOOL FOR EXPERIMENTAL ANALYSIS OF SURFACE PHENOMENA

V.I.ANISIMKIN[1)], M.PENZA[2)], E.VERONA[3)]
[1)]RAS-Institute of Radioengineering and Electronics,
Mohhovaya str., 11, 103907, Moscow, Russia
[2)]PASTIS-CNRSM, SS 7 Appia Km 7+300, 72100, Brindisi, Italy
[3)]Istituto di Acustica O.M.Corbino, CNR, via del Fosso del Cavaliere-
00133, Rome, Italy

Abstract

Propagation of Rayleigh acoustic waves below the surface of piezoelectric solids, coated with a thin layer, approximately 0.1-10 μm thick, is suggested as a tool for the experimental analysis of surface phenomena such as gas phase adsorption, desorption, diffusion, corrosion, etc. Facilities of the tool are demonstrated on considering hydrogen and water vapor adsorption on porous (polyvinil alcohol, palladium, palladium-nickel films) and monolithic (quartz, $LiNbO_3$) materials. The number of adsorbed particles, the changes in temperature, mass, density, elastic constants and electric conductivity of the layer are evaluated both for steady-state and kinetic conditions.

1. Introduction

Novel tools for the analysis of physical and chemical phenomena on the surface of solids are of fundamental importance in many areas, for example, clean electronic materials, deep vacuum apparatus, ionic engines, solar cells, etc. However, contemporary knowledge on these phenomena is restricted because of the limited amount of relevant experimental data, difficulties in reliable measurements, variety of accompanied processes, etc.

Among the different available tools, the piezoelectric quartz crystal microbalance is one of the most efficient [1]. It is based on the change in resonant frequency Δf of a piezoelectric plate produced by the change in the plate mass Δm due to gas or liquid phase adsorption. Another tool [2] is based upon the propagation of surface acoustic waves (SAWs) along the surface to be analyzed. In comparison with the quartz microbalance, operating at low frequencies (typically 10 MHz), the SAW tool has a better resolution as its typical operation frequency is in the range between 50 and 500 MHz.

C. Galassi et al. (eds.), Piezoelectric Materials: Advances in Science, Technology and Applications, 231–238.
© 2000 *Kluwer Academic Publishers. Printed in the Netherlands.*

Up to now, the performance of the SAW tool has been restricted, as solely, the change in an adsorbed mass Δm has been accounted for [2]. As shown in present paper, on accounting in addition to the balance of mass the changes in the density, elasticity, electric conductivity and temperature of an absorber, the tool provides additional information on surface phenomena. Moreover, thanks to the anisotropic nature of the SAW propagation on piezoelectric materials, the as-performed SAW tool may prove advantageous over conventional techniques known so far.

2. Theoretical background

The SAW tool is shown in Fig.1. It consists of the piezoelectric substrate 1, the test film 2, two electromechanical transducers 3 for generation and reception of SAWs, and the metal electrode 4 for screening electric fields of SAWs out of the test film. In order to analyze phenomena on the free surface of the substrate, the test film 2 and electrode 4 are absent.

Usually, SAW propagation in the basic structures of Fig.1 is studied starting from mechanical equations of motion and Maxwell's equations, together with relevant boundary conditions. However, in the most frequent case of a thin isotropic film 2 on a piezoelectric substrate 1 with arbitrary anisotropy, the perturbation approach provides a result suitable for theoretical analysis [3] :

$$
\frac{\Delta v}{v_o} = \left(\frac{\pi h}{2\lambda}\right) \cdot \left[-\frac{\Delta\rho}{\rho} \cdot A + \frac{\Delta C_{44}}{C_{44}} \cdot B + \left(\frac{\left(1 - \frac{\Delta C_{44}}{C_{44}}\right)^2}{1 - \frac{\Delta C_{11}}{C_{11}}} - 1 \right) \cdot C \right] -
$$

$$
- K^2 \cdot \frac{\Delta\sigma}{\sigma} \cdot \frac{(\sigma_s^2 / v_o^2 C_s^2)}{(\sigma_s^2 / v_o^2 C_s^2 + 1)^2} \quad + \quad TCV \cdot \Delta T \tag{1}
$$

Here, $\Delta v/v_o$ is the relative change in the SAW propagation velocity generated by changes in density $\Delta\rho/\rho$, elastic modulii $\Delta C_{ij}/C_{ij}$, sheet conductivity $\Delta\sigma/\sigma$, and temperature ΔT of film 2. h is the film thickness, λ is the acoustic wavelength; K^2, TCV, and C_s are the coupling constant, the temperature coefficient of the SAW velocity, and the capacitance per length on the surface of the substrate along the SAW path; A, B, and D are coefficients related with three mutually orthogonal mechanical displacements of SAW A_x, A_y, A_z along shear, surface normal, and propagation direction, respectively [4] (Fig.1,a).

Taking into account, that at constant volume of the film the change in film density is equal to the change in film mass ($\Delta\rho/\rho = \Delta m/m$), the number N of species adsorbed into the film is evaluated as [5]

$$N = N_{_A} \frac{\rho S b}{M} \frac{\Delta \rho}{\rho} \qquad (2)$$

Here, N_A is Avogadro's number, M is the atomic mass of a gas analyte, and S is the surface of the test film.

Finally, taking into account that adsorbed gas species produce a thin layer and, thereby, perturb the SAW propagation on the free surface of the bar substrate [6], the total number N of species adsorbed on the free surface is evaluated as [5]:

$$N = N_A \frac{2S\lambda}{\pi V^2 M (A_x^2 + A_y^2 + A_z^2)} \frac{\Delta V}{V} \qquad (3)$$

Eq.1 describes the following important result: the SAW response $\Delta v/v_0$ towards different aspects of a given surface process ($\Delta\rho/\rho$, $\Delta C_{ij}/C_{ij}$, $\Delta\sigma/\sigma$, and ΔT) can be

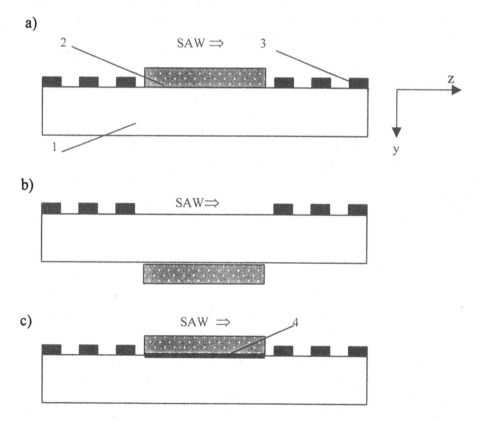

Figure 1. Test samples used for acoustic characterization of gas phase adsorption on thin solid films. (1 - piezoelectric substrate, 2 - test film, 3 - SAW transducers, 4 - metallic electrode).

enhanced or rejected by a proper selection of an acoustic substrate material, its crystallographic orientation and/or SAW propagation direction (i.e. by v_o, K^2, TCV, A_x, A_y, and A_z). This property originates from the anisotropic behavior of SAW propagation on piezoelectric single crystals making it possible to study different aspects either one by one, or in a combination or all together, even though all of them, apparently, reveal simultaneously.

Eq.1 can be applied both to steady-state (equilibrium) and kinetic (non equilibrium) conditions as it is valid for any step of a surface phenomenon. The only restriction of the SAW tool is that the surface of the solid has to support the propagation of SAW, otherwise (e.g. for liquid/solid interfaces), the application of the tool is impossible.

3. Experimental procedure

The operation of the SAW tool can be accomplished by two different ways. The first of them is based upon measurements of the SAW response $\Delta v/v_o$ for the same test film under five different experimental conditions (different propagation directions with small beam steering). Substituting the as measured $\Delta v/v_o$ into Eq.1, one gets a system of five independent equations with five unknowns $\Delta\rho/\rho$, $\Delta C_{11}/C_{11}$, $\Delta C_{44}/C_{44}$, $\Delta\sigma/\sigma$, and ΔT with v_o, A_x, , A_y, A_z, K^2, C_s, and TCV as parameters (these parameters are reported for many different materials and propagation directions in [6]). Solution of the system gives five characteristics of the surface phenomena that are unique.

Another approach is based upon concurrent analysis of different aspects of a surface process by proper selection of the crystal material, the crystal orientation, and the direction of the SAW propagation. This approach is utilized in the present paper.

In the first step, a test film is deposited on an acoustic substrate with large TCV (e.g. $Bi_{12}GeO_{20}$) (Fig.1,b). Since the SAW propagation in the structure is out of the test film, the first two terms in Eq.1 are zero, the SAW response $\Delta v/v_o$ depends solely on the change in temperature ΔT produced by a gas-film interaction, $\Delta T = \Delta v/v_o \bullet (TCV)^{-1}$, where $\Delta v/v_o$ is measured as described in [4] and TCV is taken from [6].

In the second step, a test film is deposited on three different piezoelectric substrates with metal coatings (Fig.1,c). Screening electric fields accompanied SAW propagation on piezoelectric material out of the test film allows to study separately elastic variations into the film. In order to deduce these variations, the change in temperature ΔT evaluated in the first step and the SAW response $\Delta v/v_o$ measured for the same test film under three different experimental conditions (different substrate materials) are substituted in Eq.1, getting a system of three independent equations in three unknowns $\Delta\rho/\rho$, $\Delta C_{11}/C_{11}$, and $\Delta C_{44}/C_{44}$.

Finally, on depositing the test film on free (Fig.1,a) and metallized (Fig.1,c) surfaces of a piezoelectric substrate with large coupling constant K^2 (e.g. $LiNbO_3$), one can deduce the change in film conductivity $\Delta\sigma/\sigma$. In order to do this, the measured $\Delta v/v_o$, ΔT, $\Delta\rho/\rho$, $\Delta C_{11}/C_{11}$, and $\Delta C_{44}/C_{44}$ together with tabulated v_o, A_x, A_y, A_z, K^2, C_s and TCV [6] are substituted into Eq.1 and solved towards the unknown $\Delta\sigma/\sigma$.

After that, the number of adsorbed species N for coated and uncoated piezoelectric crystals are evaluated from Eqs.2 and 3, respectively.

Table 1. Relative changes in density and elastic modulii of the not annealed Pd film due to hydrogen adsorption at 20 °C.

Gas mixture	$\Delta\rho/\rho$, 10^{-2}	$\Delta C_{11}/C_{11}$, 10^{-2}	$\Delta C_{44}/C_{44}$, 10^{-2}
$0,1\% H_2 + N_2$	+0,17	+12,7	-5
$0,5\% H_2 + N_2$	+0.33	+33,4	-30
$1,0\% H_2 + N_2$	+0.26	+35	-38

Table 2. The same for a Pd film annealed at 10^{-5}Pa for 15 hours.

Gas mixture	$\Delta\rho/\rho$, 10^{-2}	$\Delta C_{11}/C_{11}$, 10^{-2}	$\Delta C_{44}/C_{44}$, 10^{-2}
$0,1\% H_2 + N_2$	+0,33	+1,0	+1,0
$0,5\% H_2 + N_2$	+0,65	+1,75	+2,05
$1,0\% H_2 + N_2$	+0,65	+1,95	+2,05

Table 3. Comparison of the not annealed Pd and $Pd_{0.97}Ni_{0.03}$ films subjected to 1% H_2+ N_2 gas mixture at 20°C.

Gas mixture	$\Delta\rho/\rho$, 10^{-2}	$\Delta C_{11}/C_{11}$, 10^{-2}	$\Delta C_{44}/C_{44}$, 10^{-2}
Pd	+ 0,26	+ 35	- 38
$Pd_{0,97}Ni_{0,03}$	- 4,1	- 26,1	- 8,1

The outlined procedure is applicable both for steady-state and temporal conditions as it is valid for any time interval starting form the beginning of the process.

4. Results and discussion

Tables 1-3 show the changes in elastic properties of polycrystalline Pd and Pd:Ni films under hydrogen adsorption. In accordance with available knowledge on Pd-hydrogen and Pd:Ni-hydrogen interactions [7-10], the relative changes in density and elastic modulii of the films depend on the film treatment and on Ni content in Pd matrixes : after annealing, the Pd film loses some strongly-coupled pre-adsorbed oxygen responsible for $\Delta C_{ij}/C_{ij}$ and traps weakly-coupled oxygen responsible for $\Delta\rho/\rho$. The addition of small amounts of Ni decreases the catalytic activity of Pd and makes the changes in elastic modulii of Pd:Ni films less remarkable than those for Pd.

As compared with Pd, the Pd:Ni film is less sensitive towards humidity (Fig.2). On the other hand, both films adsorb more water vapors than monolithic quartz samples with polished surface. The difference in the number of adsorbed species N for different materials is probably due to various atomic forces at the surfaces of the different materials and to the much larger total area of porous films in comparison to monolithic quartz samples. On considering that one layer of tightly packed H_2O molecules contains $0.48 \cdot 10^{15}$ species per 1 cm^2, the amounts of adsorbed mono-layers are estimated to be 1000, 160 and 6 for Pd, Pd:Ni and SiO_2, respectively. The result for quartz, as obtained by the SAW technique, is in good agreement with that measured with the ellipsoid conductivity method [11]. On considering the most common orientations of weak (quartz) and strong ($LiNbO_3$) piezoelectric crystals, the lowest threshold value of the relative humidity $(RH)_{min}$ is measured for bare 128°y,x-$LiNbO_3$ (Table 4). This result can be attributed to stronger atomic forces at the surface of $LiNbO_3$, allowing H_2O

236

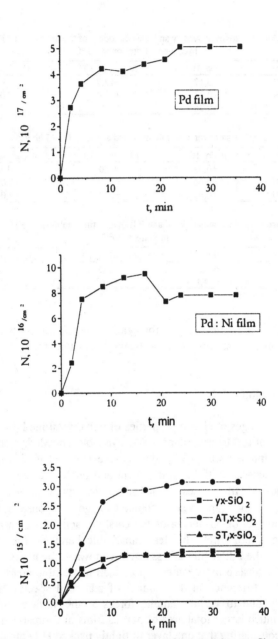

Figure 2. Number N of adsorbed water species vs. time for annealed porous films and uncoated single crystals

molecules to be efficiently attracted from the ambient. For PVA/LiNbO$_3$ structure, the SAW response $\Delta v/v_0$ depends on the electric conditions at the film/substrate interface (Fig.3). When the SAW fields are screened out of the film, the change $\Delta \sigma/\sigma$ in PVA

Table 4. Threshold RH values for different test structures.

Piezoelectric single crystal	$(RH)_{min}$, %
ST,x-SiO$_2$	3.3
y,x-SiO$_2$	2.9
AT,x-SiO$_2$	2.3
y,z-LiNbO$_2$	1.4
128°,x-LiNbO$_3$	0.5

conductivity becomes zero and the SAW responds only towards elastic and temperature variations of the film. Without screening, $\Delta\sigma/\sigma \neq 0$, the SAW response $\Delta v/v_0$ depends on the additional σ-term, and the value of the response becomes larger. The change in film conductivity increases with RH : The temperature variation ΔT of the PVA film increases with RH as well (Fig.4). Some disagreement between experimental data as measured by the SAW tool and with a thermocouple can be attributed to different experimental conditions: while the thermocouple detects the temperature at the surface of the film, the SAW tool integrates the temperature within the SAW penetration depth of ~ 100 μm.

5. Conclusions

The novel tool provides experimental evaluation of the steady-state and kinetic properties of the 6 physical parameters (N, $\Delta\rho/\rho$, $\Delta C_{11}/C_{11}$, $\Delta C_{44}/C_{44}$, $\Delta\sigma/\sigma$, ΔT), responsible for different aspects of surface phenomena. The properties of the aspects

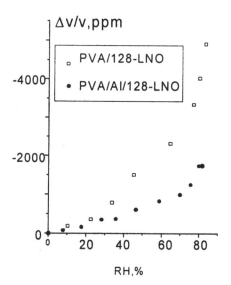

Figure 3. Calibration curves for devices using PVA/LiNbO$_3$ with or without an Al electrode at the interface.

238

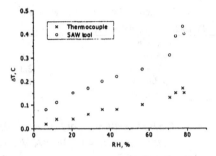

Figure 4. Temperature variations in a PVA film due to the film interaction with humid air (20 °C).

can be studied one by one, in a combination, or all together. At a given threshold value of the tool response, relevant threshold values of the measured physical parameters can be decreased by increasing the length of the test film along the propagation path.

The tool is not applicable to liquid phases since the Rayleigh mode suffers high attenuation at a solid/liquid interface. Also, it is not applicable to test films deposited onto isotropic solids, as elastic anisotropy and piezoelectricity of solids is inherently necessary for performance.

The influence of the SAW propagation (in particular, large beam steering) on the phenomena under the test should be studied in more details.

6. References

1. Mecea V.M.(1994) Loaded vibrating quartz sensors, *Sensors/Actuators*, **A40**, 1-27.
2. Martin S.J., Ricco A.J., Ginley D.S., Zipperian T.E. (1987) Isothermal measurements and thermal deposition of organic vapors using SAW devices, *IEEE Transactions* **UFFC-34**, 142-147.
3. Anisimkin V.I., Maksimov S.A. (1999) Surface acoustic wave method for analysis of the surface phenomena, *Russian Phys.: Surface Science* **8** (in press).
4. Anisimkin V.I. Kotelyanskii I.M., Verardi P., Verona E. (1995) Elastic properties of thin-film palladium for SAW sensors, *Sensors/Actuators*, **B23**, 203-208.
5. Anisimkin V.I., Maksimov S.A., Verardi P., Verona E. (1997) Effect of humidity on SAW devices, *Proc.IEEE Ultrason.Symp.* **1**, 409-413.
6. Auld B.A. (1973) *Acoustic fields and waves*, Wiley-Interscience Publication, New York, London, Sidney.
7. Yamamoto N., Tonomura S., Matsuoka T., Tsubomura H. (1980) A study on a palladium-titanium oxide Schottky diode as a detector for gaseous components, *Surface Science* **92**, 400-406.
8. Lundstrom I. (1996) Why bother about gas-sensitive field-effect devices, *Sensors/Actuators* **A56**, 75-79.
9. Chen Y., Li Y., Lisi D., Wang W.M. (1996) Preparation and characterization of Pd/Ni thin films for hydrogen sensing, *Sensors/Actuators* **B30**, 11-15.
10. Anisimkin V.I.,Verona E.(1998) New properties of SAW gas sensing, *IEEE Transaction* **UFFC-45**,1347-54.
11. Ershova G.F., Zorin Z.M., Churaev N.V. (1975) Temperature dependence of the thickness of polymolecular adsorbed water film on quartz, *Colloidal Journal* **37**, 491- 495.

THE INFLUENCE OF THE MATERIAL CONSTANTS AND MASS-LOADING ON COUPLING COEFFICIENT OF MONOLITHIC FILTERS

IRINA MATEESCU, GABRIELA POP
National Institute of Materials Physics
PO Box MG-7 76900 Bucharest-Magurele, Romania

Abstract

Based on Tiersten's analysis an accurate calculation of the coupling coefficient dependence on geometrical dimensions and mass loading for quartz and langasite monolithic structures was performed (1).

In this paper we have studied the influence of the material constants on the attenuation characteristics of two-pole langasite monolithic filters by calculation of the dependence between the coupling coefficient and elastic constant c_{66}, piezoelectric constant e_{26} and Voigt's elastic constant γ_{11}. We investigated also the effect of mass loading on the coupling coefficient of langasite and quartz monolithic filters and we have found that the effect on langasite filters is lower than on quartz filters.

The relation between the coupling coefficient and the material constants is discussed on crystals grown by various laboratories.

1. Introduction

Among the new members of the piezoelectric crystal class, the LGS crystal is one of the most promising material for applications in piezoelectric devices due to its particular properties: the absence of phase transitions up to the melting point, the absence of twins, low acoustic wave propagation losses, high electromechanical coupling coefficients (1), (2). For Y-cut langasite the particular advantages related to the monolithic filters are good thermal stability, low values of the equivalent inductance and series resistance (1, 3, and 4).

The use of langasite crystals, which exhibit coupling coefficient situated between those of quartz and of lithium tantalate, can cover the average range of filter bandwidths. While the monolithic band-pass filters using AT-cut quartz crystal show bandwidths between 0.01% and 0.35% of the central frequency and the Y-163° cut lithium tantalate filters show bandwidths of $(0.4\% \div 6\%)$ f_0, the monolithic filters on Y-cut langasite crystal exhibit pass-bandwidths of 0.3 to 0.8% of the central frequency (5, 6). Our experience in quartz and lithium tantalate monolithic filters (7, 8) allowed us to improve the design method by using an accurate electrical and geometrical calculation of quartz and langasite filters.

C. Galassi et al. (eds.), Piezoelectric Materials: Advances in Science, Technology and Applications, 239–247.
© 2000 *Kluwer Academic Publishers. Printed in the Netherlands.*

The last step of monolithic filter design is geometrical calculation of two-pole monolithic structure using the parameters of electrical equivalent circuit. In order to get a better agreement between theoretical and experimental results than that obtained with the currently used relations (5, 9), we developed (10, 11, 12, 13) a theoretical approach based on Tiersten's analysis (14), for quartz and langasite two-pole monolithic structures. The change of the coupling coefficient as a function of the electrode spacing and the plate thickness ratio for various electrode thickness of two-pole quartz and langasite filters pointed out the difference between the rigorously calculated coupling coefficient and that obtained with the approximate relations.

Based on this new theoretical calculation of the coupling coefficient dependence on structure geometry and mass loading, in this paper has been evaluated the influence of several material constants (elastic constant c_{66}, piezoelectric constant e_{26} and Voigt's elastic constant γ_{11}) and of mass loading on coupling coefficient. The results show how a small dispersion of material constant values for langasite crystals grown in various laboratories strongly modifies the attenuation characteristics of monolithic filters. We have concluded also that the effect of mass loading on the coupling coefficient of langasite filters is much smaller than that of the quartz filters. This fact is an important feature of langasite crystal compared with quartz crystal.

2. Filter geometry calculation

In order to realize a pass-band two-pole monolithic filter, the vibration mode structure of the plate should be determined by the simple thickness-shear mode excited in the constituent resonators.

In this work we discuss the filters with 10.7 MHz central frequency realized on AT - cut quartz and Y - cut langasite plates.

From the relations between the elements of the equivalent electrical circuit and the physical properties of the monolithic structure it is possible to calculate the required geometry of the individual two-pole monolithic filter (Figure 1), namely, the plate thickness t, electrode dimensions l_x, l_z and the electrode spacing d.

The resonance frequency of the plate and the plate thickness are related by

$$f_0 = N_f/t \qquad (1)$$

where f_0 is the resonance frequency of the plate (in MHz), N_f is the frequency constant of the fundamental thickness - shear mode (1680 MHz mm for quartz AT - cut and 1380 MHz mm for langasite Y - cut) and t is the plate thickness (in mm).

Using the relation (1) we find that the required thickness corresponding to 10.7 MHz central frequency are 0.158 mm for quartz plates and 0.129 mm for langasite plates. Both plates have 8.2 mm diameters. Electrode areas are calculated from the expressions of the resonator motional inductance L.

$$L = N_L / (f^3 A) \qquad (2)$$

$$N_L = N_f \cdot \frac{\rho}{8 \cdot e_{26}^2} \cdot \frac{c_{66}}{\overline{c}_{66}} \qquad (3)$$

In the relation (3) the N_f, ρ, c_{66}, c_{66}. e_{26}, constants are specific values for quartz and langasite crystals.

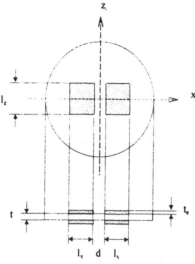

Figure 1. Two-pole monolithic structure

The conditions for separating the unwanted modes in both type of filters, for X coupling direction, are:

$l_x/t \approx 14$ for quartz filter; $l_x/t \approx 9$ for langasite filter

The sizes of geometrical configurations (plate thickness, electrode spacing, electrode thickness) are functions of specific, electrical characteristics (central frequency, bandwidth) of each type of filter.

For example, in quartz filters are used the Ag electrodes with thickness between 100 nm and 300 nm, while langasite filters use Al electrodes with thickness from 200 nm to 600 nm.

The relation currently used for the inter-resonators coupling along X direction is:

- for quartz two-pole filter:

$$k_x = 5.722 \cdot \left(\frac{2l_x}{t} + \frac{d}{t} + 2.1\right)^{-2} \exp\left(-2.28\sqrt{\frac{R}{2}} \cdot \frac{d}{t}\right) \quad (4)$$

- for langasite two-pole filter:

$$k_x = \alpha \cdot \left(\frac{2l_x + d}{t}\right)^{-2} \exp\left(-\beta\sqrt{\frac{\Delta}{f_0}} \cdot \frac{d}{t}\right) \quad (5)$$

$$\frac{\Delta}{f_0} = \frac{4}{\pi^2} \cdot k_{26}^2 + \frac{c_{11}}{c_{66}} \cdot \frac{\rho_e \cdot t_e}{\rho \cdot t} \quad (6)$$

and $\quad R = 2\,\rho_e\,t_e\,/\,\rho\,t \quad (7)$

ρ_e, ρ - electrode and substrate densities;
t_e, t - electrode and substrate thickness

The coupling coefficient being known from synthesis, it is possible to calculate the distance between electrodes of two component resonators for electrode thickness arbitrary chosen in specific range of each type of filters.

To reduce the differences between input data and the band pass of the monolithic filters we performed an accurate calculation of the coupling coefficient dependence on geometrical dimensions and the mass-loading using a theoretical approach based on Tiersten's analysis for quartz and langasite monolithic two-pole structure.

In the method described by Tiersten the three-dimensional equations of elasticity are converted into a series of two-dimensional equations by expanding the mechanical displacement and electrical displacement in a series of power of the thickness of the plate. For a narrow frequency range near the cut - off fundamental thickness - shear only the contributions of the fundamental thickness-shear and fundamental flexure modes are retained. The inter-resonator coupling coefficient was computed from the relation

$$k = 2 (\Omega_2 - \Omega_1) / (\Omega_1 + \Omega_2) \qquad\qquad (8)$$

where the Ω_1 and Ω_2 are the first two roots of the equation (1) from paper (12).

3. The influence of material constants on the LGS coupling coefficient behavior

The calculations based on Tiersten's analysis, in which was retained only thickness shear and flexure modes of vibration, shown that the influence of elastic constant c_{66}, piezoelectric constant e_{26} and Voigt's elastic constant γ_{11} is dominant on coupling coefficient behavior.

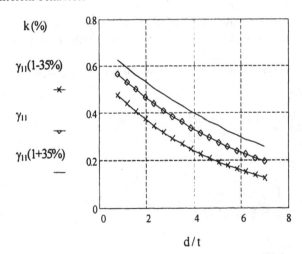

Figure 2. The coupling coefficient versus d / t for large variation
of the Voigt's elastic constant

The effects of these material constants were estimated through three sensitivity coefficients defined as: $(\Delta k/k)/(\Delta c_{66}/c_{66})$, $(\Delta k/k)/(\Delta \gamma_{11}/\gamma_{11})$, $(\Delta k/k)/(\Delta e_{26}/e_{26})$. These coefficients are calculated for various electrode spacing of langasite two-pole structure with parameters R=0.0015, $l_x = 9t$.

From these coefficients we can conclude that the most sensitive material parameter for filter design is the elastic constant γ_{11} and the piezoelectric constant e_{26} while the filter design is weakly dependent on the elastic constant c_{66}.

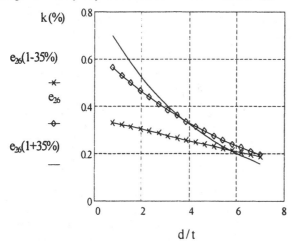

Figure 3. The coupling coefficient versus d / t
for large variation of the piezoelectric constant

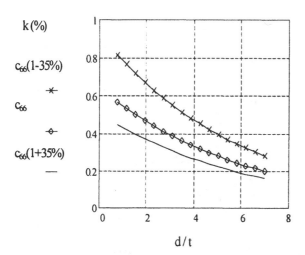

Figure 4. The coupling coefficient versus d / t
for large variation of the elastic constant

In the previous papers (12), (13) the dependence of k function on d / t for an arbitrary chosen variation (± 35 %) of material constants were evaluated. Figures 2, 3, 4 show that the coupling coefficient increases with the piezoelectric constant e_{26} and with Voigt's elastic constant γ_{11} and decreases with the elastic constant c_{66}.

In the literature could be found different values of the c_{66}, e_{26} and γ_{11} probably due to the in-homogeneity produced during the langasite growth process.

The electro-mechanical parameters of the devices based on a quaternary oxide compound (in our case LGS) are usually influenced by the structural properties of the material: stoicheiometry, lattice defects (dislocations, cracks, holes, etc) and impurities.

TABLE 1. The material constants of langasite samples

	c_{66}	e_{26}	γ_{11}
	$(10^{10}\,N/m^2)$	(C/m^2)	$(10^{10}\,N/m^2)$
1	4.35	-0.41	12.3027
2	4.2	-0.45	12.0823
3	4.216	-0.431	12.0829
4	4.32	-0.45	12.0769
5	4.22	-0.44	12.1633
6	4.2	-0.44	12.1425
7	4.235	-0.45	12.0743

In this respect it is necessary to mention that dissociation of any volatile component (O2, Ga2O3) from the compound during the synthesis and crystal growth processes can affect the stoicheiometry. Also, the thermal conditions, which characterize the development of all these experimental steps, could be responsible for the lattice damage.

We selected from the literature values of the c_{66}, e_{26} and γ_{11} constants for seven samples grown in different laboratories which are presented in table 1.

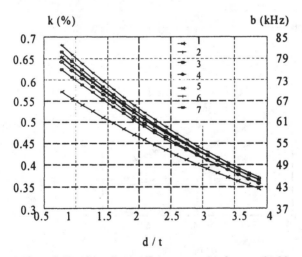

Figure 5. The variation of coupling coefficient versus d / t for several LGS samples

One observes that for c_{66} the maximum difference between the minimum and maximum values for c_{66} is ±2%, for e_{26} is 2.5% and for γ_{11} ±1.5 %. Using the accurate method based on Tiersten's analysis has been calculated the relation between the coupling coefficient and diameter / plate thickness ratio for langasite two-pole monolithic structures taking into account the constants values of these seven samples (figure5). Figure shows that the maximum difference of the coupling coefficient (11 %) is obtained for wide bandwidths (small values of d / t); for narrow bandwidths the difference between coupling coefficients corresponding to minimum and maximum constant values is about 4 %.

4. Mass-loading effect on coupling coefficient behavior

The dependence of coupling coefficient on the ratio between electrode spacing and plate thickness for some values of mass-loading has been calculated by the same accurate method previously mentioned for two-pole quartz and langasite filters.

TABLE 2. The electrode thickness t_e and the corresponding values of mass-loading R

Quartz		Langasite	
t_e (nm)	R (%)	t_e (nm)	R (%)
100	0.500	206	0.150
150	0.750	309	0.225
195	1.000	412	0.300
240	1.250	515	0.375
290	1.500	618	0.450

In table 2 are presented the mass-loading R values and corresponding electrode thickness t_e for quartz (Ag electrodes) and langasite (Al electrodes).

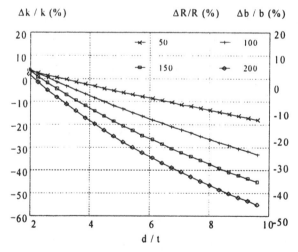

Figure 6. The relative variation of coupling coefficient and bandwidth versus d / t for quartz structure

246

The relative bandwidth change Δb / b due to the relative mass-loading variation with 50 %, 100 %, 150 % and 200 % are presented for quartz filter in figures 6 and for langasite filter in figure 7.

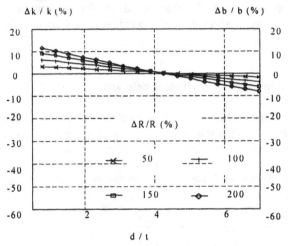

Figure 7. The relative variation of coupling coefficient and bandwidth
versus d / t for LGS structure

These figures show that the variation of the coupling coefficient (and in consequence of the bandwidth) with mass loading is quite different for quartz filter compared to langasite filter.

In the case of quartz filter the bandwidth variation increases with mass loading

(ΔR / R = 50%; 100%; 150%; 200%) and this variation is strongly amplified for high d/t ratios.

For the langasite filter the bandwidth variation is particular in the sense that the bandwidth variation with mass loading is zero around d/t=4 and shows variations in the opposite directions for values below and above d/t=4. Moreover, the mean variation with mass loading is lower for langasite than for quartz filter.

The quasi-independence of the coupling coefficient and, in consequence, of the bandwidth on mass loading seems to be one of the most important features of the langasite crystal.

5. Conclusions

An important difference between the attenuation characteristics obtained from the theoretical rigorous calculations and that obtained with usual approximate relations.

The use of the accurate design method for calculation of geometrical dimensions allows obtaining a better agreement between input data and experimental attenuation characteristics of quartz and langasite filters with 10.7 MHz.

The electrical characteristics of monolithic filters are sensitive to the material constants (Voight's elastic constant, piezoelectric constant, elastic constant). A $\pm 2.5\%$ dispersion of values of these constants could change filter's bandwidth with $\pm 20\%$.

Mass-loading effect on the coupling coefficient behavior on langasite filter is smaller than that of quartz. In the case of the langasite filter, the variation of k with the electrode thickness is not significant in the range of d / t between 3 and 5. This is an important advantage of langasite crystal over quartz crystal.

6. References

1. Gotalskaya A , Drezin D.I , Bezdelkin V.V, Stassevich V.N (1993), "Peculiarities of technology, physical properties and applications of new piezoelectric material Langasite ($La_3Ga_5SiO_{14}$)", *Proc. of the 1993 IEEE International Frequency Control Symposium* , pp. 339- 344

2. Shimamura Kiyoshi, Takeda Hiroaki, Kohno Takuya, Fukuda Tsuguo (1996), "Growth and characterization of lanthanum gallium silicate $La_3Ga_5SiO_{14}$ single crystal for piezoelectric applications", *Journal of Crystal Growth,* 163, pp. 388-392

3. Sakharov S.A, Medvedev A.V, Buzanov O. (1997), "HF langasite monolithic filters for GSM standard", *Proc. of the 11th European Frequency and Time Forum,* pp. 239-242

4. Mansfeld G.D (1998), "Langasite as material for piezoelectric devices", *Proc. of the 11th European Frequency and Time Forum,* in print

5. Sakharov S.A., Larionov I.M., Medvedev A.V (1992), "Application of langasite crystals in monolithic filters operating in shear modes", *Proc. of IEEE Frequency Control Symposium,* pp. 713-723

6. Silvestrova I. M., Pisarevsky Yu.V., Senyushenkov P.A., Krupny A.I. (1986), "Temperature dependence elastic properties $La_3Ga_5SiO_{14}$", *Sov. Phys. Solid.Stat,* v. 28, 9, pp. 2875-2878

7. Mateescu Irina and Kosinski John (1993) "Eight-pole monolithic filters using lithium tantalate", *Proc. of the 1993 IEEE International Frequency Control Symposium,* pp. 620-625

8. Zelenka Jiri, Mateescu Irina and Kosinski John (1994), "The effective coupling of the elastically coupled Y-cut $LiTaO_3$ two and three resonator structure", *Proc. of the 8th European Frequency and Time Forum,* pp. 786-791

9. Beaver W.D. (1968), "Analysis of elastically coupled piezoelectric resonators", *Journ. of Acoustic Soc. Amer.,* v.43 May, pp. 972-981

10. Mateescu Irina, Zelenka Jiri and Şerbănescu Alexandru (1996), "Contribution to the design of the wave filters composed from the elastically coupled two resonators structure", *Proc. of the 10th European Frequency and Time Forum,* pp. 132-137

11. Mateescu Irina, Şerbănescu Alexandru, Pop Gabriela, and Ghiţă Christina (1997), "The influence of the piezoelectric crystal properties on the monolithic filters design", *Analles de Chimie,* vol. 22, nr.8, pp. 713-716

12. Mateescu Irina, Pop Gabriela, Manea Adrian, Lazarescu Mihai and Ghiţă Christina (1998), "Investigation on behaviour of coupling coefficient of monolithic structure", *Proc. of the 11th European Frequency and Time Forum,* in print

13. Mateescu Irina, Pop Gabriela and Ghiţă Christina (1998), "A new theoretical approach for the coupling coefficient in the monolithic structures", *Proc. of the 1998 IEEE International Frequency Control Symposium,* pp.

14. Tiersten H. F. (1969), "Linear piezoelectric plate vibrations", *Plenum Press,* N.Y.

THICK AND COMPOSITE PIEZOELECTRIC COATINGS FOR BIOMEDICAL AND HIGH TEMPERATURE ULTRASOUND

M. SAYER, M. LUKACS, G. PANG, L. ZOU, Y.CHEN
Department of Physics, Queen's University, Kingston, Ontario, K7L 3N6 Canada
C.K. JEN
Industrial Materials Institute, National Research Council of Canada 75 Blvd de Mortagne, Boucherville, PQ, J4B 6Y4 Canada

ABSTRACT: Piezoelectric sol gel composites involve the coating and firing of dispersions of piezoelectric powders within a ferroelectric or piezoelectric sol gel. They provide a means to fabricate piezoelectric coatings of thickness 3-200μm. Such coatings have been used to implement single element and array pulsed ultrasonic transducers at frequencies in the range 20–160 MHz for medical imaging, and at lower frequencies as high temperature pulse echo transducers for industrial process control. The factors that lead to successful piezoelectric devices are discussed along with methods of piezoelectric characterization.

1. Introduction

Minimally invasive surgical procedures using catheter and probe based surgical tools require the parallel development of methods of imaging structures at the site of interest. Focussed high frequency ultrasound in the frequency range 30-200 MHz provides high resolution imaging over depths up to several millimeters [1]. Imaging modalities based on electronically scanned high frequency pulsed ultrasound require multi-element piezoelectric transducer arrays in one or two dimensions [2,3]. Such arrays require no mechanical scanning motions. Biomicroscopy has been used for eye, skin and intravascular imaging [4-6]. Other applications are found in general-purpose endoscopy and in high resolution non-destructive testing.

For operation at frequencies from 15-60MHz, individual elements of 64-256 element piezoelectric transducer arrays must have a strong piezoelectric response within an element of thickness of 100-25 μm, an element width of 15-40 μm and an element spacing of 40-10 μm. Other challenges inherent in the design of such arrays include acoustic and electrical impedance matching of the transducer elements to the propagating medium and power source. Serial or parallel processing of signals from all array elements is required. Packaging and the electrical interconnection of the array to signal processing electronics must be compatible with the surgical application. For the

C. Galassi et al. (eds.), Piezoelectric Materials: Advances in Science, Technology and Applications, 249–260.
© 2000 *Kluwer Academic Publishers. Printed in the Netherlands.*

technology to be economically successful, it is preferable that arrays be manufactured by batch processing methods compatible with semiconductor manufacturing.

Piezoelectric ultrasound transducers function by driving a piezoelectric structure at its fundamental thickness resonance. The thickness of the piezoelectric is inversely related to the frequency of oscillation and a device operating at a frequency > 40 MHz typically requires a piezoelectric layer < 50μm in thickness. The higher the frequency, the thinner the piezoelectric and the more difficult the fabrication of the transducer. In a phased array, the pitch of the elements must be one half the wavelength of sound in the propagating medium, and the width to height ratio must be ≤ 0.6 [7]. This can be relaxed up to a full wavelength for a linear array. For a one-wavelength pitch at 40 MHz in water, the array elements are 50μm thick, 20-25μm wide with a spacing of 15μm. Image formation from a one-dimensional linear array requires a minimum of 64 elements, with the preference being > 256 elements. The element count rises equivalently for two dimensions.

Transducer fabrication on this dimensional scale is not trivial. 64 element linear arrays to 20 MHz have been fabricated by mechanically dicing and assembling bulk piezoelectric ceramics, but the labor and precision involved in this assembly is excessive. The large number of electrical connections also becomes prohibitively difficult as the frequency and the number of elements increases [8]. It becomes attractive to consider methods of additive or subtractive patterning of active piezoelectric coatings fabricated with the required thickness, and then to use semiconductor packaging methods for array interconnects. This is part of the rapidly developing field of microsystems technology (MST) [9]. In this paper we demonstrate two applications of this composite technology– transducers for high frequency imaging and coatings for high temperature transducer design for industrial process control.

2. Sol Gel Composites

2.1 FABRICATION

Conventional sol gel processing or metallorganic decomposition of piezoelectric ceramics such as lead zirconate titanate is well established [10]. Multi-layer spin coating methods can achieve a film thickness close to 10μm [11,12], but the process is laborious at 0.2-0.5μm/coat and does not achieve the 50μm thickness required for 40 MHz ultrasonic transducer design. A method of achieving such a thickness is the formation of a sol gel composite. A suspension of a piezoelectric powder within an acid based sol gel [11] is spin, spray or dip coated onto a surface [13,14]. Substrates may include platinum-coated silicon or alumina, bulk aluminum or stainless steel. The coating is dried and then fired on a hotplate at < 500°C for 2mins to create a coating 3-4μm in thickness. This process is repeated for multiple layers until the required thickness is exceeded. The completed multi-layer is annealed in air in a furnace at 600-650°C for 1 hour and is then lapped to the desired thickness. Since sol gel composites may be prepared from different sol gel and powder phases, the two components will be written, for example, PZT/PZT or PZT/LiTaO$_3$.

The formation of a successful composite has to take a number of factors into account. The powder phase should pack as densely as possible, while the sol phase should percolate the entire matrix with the smallest possible shrinkage during processing. The suspension must have a viscosity low enough to allow an even distribution of material during spin or spray coating. To a large extent these requirements are incompatible, but a successful coating can be achieved if the particle size of the powder is uniform and has a sub-micron dimension (<0.8μm). For example, densities >90% theoretical are achieved in zirconia/zirconia sol gels using powders <0.3μm mean diameter. Figure 1 shows the powder morphology of two PZT powders used in composite preparation. The large grain commercial powder with a broad size distribution shown in Figure 1a) does not form an acceptable sol gel composite. However, a selected powder (Pz29) from Ferroperm Inc. [15] (Figure 1b) which has a particle size distribution of 0.8 ± 0.3 μm gives a composite of density in the range of 85% (6350 kg/m^3) which has reasonable characteristics.

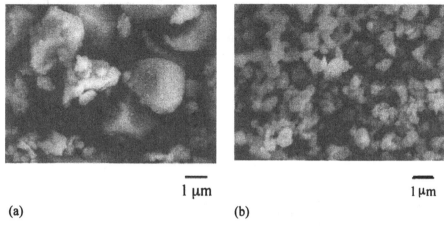

1 μm   1 μm

(a)   (b)

Figure 1. SEM images of PZT powders used in sol gel composites
a) commercial 5H, (b) Ferroperm Pz29.

The fired sol gel phase and the powder within the composite must be in intimate physical contact for both mechanical adhesion and continuity of electric field between the two phases. Acid based sols have been the most successful in creating the strongest chemical and mechanical bond within the composite. Representative gel recipes for PZT are based on zirconium and titanium propoxides [11] or butoxides [16], with lead acetate as the lead salt, and acetic acid and methanol as solvents. For high frequency piezoelectric purposes, a sol gel composite has a favorable structure since the field distribution between the powder and the sol gel derived phase is primarily set by considerations of capacitance. The sol gel derived phase can have a relative permittivity that is comparable to or may exceed that of the powder. Under ac or pulse excitation, the proportion of the electric field applied to the powder is therefore enhanced, and the piezoelectric response attributed to the powder is increased. Applications of this fact will be shown later using PZT/lithium tantalate sol gel composites for high temperature pulse-echo applications. Because of the complex bi-phasic structure and low firing temperature, the mechanical properties of a two-phase composite are not those of a sintered ceramic. However the low mechanical quality factor characteristic of a ceramic gives rise to electro-mechanical vibrations with a broad frequency bandwidth. This is desirable for pulse-echo ultrasonics where well-defined timing pulses and a lack of long term 'ringing' benefits image formation. In order to achieve improved piezoelectric performance for specific purposes, the multi-layer fabrication process allows the effective density of the coating to be improved if thinner layers of pure sol gel are alternated with layers of composite.

2.2   CHARACTERIZATION

Measurement of the piezoelectric activity of a composite coating or a thin sintered ceramic firmly attached to a surface has to take into account the presence of the underlying substrate. At low frequencies, the piezoelectric coefficients d_{33} and d_{31} have been measured by optical and cantilever methods [17]. At higher frequencies, the thickness mode parameters (in the 33 direction) of the piezoelectric response can be determined from measurements of the complex impedance of the film on the substrate as a function of frequency from 1kHz to 1GHz [18]. The upper frequency limit is set by the condition that the impedance of the sample is measurable and that the area of the sample ensures a one-dimensional response.

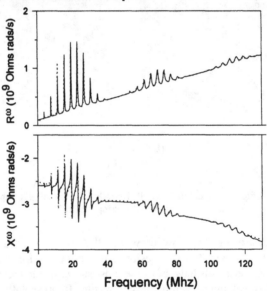

Figure 2   The experimental (solid line) and best-fit (dotted line) impedance spectra for a 70.5μm PZT composite sol gel layer of density 6340 kg/m³ on an 0.8 mm aluminium substrate of density 2700 kg/m³ with a top electrode diameter of 1mm [18]. The first three thickness mode piezoelectric resonances are clearly visible. The best-fit spectrum takes into account the frequency dispersion of the material parameters.

Figure 2 shows three piezoelectric resonances induced by a 70.5μm thick sol gel composite PZT coating on a 0.82mm thick aluminium substrate.   The discrete resonances within each envelope correspond to thickness vibrational modes of the substrate. Using a one-dimensional resonator model [19] and a software fitting routine [20], the piezoelectric and material constants of the coating can be derived. These include the open-circuit elastic stiffness c^D_{33}, the clamped permittivity ε^S_{33} and the h_{33} piezoelectric constant of the piezoelectric layer and the elastic stiffness c^Ds of the substrate. The properties are expressed as complex variables in order to account for losses. The technique allows measurement of a parameter that is otherwise difficult to assess for thin films. The electromechanical coupling constant k_t can be calculated from $k_t^2 = (h^2_{33} \varepsilon^S_{33})/c^D_{33}$. Under conditions in which multiple resonances due to the coating are observed, the frequency dispersion of the real and imaginary parts of the parameters may be calculated [18]. In general, for sol gel composites, an assessment of the properties of PZT sol gel coatings up to 30μm thick on alumina and stainless steel substrates demonstrate an electromechanical coupling coefficient of up to 0.35. k_t for PZT/lithium tantalate sol gel composites is about 0.17. These parameter values are consistent with qualitative expectations based on extrapolation between the relative properties of well sintered ceramic and a composite.

3.   Piezoelectric Array Structures

Piezoelectric arrays of the dimensions discussed above serve as a useful demonstration of batch fabricated microstructures created from sol gel composites.   Because the material is created from a liquid suspension both subtractive and additive patterning can be used.

3.1   SUBTRACTIVE PATTERNING

Subtractive patterning implies the formation of mechanical structures by cutting or etching into a previously prepared matrix. In this work subtractive patterning has been achieved by laser machining. A PZT sol gel composite coating which has the required thickness is formed on a platinized alumina or silicon substrate. The final thickness and polish necessary to prepare effective electrodes is achieved by diamond polishing. An important issue in the batch fabrication of active structures of small dimensions is the means of achieving connections to the electrodes. If connections are made through bonds on the surface of a piezoelectric or ferroelectric layer, parasitic capacitance or unwanted piezoelectric resonances may severely degrade the performance of the device.   To minimize these effects, prior to the deposition of the top electrode metallization, strips of insulating polyimide or photoresist are patterned on the surface. The electrode areas between the strips are in direct contact with the ceramic coating, and are poled by applying a field of 9V/μm at 180°C. The areas under the polymer are not piezoelectric and the series capacitance of the low permittivity insulator minimizes the parasitic capacitance. When the active area of the subsequent array pattern is cut it is aligned with the poled, active areas with the bond pads and connecting metal interconnects lying over the polymer. For fine line piezoelectric structures it is desirable to minimize temperature rise and consequent depoling within the ceramic using strongly absorbed laser light with a wavelength shorter than the absorption edge for PZT. The system employed in current work is a Lumonics PM-844 KrF excimer laser with an emission wavelength at 248nm and pulse length of 10ns. The FWHM dimensions of the raw laser beam is 10mm high by 25 mm wide. This is optically homogenized to a plane beam 5 mm by 30mm in size. In order to achieve batch fabrication of numerous array transducers, the beam is passed through a photolithographically patterned chrome-quartz mask and then demagnified by a factor of x10 to create an ablation image on the sample. This achieves the threshold necessary for ablation of the sample without damage to the mask. For masks which are larger than the area of the homogenised beam, movement of both mask and sample is implemented. Figure 3 shows laser machined patterns and cuts in 30μm PZT.

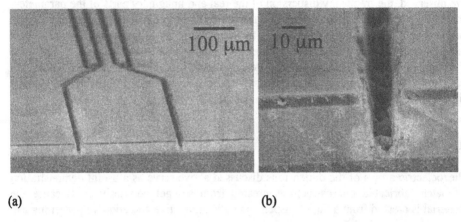

(a)   (b)

Figure 3   a) Angled SEM view of a section of a 3-element array showing the definition of the individual elements b) Laser cuts <15μm wide in a sol gel composite/metallisation layer structure. The PZT is approximately 30μm thick.

The depth of the laser cut is set by the number of laser pulses. In PZT sol gel composite the etch rate is 0.2μm/pulse at a fluence of 6.5 J/cm². Figure 4 a) shows multiple arrays fabricated on a 2.54 cm square alumina substrate. An individual array is shown in Figure 4 b).

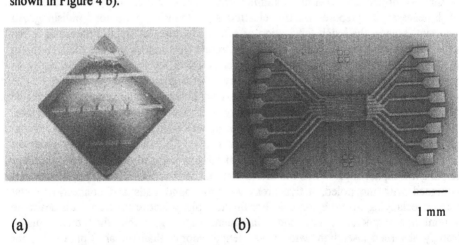

1 mm

(a)   (b)

Figure 4   a) An optical photograph of a set of 16-element arrays manufactured on an alumina substrate 2.54 cm x 2.54 cm.. The light strips show the active regions which do not have polymer under the gold electrode. b) detail of the individual arrays.

3.2   ADDITIVE PATTERNING

Additive patterning implies the firing and annealing of the ceramic microstructure in an appropriate mold. Although the sol gel composite is fired at the comparatively low temperature of 600-650°C compared to the sintering temperature of a normal ceramic, this still implies that high temperature metal molds must be created as an intermediate step between deep organic photolithography and the firing of the ceramic structure.

As an adaptation of a LIGA based micromolding technique [9], either AZ4000 or SU-8 deep photoresist is patterned by optical methods to the form and depth of the mold. Nickel is pulse electroplated into the pattern to form a metal mold and the photoresist is removed by dissolution in organic solvents or by thermal pyrolysis in an oxidising atmosphere. The sol gel composite solution can then be filled into the mold with compaction being assisted by ultrasonic agitation. In a similar manner to the formation of a coating, the sol gel composite is dried and fired at 400°C, with successive applications of the sol gel composite being made to complete the filling of the mold. A final firing at 550-600°C for 1 hour completes the microstructure. The nickel mold is then chemically removed by an acid etch. Figure 5 shows a planar view and cross-section of structures formed in this manner. The planar dimensions of this preliminary test mask are larger and the thickness is smaller than required for an ultrasonic device. In this technique the upper electrode must be deposited and patterned in subsequent steps.

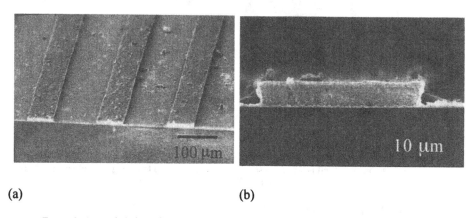

(a)   (b)

Figure 5: a) a angled view of sol gel composite PZT linear structures 10μm thick deposited by micromolding, b) SEM cross-section.

3.3   INTERCONNECTION TECHNOLOGIES

For imaging purposes, linear arrays of >64 elements are desirable with 256 elements being a commercial objective. In order to minimize acoustic damping of the response of individual elements, the upper gold electrode metallisations are < 0.1μm thick. This makes them fragile as the basis for interconnects. This implies that additional bonding and interconnect structures must be made by semiconductor techniques. Techniques under consideration include photolithographic definition of the bond pad regions followed by the deposition of 2-3μm thick electroless gold bumps to facilitate wire or solder-bumped flip chip [21] bonding to an appropriate header or flexible circuitry [8]. Following procedures developed for single element transducers prepared from sol gel composites [22], an on-going objective is to remove the substrate required for support during array fabrication. The choice of substrate is primarily determined by the need to withstand the thermal budget during processing. This may not lead to optimal choice as an acoustic backing layer. A further objective is to achieve an upper planar metal electrode that acts as an electrical ground plane and serves to provide some environmental protection to the device. In the long run, for practical multi-element devices for medical applications, multiplexing and beam forming delay circuits will need to fabricated in silicon technology and be packaged close to the array. This constitutes a significant area of research and development for the immediate future.

3.4 PULSE-ECHO RESPONSE

The piezoelectric pulse-echo response of an individual element of a 16-element sol gel composite array operated in water is shown in Figure 6.

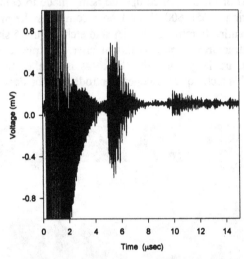

Figure 6: Pulse-echo response for a PZT sol gel composite array element averaged 20 times. The large signal on the left is the response to the driving pulse and following two pulses are echoes from a quartz flat immersed in water [23].

The amplitude of the first echo for a single pulse is about 2 mV peak to peak but when the pulse is signal averaged twenty times the amplitude dropped by a factor of 2. However, the averaged response remains significantly higher than the background noise. The impedance of the array element is 1.3 kΩ at 18 MHz and is not well matched to the driving circuit. No electrical or mechanical matching has been performed. The narrow band response is due to the effects of the alumina substrate. This aspect of array electronics requires further attention through materials and structural design.

4. High Temperature Composite Transducers

Ultrasonic pulse-echo methods have important applications in industrial process control. Echoes received from interfaces in multi-layer systems can provide information about the media contained within a space or of undesirable defect structures [24]. The requirements for high-resolution timing suggests that high frequency, broad bandwidth transducers are an important part of the equipment infrastructure of such measurements. For process control involving high temperatures, for example, the evaluation of mold filling in polymer extrusion (250-350°C) or metal die casting (600-800°C), the transducers involved must not only operate at these temperatures, but degradation of the performance of the transducer and of the couplant between the transducer and the mold with time must be avoided. While bulk ceramic transducers such as aluminum nitride AlN have acceptable Curie temperatures, most organic based coupling methods degrade with time. In order to obviate this problem sol gel composite transducers 50-100μm thick can be manufactured with no intervening couplant directly on a appropriate metal guide rod or even on the surface of the mold itself.

4.1 HIGH TEMPERATURE SOL GEL COMPOSITES

Lithium tantalate $LiTaO_3$ (LT) and lithium niobate $LiNbO_3$ (LN) powders have Curie temperatures of $620^\circ C$ and $>1000^\circ C$ respectively. Both crystalline structures are piezoelectric with relative permittivity at 300K of 50 and 39 respectively. In polycrystalline form it is difficult to pole these materials at temperatures approaching their Curie temperatures because of dielectric breakdown. $LiTaO_3$ powder with an average diameter of several microns was obtained from Superconductive Components Inc. Initial efforts to create sol gel composite structures used lithium tantalate (LT) and lithium niobate sols with 2-methoxyethanol as a solvent. These sols have relatively low viscosity and are stable only at relatively low molarity. The combination of relatively large particle size with low viscosity sols create composites having large porosity within the microstructure means that successfully poled LT/LT or LN/LN composites have not yet been achieved .

An alternative approach based on the discussion of Section 2 in which the ac field across the powder is enhanced by using a sol gel phase that has a higher relative permittivity than the powder has been successful [25]. A PZT/LT sol gel suspension was prepared by blending 0.62g $LiTaO_3$ into 1g of an acetic acid based PZT sol [11]. The sol gel derived PZT phase is expected to have a relative permittivity of >500 [16].. After ultrasonic agitation, the coating was deposited onto a flat stainless steel substrate by spin coating at 2000 rpm for 20 s. The organic material was removed by firing at $420^\circ C$ for 2 min. The coating/ heating process was repeated to build up a multi-layered coating of thickness 50-100μm. The sample was then annealed at $650^\circ C$ for 1 hr and the surface was finally polished mechanically. A top electrode of ~400 nm Pt was sputtered in argon and the ceramic was poled at 380 °C for 30 min with an electric field of 3V/μm. At the conclusion of this fabrication process the sample showed no evidence of delamination from the substrate.

4.2 COMPOSITE CHARACTERISATION

Characterization at 300K using the high frequency impedance showed piezoelectric activity with a real component of the electromechanical coupling coefficient $k_t = 0.17$. The primary origin of the piezoelectric activity was assessed from dielectric measurements of the composite as a function of temperature. Figure 7 shows the relative permittivity and dissipation factor for the composite as a fucntion of temperature and frequency. Both the relative permittivity and the dissipation factor increase with increasing temperature. However, the relative permittivity shows a small peak within a temperature range of 300 to 400°C at 100 kHz. This can be explained by the contribution of the dielectric relaxation of the PZT phase in the composite near its Curie temperature

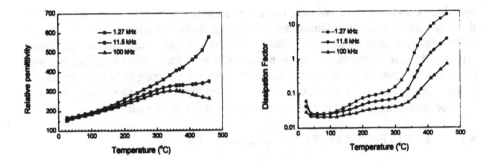

Figure 7: a) the relative permittivity and b) the dissipation factor as a function of temperature for a PZT/LT composite at 1.27 kHz, 11.5 kHz and 100 kHz..

The high temperature ultrasonic transducer properties were measured by a pulse-echo technique, transmitting and receiving ultrasound within the stainless steel substrate. In order to simulate the eventual conditions of application the 56μm thick composite sample on 2.9 mm thick stainless steel was cycled continuously from 20 to 370°C Figure 8 shows the pulse-echo responses on the fourth cycle.

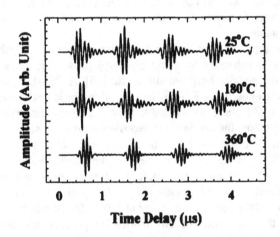

Figure 8: Pulse–echo response from a 56μm PZT/LT composite on 2.9 mm stainless steel. reflected signals on the 4th thermal cycle.

The pulse echo response is stable upon cycling up to 368°C with an initial decrease on successive cycling being followed by a subsequent increase in performance. The latter is not yet fully understood but may be due to changes at the PZT/LiTaO$_3$ interface which lead to a decrease in the proportion of the field distributed across this region. It is not yet established to what extent polarization of LaTiO$_3$ occurs within the composite, but the combination of materials appears to utilise the properties of the PZT phase to the fullest extent.

5. Summary and Conclusions

Sol gel composites provide a means of creating piezoelectric coatings in the thickness range 3-200μm by a process which is compatible with conventional semiconductor processing. The process involves a lower thermal budget than conventional sintering. If it is recognized that the materials properties differ from those of the comparable ceramic, the technology provides a significant new tool for the design of active piezoelectric devices.

6. Acknowledgements

The work was supported by Materials and Manufacturing Ontario, the Natural Sciences and Engineering Research Council of Canada, the Medical Research Council of Canada, Gennum Corporation, Photonics Research Ontario and Datec Coating Corporation. Discussions with and technical assistance from S.Foster and M.Watt are acknowledged. Access to impedance analysis facilities at the Royal Military College of Canada is appreciated.

7. References

1. Foster, F.S., Pavlin, C.J., Lockwood, G.R., Ryan, L.K., Harasiewicz, K.A., Berube, L. and Rauth, A.M. (1993), Principles and Applications of Ultrasound Backscatter Microscopy, *IEEE Trans. Ultrasonics, Ferroelectrics and Frequency Control* 40, 608-617
2. Smith, S.W., Trahey, G.E. and von Ramm, O.T. (1992), Two-Dimensional Arrays for Medical Ultrasound, *Ultrasonic Imaging* 14, 213-219
3. Angelsen, B.A.J., Torp, H, Holm, S, Kristoffersen, K and Whittingham, T.A., (1995), " Which Transducer Array is Best", *Eur. J.Ultrasound* 2, 151-164
4. Pavlin, C.J. and Foster, F.S. (1994), High Frequency Ultrasound Biomicroscopy: Application to the Eye, *Imaging* 7, 509-515
5. Yano, T., Fukukita, H., Uneo, S. and Fukumoto, A. (1987), High Frequency Ultrasonic Diagnostic System for Dermatological Examination, *1987 IEEE Ultrasonics Symposium Proceedings* 875-879
6. O'Donnell, M., Eberle, M.J., Stephens, D.N., Lizza, J.L., Shapo, B.M., Crowe, J.R., Choi, C.D., Chen, J.J., Muller, D.M.W., Kovach, J.A., Lederman, R.L., Ziegenbein, R.C., Wu, C.C., San Vincente, K and Bleam, D. (1997) Catheter Arrays: Can Intravascular Ultrasound Make a Difference in Managing Coronary Artery Disease, *1997 IEEE Ultrasonics Symposium Proceedings,* 147-151
7. Sato, J., Kawabuchi, M. and Fukumoto, A. (1979) Width to Thickness Ratio for Elements in Electronically Scanned Ultrasound Diagnostic Systems, *J.Acoust. Soc. Amer.* 66, 1609
8. Davidsen, R.E. and S.W.Smith (1998) Two Dimensional Arrays for Medical Ultrasound using Multi-layer Flexible Circuit Interconnection, *IEEE Trans. Ultrasonics, Ferroelectrics and Frequency Control* 45, 338-348
9. Madou, M. (1997) *Fundamentals of Microfabrication*, CRC Press, Boca Raton.
10. Sayer, M. and Sedlar, M. (1995), "Comparative Sol Gel Processing of PZT Thin Films", *Integ. Ferroelect.*, 7, 247-258
11. G.Yi, Z.Wu and M.Sayer (1988) Preparation of Pb(Zr,Ti)O₃ Thin Films by Sol-Gel Processing: Electrical, Optical and Electroptic Properties, *J.Appl.Phys.* 64, 2717-2723
12. Haertling, G.H. (1997) Thickness Dependent Properties of Acetate-Derived PLZT Films, *Integ. Ferroelect.* 14, 219-224
13. Barrow, D.A., Petroff, T.E. and Sayer, M. (1995) Thick Ceramic Coatings Using a Sol Gel Based Ceramic-Ceramic 0-3 Composite, *Surf. and Coat. Techn.* 76-77, 113-118
14. Barrow, D.A., Petroff, T.E., Tandon, R. and Sayer, M. (1997) Characterisation of Thick Lead Zirconate Titanate Films Fabrication using a New Sol Gel Based Process. *J.Appl.Phys.* 81, 876-881
15. Ferroperm Ltd (1995) Ferroperm Piezoelectric Materials Data Sheet, Hejreskovvej 6, DK-3490 Kvistgard, Denmark.

16. Olding, T.R. Leclerc, B. and Sayer, M. (1999), Processing of Multilayer PZT Coatings for Device Purposes, *Integ. Ferroelect.* (in press May 1999)
17. Gardeniers, J.G.E., Verholen, A.G.B.J, Tas, N.R. and M.Elwenspoek (1998) Direct Measurement of Piezoelectric Properties of Sol Gel PZT, IMF-9, *J.Korean Phys.Soc.* **32**, S1573-1537
18. Lukacs, M, Olding, T., Sayer, M., Tasker, R. and Sherrit, S. (1999) Thickness Mode Material Constants of a Supported Piezoelectric Film, *J.Appl.Phys.* **85**, 2835-2843
19. Lukacs, M (1999) Single Element and Linear Array High Frequency Ultrasonic Transducers Using PZT Sol Gel Composites, *Ph.D Thesis*, Queen's University, Kingston, ON K7L 3N6
20. Piezoelectric Resonance Analysis Program, http//www.canlink.com/tasi/tasi.html, TASI Technical Software, Kingston, ON, Canada (1998)
21. Lau. J.H., Flip Chip Technologies, McGraw-Hill, New York 1996
22. Lukacs, M, Sayer, M. and Foster S (1999) Single Element High Frequency (>50MHz) PZT Sol Gel Composite Ultrasound Transducers, *IEEE Trans. Ultrasonics, Ferroelectrics and Frequency Control* (in press, 1999)
23. Lukacs, M., Sayer, M. and Foster, S. (1999) High Frequency Ultrasonics Using PZT Sol Gel Composites, *Integ. Ferroelect.* (in press - May 1999)
24. Jen, C.K., Cao, B, Ngyuen, K.T., Loong, C.A. and Legoux, J-G (1997) On-line Monitoring of a Die-Casting Process using Buffer Rods, *Ultrasonics* **35**, 335-344
25. Chen, Y., Sayer, M., Zou L and Jen, C.K. (1999), Lithium Tantalate/Lead Zirconate Titanate Composite Ultrasonic Transducers, *Appl.Phys.Lett.* **74**, (in press 1999)

PULSED-LASER DEPOSITION AND CHARACTERIZATION OF THIN FILMS

D. Bäuerle, M. Dinescu, R. Dinu, J. Pedarnig, J. Heitz,

R. Schwödiauer, S. Bauer, S. Bauer-Gogonea

Angewandte Physik, Johannes-Kepler-Universität Linz,

A-4040 Linz, Austria

Abstract:

Pulsed-laser deposition (PLD) has become a very popular technique for the fabrication of thin films of multicomponent materials. Among the materials studied in most detail are compound semiconductors, dielectrics, ferroelectrics, electrooptic and giant magnetoresistance oxides, high-temperature superconductors, polymers, and various types of heterostructures. PLD is very reliable, offers great experimental versatility, it is fairly simple and fast - as long as small-area films of up to several square-centimeters are to be fabricated. For these reasons, PLD is particularly suitable in materials research and development. The strong non-equilibrium conditions in PLD allow, however, some *unique* applications: Among those are *metastable* materials that cannot be synthesized by standard techniques and the fabrication of films from species that are generated only during pulsed-laser ablation. With certain systems, the physical properties of such films are superior to those fabricated by standard evaporation, electron-beam evaporation, etc.

This paper gives an overview on the fundamentals and applications of PLD with special emphasis on dielectric and piezoelectric thin films.

1. Introduction

Lasers can be used to fabricate thin extended films by condensing on a substrate surface the material that is ablated from a target under the action of laser light [1]. Depending on the specific laser and material parameters, ablation takes place under quasi-equilibrium conditions, as in laser-induced thermal vaporization, or far from equilibrium, as in pulsed-laser ablation. In the latter case, thin-film formation is termed pulsed-laser deposition (PLD). Pulsed-laser deposition is of particular interest

261

C. Galassi et al. (eds.), Piezoelectric Materials: Advances in Science, Technology and Applications, 261–271.

262

because it enables one to fabricate multicomponent stoichiometric films from a *single* target.

A typical setup employed for film deposition is schematically shown in Fig. 1. It essentially consists of a laser, a reaction chamber, a target, and a substrate. The material ablated from the target is condensed on the substrate and forms a thin film. Ablation can take place in either a vacuum or a low pressure inert or reactive atmosphere. The latter technique is termed *reactive* laser ablation (reactive laser sputtering). The targets mainly employed in PLD are ceramics and, in special cases, liquids. To a lower extent, targets in single crystalline, polycrystalline, powdery, or amorphous form are used as well. For uniform ablation, the target is rotated and scanned with respect to the laser beam (Fig.2). By this means, surface roughening and structure formation can be significantly suppressed. As a consequence, the density of particulates and displacements of the plasma plume direction are minimized and the ablation rate and profile of the deposited film remain almost constant with respect to the number of laser pulses.

Pulsed-laser deposition is particularly suitable in materials research and development. The strong non-equilibrium conditions in PLD allow, however, some *unique* applications:

- The synthesis of metastable materials that cannot be produced by standard techniques.

- The fabrication of films from species that are generated only during pulsed-laser ablation. With certain systems, the physical properties (microstructure, morphology, adhesion, optical, electrical, etc.) of such films are superior to those fabricated by standard evaporation, electron-beam evaporation, etc.

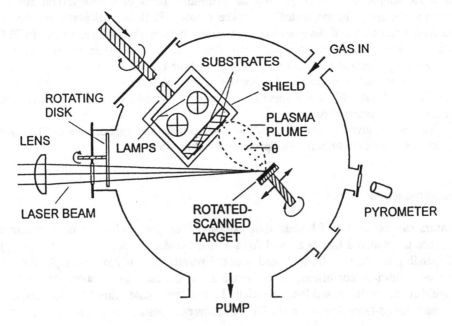

Figure 1. Schematic picture of an experimental setup employed in PLD [1].

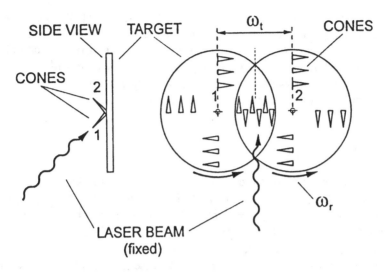

Figure 2. Surface structures which align with the incident laser beam, e.g. columnar structures (cones), can be suppressed by simultaneous rotation and translation of the substrate with incommensurate frequencies ω_r and ω_t, respectively. By this means, each target site is ablated from opposing incident angles [2].

Besides restrictions with respect to film areas, the major disadvantage of PLD is related to particulates that appear on the substrate and film surface.

Pulsed-laser ablation can be based on thermal, photophysical, and photochemical mechanisms [1]. From the aspect of film formation, however, the detailed ablation mechanisms are of minor relevance. It is only important that ablation takes place on a time scale that is short enough to suppress the dissipation of the excitation energy beyond the volume ablated during the pulse. Only with this condition, can damage of the remaining target and its segregation into different components be largely avoided. In this regime of interactions, the relative concentrations of species within the plasma plume remain almost unchanged for successive laser pulses and they are almost equal to those within the target material. This is the main reason why pulsed-laser deposition has been found to be useful, in particular for the deposition of thin films with complex stoichiometry. Among the materials studied in most detail are compound semiconductors, dielectrics, ferroelectrics, electrooptic and giant magnetoresistance oxides, high-temperature superconductors, polymers, and various types of heterostructures. For the fabrication of smooth and stoichiometric films, excimer lasers and higher harmonics of Nd:YAG lasers together with dense ceramic targets in the form of sintered pellets are most appropriate [1].

2. Dielectric materials

Among the dielectric materials deposited as thin films are diamond-like carbon [1,3], oxides like SiO_2, ZrO_2, Al_2O_3, etc. [1], many fluorides such as MgF_2, CaF_2, SrF_2, etc.

[1], and nitrides like BCN [4], BN, C_xN_y [5], Si_3N_4 [6,7] etc. [1]. The loss of oxygen observed with many oxides can be compensated by reactive ablation in O_2 at pressures between 10^{-4} and 1 mbar. Nitride films were synthesized by reactive ablation in NH_3 or N_2 atmosphere. Hybrid techniques using, e.g., PLD in combination with an RF-source [7] or in an ion beam [8] have been employed as well [7]. The deposition rates achieved with these materials are, typically, between 0.5 Å/s and 50 Å/s.

3. Ferroelectric materials

Among the ferroelectric films fabricated by PLD are oxidic perovskites such as $BaTiO_3$ [9], $KTa_{1-x}Nb_xO_3$ (KTN), $PbTi_{1-x}Zr_xO_3$ (PZT) [9-11], $Pb_{1-y}La_yTi_{1-x}Zr_xO_3$ (PLZT) [12], $Sr_xBa_{1-x}TiO_3$ [13,14], incipient ferroelectrics [15] such as $SrTiO_3$ and $KTaO_3$, and perovskite-like materials such as $Bi_4Ti_3O_{12}$ (BTO), $LiNbO_3$, $Pb(Mg_{1/3}Nb_{2/3})O_3$ (PMN) [16], $SrBi_2Ta_2O_9$ (SBT) [17,18], etc. [1].

Epitaxial and oriented *polycrystalline* films of these materials have been grown on different types of substrates such as $SrTiO_3$, $LaAlO_3$, MgO, fused quartz (a-SiO_2), Si, etc. With Si substrates, buffer layers, mainly of YSZ and a-SiO_2 are employed. The structural, morphological, and ferroelectric properties of films depend strongly on the laser parameters, the type of substrate material, and the substrate temperature. For example, with crystalline films of $PbTi_{1-x}Zr_xO_3$ ($0.48 \leq x \leq 0.55$) deposited by means of 248 nm KrF-laser radiation on bare and Pt or Au coated Si substrates at temperatures $T_s \approx 600°C$, one finds a dielectric constant $\varepsilon' \approx 850$, a remanent polarization $P_r \approx 22$ $\mu C/cm^2$, a coercive field $E_c \approx 40$ kV/cm, and a resistivity $\rho \approx 10^{13}$ Ωcm [19]. With the *same* material and substrate, but with Nd-YAG laser radiation and $T_s \approx 375°C$ the values are $\varepsilon' \approx 500$, $P_r \approx 15$ $\mu C/cm^2$, $E_c \approx 100$ kV/cm [20].

The fabrication of multilayer structures has been demonstrated for many material combinations:

- Ferroelectric/dielectric structures such as SBT/Bi_2O_3 [17], $Pb(Mg_{1/3}Nb_{2/3})O_3$ – $PbTiO_3$ (PMN-PT)/$SrRuO_3$ [12].
- Ferroelectric/high-temperature superconductor structures as, e.g., BTO/$YBa_2Cu_3O_7$(YBCO) [21], PZT/YBCO [11,22], YBCO/PZT/YBCO [20,23].
- Ferroelectric/giant magneto-resistive (GMR or CMR) structures like PZT/$La_xCa_{1-x}MnO_3$ [24].

Such structures are of interest for various applications, such as microwave signal processing devices, magneto-sensitive memory cells, high-density dynamic random access memories (DRAMs), nonvolatile ferroelectric RAMs, etc. Among the most promising compound for the latter application is SBT, because of its very small polarization fatigue of up to 10^{12} cycles, low imprint tendency, and low leakage currents [25]. Subsequently, we discuss some recent results on the PLD of SBT films [17].

3.1. SrBi$_2$Ta$_2$O$_9$ (SBT) FILMS

For the fabrication of SrBi$_2$Ta$_2$O$_9$ films and Bi$_2$O$_3$/SrBi$_2$Ta$_2$O$_9$/Bi$_2$O$_3$ multilayers on Pt/Si/SiO$_2$ coated Si(100) substrates, Bi-enriched targets (SrBi$_{2.2}$Ta$_2$O$_9$) have been used in order to compensate for the Bi loss during film deposition [17]. SBT films were grown at substrate temperatures of 500 $\leq T_s \leq$ 800°C and oxygen pressures of p(O$_2$)≈0.4mbar by means of KrF-laser radiation (λ=248nm, τ_{FWHM}≈30ns, Φ=2-4J/cm^2, repetition rate 5Hz). With these parameters, the deposition rates were, typically, 0.6Å/pulse. The film thicknesses employed were around 450nm. At a fluence of 2J/cm^2, the best film structure, consisting only of the ferroelectric phase with a dominant (115) crystallite orientation was obtained at a substrate temperature of T$_s$=750°C. The formation of the ferroelectric phase was favored at a lower temperature, T$_s$=650°C, for Bi$_2$O$_3$/SrBi$_2$Ta$_2$O$_9$/Bi$_2$O$_3$ multilayer structures.

The SBT films showed hysteresis loops with a maximum remanent polarization of P$_r$=6.5µC/cm^2 (Fig. 3), and a coercive field of E$_c$=35kV/cm. With Φ=2.5J/cm^2, the maximum remanent polarization has been obtained for films deposited at T$_s$=700°C. The remanent polarization agrees well with measurements on liquid-phase deposited films prepared at 800°C [26] and is slightly above the value of bulk SBT (P$_r$=6µC/cm^2 [27]). Fig. 4 shows the dielectric constant of films deposited at T$_s$=700, 750 and 800°C

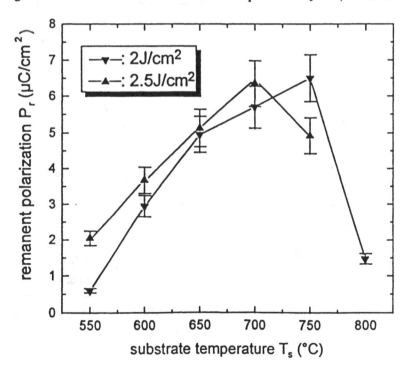

Figure 3. Remanent polarization P$_r$ for SBT films deposited at different substrate temperatures T$_s$ (248 nm KrF, Φ=2 and 2.5J/cm^2; after [17]).

with $\Phi=2J/cm^2$. In agreement with XRD investigations and the results in Fig. 3, the strength of the dielectric anomaly is largest for films deposited at $T_s=750°C$. For temperatures well below $T_{max} \equiv T(\varepsilon'_{max})$ the dielectric constant of a ferroelectric material with a second order phase transition can be described by a Curie-Weiss (CW) law with a Curie-Weiss constant twice that of the paraelectric phase. In the region $\pm40°C$ around T_{max} we find a relaxor-like behavior. The dependence over the whole temperature range can be approximated by: $1/\varepsilon' - 1/\varepsilon'_{max} = B(T-T_{max})^m$, with m=2 for the relaxor state, and with m=1 and $\varepsilon'_{max} \to \infty$ within the CW-region. The fit in Fig. 4 in the range $\pm40°C$ around T_{max} yields m=1.8 and m=2 for films deposited at T_s =700 and 750°C, respectively. It confirms the tendency to relaxor-like behavior around T_{max}. At temperatures below 250°C the deviation of ε' from the CW fit has been ascribed to domain-wall motions. The SBT films revealed no significant fatigue after $4\cdot10^{10}$ switching cycles at 120°C. Domain pinning led to slight fatigue at room temperature. However, recovery of the non-switchable polarization was possible by keeping the film under field at 90°C. Leakage currents of $5\cdot10^{-8}A/cm^2$ for fields up to 100kV/cm have been determined. This is sufficient for applications in dynamic random access memory devices.

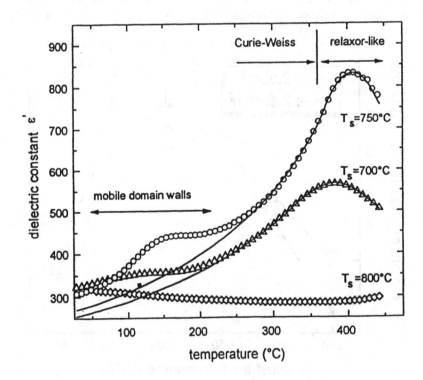

Figure 4. Dielectric constant ε' vs. temperature for SBT films deposited at T_s=750°C, 700°C, and 800°C. The full curves have been calculated (after [17]).

4. Teflon (PTFE) films

The good mechanical, thermal, and chemical stability of polytetrafluoroethylene (PTFE) together with its low surface adhesion, frictional resistance, and low dielectric constant, makes this material unique for numerous applications in mechanics, microelectronics, chemistry, medicine, and bioscience. For certain applications, it is desirable to fabricate PTFE in the form of thin films. Subsequently, we discuss the fabrication, characterization and electret properties of pulsed-laser deposited polytetrafluoroethylene (PLD-PTFE).

PTFE films were deposited by means of KrF-laser radiation from pressed powder targets (grain sizes between 6-9μm) [28]. The films are highly crystalline, as shown in the optical polarization micrographs of Fig. 5a,b. The size of the spherulites could be enlarged to diameters up to 500μm by postannealing at 550°C. The high crystallinity of the PLD-PTFE films is demonstrated in IR transmission spectra (Fig. 6) [28], where all bands related to amorphous PTFE are missing in the PLD films. PLD-PTFE films also show the characteristic first-order structural phase transitions from triclinic to hexagonal and pseudohexagonal at 20 and 30°C, respectively. This is shown in Fig. 7 where we have plotted the dielectric constant vs. temperature [29]. During film formation, PTFE grains are laser-transferred from the target to the substrate with subsequent melting and crystallization. Interactions between particulates and the plasma plume seem to play an important role in the film formation process [30].

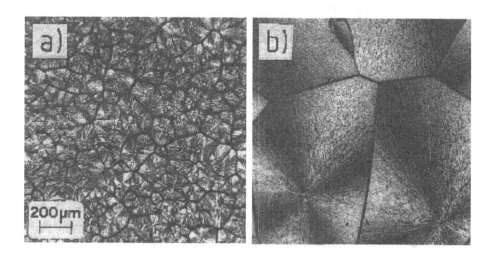

Figure 5. Optical polarization micrographs showing the surface morphology of KrF-laser-deposited PTFE-Teflon films on Si substrates (λ = 248 nm, ϕ = 2.5 J/cm^2, N_l = 2500 pulses, 2 Hz). The magnification is the same with both pictures. (a) T$_s$ = 400 °C (b) T$_s$ = 400 °C and additional post-annealing at T$_s$ = 550 °C (after [28]).

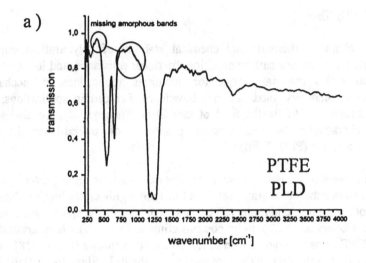

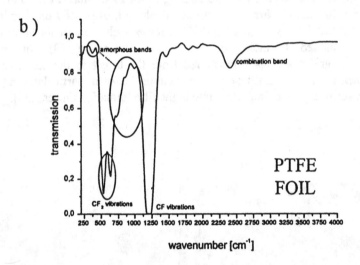

Figure 6. IR spectra for (a) a post-annealed film deposited from a pressed powder pellet at T_s= 355°C, T_a = 550 C, (b) a commercial PTFE foil (d = 25 μm) (after [28]).

PTFE is well known as an excellent electret. An electret is a dielectric exhibiting a quasi-permanent electrical charge. The electret charge may either be a "real" charge, or an inherent polarization, or a combination of both [31]. True charge electrets, such as nonpolar PTFE, FEP (tetrafluoroethylene-co-hexafluoropropylene copolymer), and PFA (tetrafluoroethylene-co-perfluoropropoxyethylene) have found widespread applications in acoustic transducers, such as electret microphones [31]. The increasing demand for miniaturized devices makes PTFE films deposited on substrates an interesting charge electret.

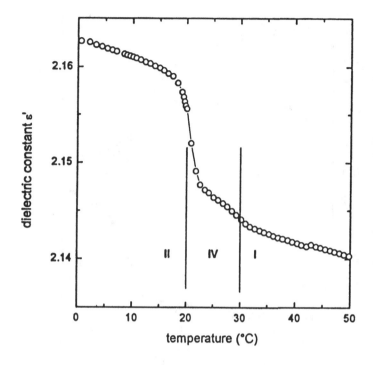

Figure 7. Dielectric constant ε' of a PLD-PTFE film capacitor versus temperature (after [29]).

The suitability of PLD-PTFE films as charge electret has been demonstrated by measuring the thermally-stimulated voltage decay of corona-charged PLD samples at a rate of 4°C/min. Details of the experiment have been reported in [32]. Fig. 8 shows the surface voltage decay of the PLD-PTFE electret. For comparison, data from commercially available PTFE foils are included. Most notable is the excellent temperature-dependent charge stability of PLD-PTFE films with large spherulites, comparable or even better than that of the PTFE foil. Thus PLD-PTFE films add a new member to the family of "Teflon" materials, interesting for miniaturized electret applications.

270

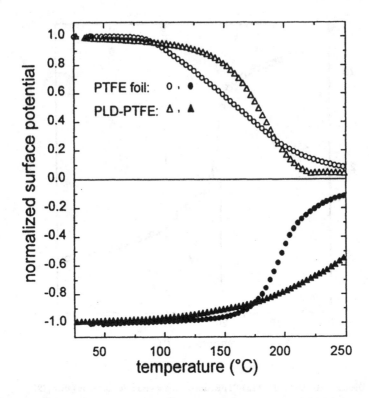

Figure 8. Thermally-stimulated surface potential decay for PTFE foils and PLD-PTFE films with large spherulites (after [32]).

References

1. Bäuerle, D. (1996) *Laser Processing and Chemistry*, Springer Verlag, Berlin, Heidelberg (the third edition will be published in 2000).
2. Arnold, N. and Bäuerle, D. (1999) Uniform target ablation in pulsed-laser deposition, *Appl. Phys. A* **68**, 363-367.
3. Yamamoto, K., Koga, Y., Fujiwara, S., Kokai, F., and Heimann, R.B. (1998) Dependence of the sp^3 bond fraction on the laser wavelength in thin carbon films prepared by pulsed laser deposition, *Appl. Phys. A* **66**, 115-117.
4. Dinescu, M., Perrone, A., Caricato, A.P., Mirenghi, L., Gerardi, C., Chica, C., and Frunza, L. (1998) Boron carbon nitride films deposited by sequential pulsed laser deposition, *Appl. Surf. Sci.* **127-129**, 692-696.
5. Vivien, C., Dinescu, M., Meheust, P., Boulmer-Leborgne, C., Caricato, A.P., and Perrière, J. (1998) Nitride-molecule synthesis in plasma produced by reactive laser ablation assisted by RF discharge for thin-film deposition, *Appl. Surf. Sci.* **127-129**, 668.
6. Mihailescu, I.N., Lita, A., Teodorescu, V.S., Gyorgy, E., Alexandrescu, R., Luches, A., Martino, M., and Barborica, A. (1996) Synthesis and deposition of silicon nitride films by laser reactive ablation of silicon in low pressure ammonia – a parametric study, *J. Vac. Sci. Technol.* A **14**, 1986-1994.
7. Leborgne, C., Dinescu, G., Vivien, C., Stanciu, C., and Dinescu, M. (1999) Appl. Surf. Sci. (in press).
8. Bae, J.H. and Cho, J. (1999) Structural, morphological, and mechanical properties of amorphous carbon and carbon nitride thin films deposited by reactive and ion beam assisted laser ablation, *J. Materials Science* **34**, 1093-1098.

9. Lin, Y., Zhao, B.R., Peng, H.B., Xu, B., Chen, H., Wu, F., Tao, H.J., Zhao, Z.X., and Chen, J.S. (1998) Growth and polarization features of highly (100) oriented $Pb(Zr_{0.53}Ti_{0.47})O_3$ films on Si with ultrathin SiO_2 buffer layer, *Appl. Phys. Lett.* **73**, 2781-2783.

10. Lappalainen J., Frantti, J., and Lantto, V. (1997) Electrical and mechanical properties of ferroelectric thin films laser ablated from a $Pb_{0.97}Nd_{0.02}(Zr_{0.55}Ti_{0.45})$ O_3 target, *J. Appl. Phys.* **82**, 3469-3477.

11. Grishin A.M., Yamazato, M., Yamagata, Y., and Ebihara, K. (1998) Structure and polarization in epitaxial ferroelectric $PbZr_{0.52}Ti_{0.48}O_3/YBa_2Cu_3O_{7-x}/Nd:YalO_3$ thin films, *Appl. Phys. Lett.* **72**, 620-622.

12. Tyunina, M., Levoska, J., Sternberg, A., and Leppävuori, S. (1998) Relaxor behavior of pulsed laser deposited ferroelectric $(Pb_{1-x}La_x)(Zr_{0.65}Ti_{0.35})O_3$ films, *J. Appl. Phys.* **84**, 6800-6810.

13. Xiong, S.B., Ye, Z.M., Chen, X.Y., Guo, X.L., Zhu, S.N., Liu, Z.G., Lin, C.Y., and Jin, Y.S. (1998) Ferroelectric $Sr_xBa_{1-x}Nb_2O_6$ optical waveguiding thin films on Si O_2-coated Si(1000) substrates, *Appl. Phys. A* **67**, 313-316.

14. Nakano, M., Tabata, H., Tanaka, K., Katayama, Y., and Kawai, T. (1997) Fabrication and characterization of $(Sr,Ba)Nb_2O_6$ thin films by pulsed laser deposition, *Jpn. J. Appl. Phys. Part. 2 Lett.* **36** (10A), L1331-L1332.

15. Migoni, R., Bilz, H., and Bäuerle, D. (1976) Origin of Raman scattering and ferroelectricity in oxidic perovskites, *Phys. Rev. Lett.* **37**, 1155-1158

16. Maria, J.-P., Hackenberger, W., and Trolier-McKinstry, S. (1998) Phase development and electrical property analysis of pulsed laser deposited $Pb(Mg_{1/3}Nb_{2/3})O_3$-$PbTiO_3$ (70/30) epitaxial thin films, *J. Appl. Phys.* **84**, 5147-5154.

17. Dinu, R., Dinescu, M., Pedarnig, J.D., Gunasekaran, R.A., Bäuerle, D., Bauer-Gogonea, S., and Bauer, S. (1999) Film structure and ferroelectric properties of in-situ grown $SrBi_2Ta_2O_9$ films, *Appl. Phys. A* in press.

18. Fujimura, N., Thomas, D.T., Streifer, S.K., and Kingon, A.I. (1998) Preferred orientation, phase formation and the electrical properties of pulsed laser deposited $SrBi_2Ta_2O_9$ thin films, *Jpn. J. Appl. Phys.* **37**, 5185-5188.

19. Roy, D., Krupanidhi, S.B., and Dougherty, J.P. (1992) Excimer laser ablation of ferroelectric $Pb(Zr,Ti)O_3$ thin films with low pressure direct-current glow discharge, *J. Vac. Sci. Technol.* A **10**, 1827-1831.

20. Verardi, P., Dinescu, M., Craciun, F., and Sandu, V. (1997) Oriented $PbZr_xTi_{1-x}O_3$ thin films obtained at low substrate temperature by pulsed laser ablation, *Thin Solid Films* **311**, 171-176.

21. Maffei, N. and Krupanidhi, S.B (1992) Excimer laser-abtated bismuth titanate thin films, *Appl. Phys. Lett.* **60**, 781-783.

22. Liu, B.T., Hao, Z., Chen, Y.F., Xu, B., Chen, H., Wu, F., Zhao, B.R., Kislinskii, Y., and Stepantsov, E. (1999) Investigation on $Ag/Pb(Zr_{0.53}Ti_{0.47})O_3/YBa_2Cu_3O_{7-\delta}$ three-terminal system with small gate area, *Appl. Phys. Lett.* **74**, 2044-2046.

23. Ramesh, R., Chan, W.K., Wilkens, B., Gilchrist, H., Sands, T., Tarascon, J.M., Keramidas, V.G., Fork, D.K., Lee, J., and Safari, A. (1992) Fatigue and retention in ferroelectric Y-Ba-Cu-O/Pb-Zr-Ti-O/Y-Ba-Cu-O heterostructures, *Appl. Phys. Lett.* **61**, 1537-1539.

24. Grishin, A.M., Khartsev, S.I., and Johnsson, P. (1999) Epitaxial ferroelectric/giant magnetoresistive heterostructures for magnetosensitive memory cell, *Appl. Phys. Lett.* **74**, 1015-1017.

25. Paz de Araujo, C.A., Cuchiaro, J.D., McMillan, L.D., Scott, M.C., and Scott, J.F. (1995) Fatigue-free ferroelectric capacitors with platinum electrodes, *Nature* **374**, 627-629.

26. Noguchi, T., Hase, T., and Miyasaka, Y. (1996) Analysis of the Dependence of Ferroelectric Properties of Strontium Bismuth Tantalate (SBT) Thin Films on the Composition and Process Temperature, *Jpn. J. Appl. Phys.* **35**, 4900-4904.

27. Lines, M.E. and Glass, A.M. (1977) *Principles and Applications of Ferroelectrics and Related Materials*, Clarendon Press, Oxford.

28. Li, S.T., Arenholz, E., Heitz, J., and Bäuerle, D. (1998) Pulsed-laser deposition of crystalline Teflon (PTFE) films, *Appl. Surf. Sci.* **125**, 17-22.

29. Schwödiauer, R., Heitz, J., Arenholz, E., Bauer-Gogonea, S., Bauer, S., and Wirges, W. (1999) Pulsed-Laser-Deposited and Plasma-Polymerized Polytetrafluoroethylene (PTFE)-Like Thin Films: A Comparative Study on PTFE-Specific Properties, *J. Polym. Sci. B: Polym. Phys.*, in press.

30. Heitz, J. and Dickinson, J.T. (1999) Characterization of particulates accompanying laser ablation of pressed polytetrafluoroethylene (PTFE) targets, *Appl. Phys. A*, (1999) in press.

31. Sessler, G.M. (1999) *Electrets*, 3rd ed. Vol I, Laplacian Press, Morgan Hill.

32. Schwödiauer, R., Bauer-Gogonea, S., Bauer, S., Heitz, J., Arenholz, E., and Bäuerle, D. (1998) Charge stability of pulsed-laser deposited polytetrafluoroethylene film electrets, *Appl. Phys. Lett.* **73**, 2941-2943.

CORRELATED COMPOSITIONAL, MORPHOLOGICAL AND ELECTRICAL CHARACTERIZATION OF PLD PIEZOELECTRIC THIN FILMS

F. CRACIUN, P. VERARDI, M. DINESCU[1], L. MIRENGHI[2],
F. DINELLI[3]
CNR Istituto di Acustica "O.M. Corbino", Area di Ricerca Tor Vergata, Via del Fosso del Cavaliere 100, 00133 Roma, Italy
[1] *Institute of Atomic Physics, NILPRP, R76900 Bucharest, Romania*
[2] *PASTIS-CNRSM, 72100 Brindisi, Italy*
[3] *University of Oxford, Department of Materials, Oxford, UK*

Abstract

Using different correlated techniques, we addressed the problem of the uniformity of the properties of piezoelectric thin films obtained by pulsed laser deposition, with particular emphasis on lead zirconate titanate thin film. This problem is presented together with a survey of different experimental findings and in the context of existing theoretical models. The results allow us to define a uniform deposition area over which properties like film thickness, composition, crystalline structure and piezoelectric properties remain almost constant in the selected experimental conditions.

1. Introduction

Recently, pulsed laser deposition (PLD) demonstrated to be an appropriate technique for obtaining thin films with complicated composition like oxide ferroelectrics and high T_c superconductors [1]. The main advantages of PLD: congruent evaporation (which allows to preserve the stoichiometry of a multielement target in the growing thin film) and the possibility of drastically decreasing the substrate temperature and simultaneously improving the film crystallinity (due to the high energy of the particles arriving on the substrate) make this technique very attractive [1]. However, PLD suffers from few disadvantages, not evident with other film growth techniques, which have inhibited its full extension in the large scale fabrication of single element thin films and multilayers: 1) a highly forward directed plume which leads to a nonuniform thickness film [1-14]; 2) a different angular distribution of the various ejected constituents of the

273

C. Galassi et al. (eds.), Piezoelectric Materials: Advances in Science, Technology and Applications, 273–284.
© 2000 *Kluwer Academic Publishers. Printed in the Netherlands.*

plasma plume, which can produce spatial compositional nonuniformities in the film [1-4,7]; 3) the deposition of large size particulates on the film surface [1]. Moreover, the growth of multicomponent oxide ferroelectrics with volatile constituents , like e.g. Pb in Pb $(Zr,Ti)O_3$ (PZT) is even more complicated due to the growing of metastable pyroclore phases, especially in the presence of off-stoicheiometric compositions [1]. Therefore correlated compositional, morphological and electrical characterisation are necessary in order to define the homogeneous area extension and the relationship with PLD process parameters. Different piezoelectric thin films obtained by PLD have been examined [15-19], but correlated investigations have been performed mainly on PZT, due to its high interest for applications. The samples have been divided in different zones in order to verify the existence of chemical and structural differences between different regions. Few techniques like XRD, XPS, SIMS, AFM as well as local dielectric and piezoelectric measurements have been employed for characterisation. Results concerning film thickness distribution, differences in the crystallinity, orientation, chemical composition as well as piezoelectric properties measured in different deposition zones are presented in the following sections.

2. Pulsed laser deposition of PZT piezoelectric films

PZT films used as an example here were deposited by laser ablation of a bulk $Pb(Ti_{0.6}Zr_{0.4})O_3$ ceramic target in oxygen reactive atmosphere. Fig. 1 shows the experimental set-up. The deposition vacuum chamber was evacuated by a turbomolecular pump down to approximately 10^{-6} Torr and then filled with oxygen to 150 mTorr. Si (111) substrates, on which Au/Cr electrodes with thickness of about 100 nm have been previously evaporated, were attached to a heated holder. The targets were fixed parallel to the substrate on a holder, which could rotate and translate. The beam from a pulsed Nd-YAG laser (wavelength 1.06 µm, energy 300 mJ/pulse, pulse duration 10 ns, pulse frequency 10 Hz) was incident at 45^0 with respect to the target normal and focused on its surface (with a circular spot of about 1 mm^2). Deposition at different laser fluences was performed and the best value for obtaining piezoelectric films (approximately 25 J/cm^2) identified (at lower fluences, achieved by varying pulse energy, samples obtained at low substrate temperatures were slightly or not at all piezoelectric) [15]. The deposition was performed at a substrate temperature of about 375^0C for approx. 40 min. The target-substrate distance was kept at 5 cm. In these conditions, the average deposition rate was about 0.1 nm/pulse.

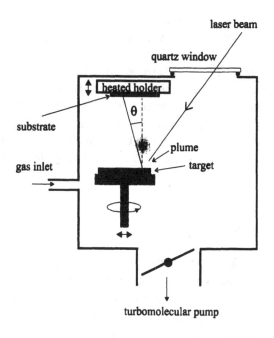

Figure 1. Schematic diagram of the PLD system.

After deposition the samples were subjected to characterisations performed on different areas at various distances from the centre of deposition; this was defined as the projection of the target point where the laser beam was incident, on the substrate (Fig. 1).

Film thickness was measured in these zones by secondary ion mass spectroscopy (SIMS).

XPS studies have been performed by a VG ESCALAB 210 Spectrometer, using a non-monochromatic Al K_α X-ray source (300 W) in the five channel hemispherical analyser. Wide scans in the binding energy scale 0-1200 eV (at 50 eV analyser pass energy) were collected both onto as received surface and after the sputter cleaning in order to put in evidence all the constituents of the film. Narrow scans of Ti 2p, Zr 3d, Pb 4f, O 1s and C 1s were also acquired at 20 eV and 0.1 eV/channel pass energy and 100 ms of dwell time in order to give a better insight into the chemical bonds form and a semiquantitative analyse of the present chemical species.

XRD studies were performed in different zones to investigate the crystallinity of the deposited films by using the CuK$_\alpha$ line (with an average wavelength value of 1.54 Å) produced by a DRON 20 diffractometer with nickel filter.

Sample surface morphology was examined by SEM and Atomic Force Microscopy (AFM) [17]. A typical AFM configuration working in contact mode and using a feedback circuit to keep the normal force constant was employed. The lateral force microscopy mode has been also used which helps in distinguishing materials of different frictional behaviour on the same surface.

The dielectric properties have been obtained by measuring the capacity of the film over which top electrodes have been deposited. The piezoelectric d_{33} constant of PZT films was measured by a direct method, in which a stress signal is applied by an ultrasonic transducer on the piezoelectric film. The proportional electric voltage produced by the tested film is measured by an oscilloscope, compared with that obtained from a thin quartz sample and used to calculate the longitudinal piezoelectric constant [16].

3. Film thickness profile characterisation

It has been found by different authors that the flux of the ablated material during PLD process can be characterised by a polar emission described by $\cos^n\theta$ (with n>>1), where θ is the angle of deposition, measured from the target normal [3,4,6,7,10]. Consequently, when thickness spatial variations are examined onto the deposited film, a $\cos^{n+3}\theta$ dependence is found. When related to film measurements, $\theta = \tan^{-1}(d/L)$, with d the distance from the centre of deposition and L the substrate-target distance. Few models have been proposed to explain the forward peaking of the polar distribution. In ref. [14], Kools used a Monte Carlo simulation method to model the interaction between the plasma cloud and a diluted gas background with pressures up to 200 mTorr. He found forward focusing to appear mainly due to geometrical effects of the collisions. By using a different approach, Anisimov, Bäuerle and Lukyanchuk [12] found that hat plume directionality can be attributed to the large differences in the pressure gradients normal and parallel to the target surface. Regarding the influence of the laser parameters, both theoretical and experimental findings agree that a reduction in the laser spot size (while keeping the fluence constant) leads to broadening of the polar distribution [2,6], while there is much less consensus on the effect of the fluence variation at constant spot size. Regarding the effects of the background gas, Lichtenwalner et al. [4] found a narrowing of PZT plume and film thickness profile when oxygen pressure was increased up to 300 mTorr, while Tyunina et al. [5] reported a gradually broadening profile of PZT plume for an increase in oxygen pressure to 100 mTorr, followed by a narrowing of the profile at higher pressures.

Fig. 2 shows the obtained thickness profile on the PZT samples prepared as described in the previous section (intermediate curve), together with other results obtained by Licthenwalner et al. [4] on PZT films deposited by using a KrF excimer laser (fluence 2.2 J/cm^2, spot size 2x9 mm^2, pulse duration 25 ns, repetition rate 3 Hz, target-substrate distance 6 cm and O$_2$ pressure 100 mTorr) (lower curve) and by Tyunina et al. [5] on PZT films produced by using a tripled Nd-YAG laser ($\lambda = 355$ nm,

$\tau = 10$ ns, spot size 2.5×1.5 mm^2, fluence 1.5-4 J/cm^2, target-substrate distance 55 mm and O$_2$ pressure 100 mTorr) (higher curve).

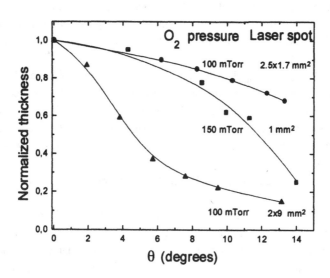

Figure 2. Normalised thickness of PZT thin films vs. the deposition angle, as reported by different authors (● – ref. [5]; ▲ – ref. [4]; ■ – this paper). The continuous lines are guides for eye.

The results are normalised to the film thickness t as determined in the centre of deposition, t_0, which in our case is approx. 2.1 μm (for the results taken from ref. [4] and represented by triangles $t_0 \approx 3200$ Å).

By comparing the different sets of results it can be observed that in all the cases the thickness varies with the distance from the centre of deposition. With respect to the results of Tyunina *et al.*, the thickness profile obtained on our samples is narrower, probably due to the higher value of the O$_2$ pressure, which increases the number of collisions inside the plume. Instead the film profile is comparatively broader than that represented by the lower curve, probably due to the much smaller spot size area which was used in our experiments. Therefore in order to reduce the forward peaking of the angular distribution, one should work with very focused laser beams and lower oxygen pressure, while keeping the laser fluence under control. But nevertheless the oxygen pressure cannot be lowered below a certain optimum value, which is necessary to preserve the stoichiometry of the samples [1].

Investigations of the film thickness profile have been made also by Hau *et al.* [3] on PZT, but in the absence of background oxygen gas.

4. Composition distribution characterisation

The characterisation of composition distribution is a key problem for films obtained by PLD from multicomponent targets [2]. It has been experimentally observed that polar distributions of the ablated material from such targets depends on the mass of the plume elements [2,3], making therefore difficult the obtaining of large deposition area with uniform stoichiometry.

By studying the angular distribution of PZT films obtained by XeCl laser deposition in vacuum (λ = 308 nm, spot size 0.5x2.2 mm^2, laser fluence 0.5-5 J/cm^2) Hau *et al.* [3] showed that the distribution of Ti and Zr is predominant in the central zone, while the distribution of Pb is much broader, and there is a considerable deposition of Pb at angles away from the centre. They attributed this result to the low melting temperature of Pb and its oxides, which favours an ejection dominated by thermal evaporation, therefore a much broader distribution of Pb with respect to those of Ti and Zr.

Fig. 3 shows XPS spectra taken on our PZT films examined in the same zones where the thickness was evaluated. Fig. 3a) presents a survey spectrum of the sample after the surface cleaning by sputtering, taken on a zone situated in the centre of the film. The peaks corresponding to Pb, Zr and Ti are shown. The O 1s peak at 531 eV is asigned to oxygen bonded in PZT.

From the Ti/Zr ratio we can deduce that our film has the $Pb(Ti_{0.6}Zr_{0.4})O_3$ composition. Some other (very small) peaks are present in both spectra, due to the contamination in different processing phases. XPS spectra were taken on two main zones: the central one with an area of about 1.5 cm^2 and the external one (at a distance of about 1.5 cm toward the border). A comparative (semiquantitative) analysis of the two zones gives interesting information concerning the type and the amount of the chemical bonds. Fig 3 b), c) and d) present respectively the Pb 4f, Ti 2p and Zr 3d ranges of the XPS spectra for the central and eccentric zones. Analyses of detailed spectra for Pb, Ti and Zr evidenced common features but also some differences for the two area. Referring to the position of the bonding energy of the Pb doublet (3d5/2 at 139.2 eV for the central and 139.3 eV for the eccentric zone) we can affirm that it is bonded in a compound of $PbAO_3$ type, where A is a metal of the structure (Ti or Zr in our case). In particular, for A=Ti ($PbTiO_3$) the value reported in literature for the Pb 3d5/2 peak is 138 eV. The value we found experimentally for the photoelectronic peak of Pb is very close to that value. The shift could be a consequence of a superposition of two effects: the alternative presence of Ti and Zr ions and the loading due to the film nature (containing oxygen). Data about the binding energy of Pb in $PbZrO_3$ structures are not available in literature. The positions we identified for Ti 2p and Zr 3d peaks are those reported in literature to correspond to the stable oxides (TiO_2 and ZrO_2). We can then suppose that the chemical neighbour being the same for both, an interchange between these two elements (Ti and Zr) in the identified structure appears.

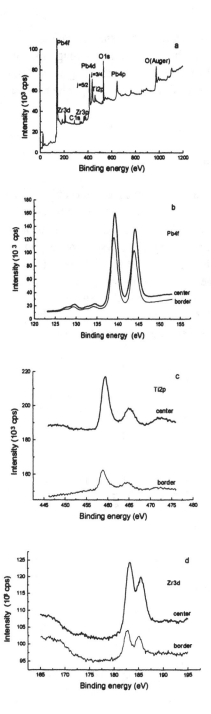

Figure 3. Survey XPS spectrum on the central zone of a PZT film (a) and XPS profiles of Pb 4f (b), Ti 2p (c) and Zr 3d (d) measured in the central zone of deposition and in an eccentric zone.

Both the Ti and Zr concentrations are decreasing from the centre to the border, but the ratio Ti/Zr is always the same, corresponding to the compound $Pb(Ti_{0.6}Zr_{0.4})O_3$ The Pb excess is found to be bonded with oxygen in PbO_2 compound. This is in agreement also with the results of Hau et al. [3], which evidenced Pb in excess on the border of PZT films.

5. Structural investigations

X-ray diffraction patterns analysis performed on central and eccentric zones of the PZT films, like in the previous sections, confirm the XPS results. The XRD spectra for the central zone and for an eccentric zone situated toward the border are presented in Fig. 4 a,b. The central zone contains only perovskite phase, with mainly (111) crystalline orientation (Fig. 4a). The pyroclore phase, if exists, is in a very small amount, not crystalline. In Fig. 4 b the corresponding spectra for a region situated toward the border shows that: 1) the film is thinner than in the central zone; 2) it contains still a quite big amount of piezoelectric PZT phase and 3) a pyroclore phase is present. Minor quantities of perovskite phase with different orientations (002) and (220) are also present.

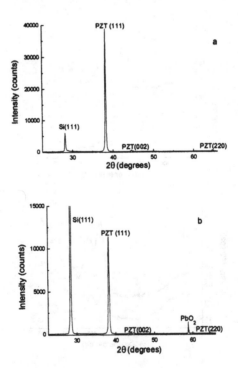

Figure 4. XRD diffraction pattern of PZT film for the central zone (a) and for an eccentric zone (b).

SEM images showed evidence of a fine microstructure with rounded grains of submicronic diameters. We did not find sensible differences between the SEM images taken on the central zone and on the eccentric one.

AFM topographic and lateral force images taken on PZT samples in different zones showed that their surface is uniformly covered with almost circular grains of PZT and it is homogeneous (as it has been evidenced in friction images). 3D islands with size ranging between few hundreds of nm and one thousand nm and with heights of about few hundreds nm have been observed. The height profile scan showed 3D islands with rounded smooth profile. The majority of islands had a height ranging between 100-300 nm. The statistical analysis gives a mean height of about 244 nm and a r.m.s. roughness of about 101.7 nm [17].

Regarding the presence of particulates on the film surface, SEM images showed an almost uniform distribution over a large deposition area. The size distribution of particulates has been also obtained from SEM images of the PZT film surface. Fig. 5 shows results for two different PZT films, obtained by PLD by using different targets. It can be observed (curve 2) that the largest number of particulates have a diameter of about 0.6 μm, but there is also a certain percentage of particulates with diameters higher than 2 μm. In a previous study [20] we showed that these particulates originate from the ablation conical features that are developed on the target during the interaction with the high energy laser beam. These large diameter particulates could be eliminated by using a special target with fine grains [20]. Indeed, the curve 1 in fig. 5 obtained for films deposited from this target shows that large diameter particulates are almost absent in this case.

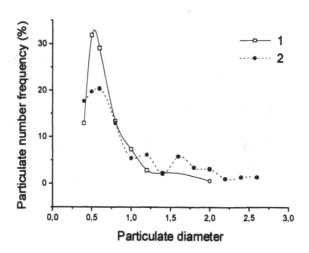

Figure 5. Size distribution of particulates found on the surface of films produced from two different targets: a) commercial PZT; b) PZT target with fine grains.

6. Piezoelectric and dielectric characterisation

The problem of uniformity of piezoelectric properties was less addressed, mainly due to the lack of local inspection techniques. We used the special device described in ref. [16] which allows to obtain local values in different zones of the film. Fig. 6 shows the variation of the longitudinal piezoelectric constant d_{33}, normalised to its highest value obtained in the centre of deposition ($d_{33}^0 \approx 42$ pC/N), measured at different distances from the centre of deposition. Its reduction with the deposition angle θ is less drastically than that of the thickness (reported for comparison on the same graph); indeed the decreasing of the piezoelectric constant should be more related to compositional or structural off-axis variations, rather than a size effect.

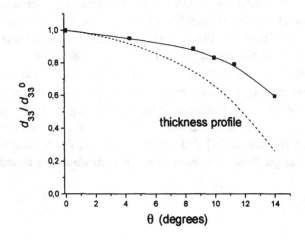

Figure 6. Normalized piezoelectric constant vs. deposition angle. The continuous line is a guide for eye. The dashed line represents the normalized film thickness vs. deposition angle.

Regarding the dielectric constant, measurements performed on different zones did not evidenced a clear spatial variation, but their accuracy was affected by both capacity and thickness determination errors. An average value of the relative dielectric constant $\varepsilon_r \approx 500$ was obtained.

7. Summary and conclusions

An important drawback of the piezoelectric films obtained by PLD is the spatial nonuniformity of their thickness and composition. By making local measurements on PZT films, we evidenced also structural and piezoelectric spatial variations.

Many applications of the piezoelectric thin films require a relatively large area with uniform properties. By using different correlated techniques like XPS, SIMS, XRD, SEM, AFM and piezoelectric measurements we defined an area of about 1.5 cm^2 (or a deposition angle of about 10°) where most of the emitted material is contained, the stoichiometry and structure are almost preserved and the piezoelectric constant variation is less than 20 %.

Nevertheless, substrate displacement devices should be incorporated in the design of future PLD industrial systems for deposition over large areas.

8. Acknowledgment

C.F, V.P. and D.M. gratefully acknowledge the support for this work from NATO-SfP Programme, SfP 97-1934.

9. References

1. Chrisey D.B. and Hubler G.K. (eds.) (1994) *Pulsed Laser Deposition of Thin Films*, J. Wiley&Sons, New York.
2. Saenger K.L. (1994) Angular Distribution of Ablated Material, in D.B. Chrisey and G.K. Hubler (eds.), *Pulsed Laser Deposition of Thin Films*, J. Wiley&Sons, New York, pp. 199-227, and references therein.
3. Hau S.K., Wong K.H., Chan P.W., Choy C.L. and Wong H.K. (1992) Angular distribution of XeCl laser deposition of Pb(Zr$_{0.48}$Ti$_{0.52}$)O$_3$ films, *J. Mater. Sci. Lett.* **11**, 1266-1268.
4. Lichtenwalner D.J., Auciello O., Dat R. and Kingon A.I. (1993) Investigation of the ablated flux characteristics during pulsed laser deposition of multicomponent oxides, *J. Appl. Phys.* **74**, 7497-7505.
5. Tyunina M., Wittborn J., Björmander C. and Rao K.V. (1998) Thickness distribution in pulsed laser deposited PZT films, *J. Vac. Sci. Technol.* **A16**, 2381-2384.
6. Weaver I. and Lewis C.L.S. (1996) Polar distribution of ablated atomic material during the pulsed laser deposition of Cu in vacuum: Dependence on focused laser spot size and power density, *J. Appl. Phys.* **79**, 7216-7222.
7. Venkatesan T., Wu X.D., Inam A. and Wachtman J.B. (1988) Observation of two distinct components during pulsed laser deposition of high T$_c$ superconducting films, *Appl. Phys. Lett.* **52**, 1193-1195.
8. Singh R.K., Holland O.W. and Narayan J. (1990) Theoretical model for deposition of superconducting thin films using pulsed laser evaporation technique, *J. Appl. Phys.* **68**, 233-247.
9. Ballesteros G.M., Afonso C.N. and Perrière J (1997) Angular distribution of oxide films deposited by multi- or single-component laser ablation, *Appl. Surf. Sci.* **109/110**, 322-326.
10. Szöreny T. and Ballesteros J.M. (1997) Dependence of the thickness profile of pulsed laser deposited bismuth films on process parameters, *Appl. Surf. Sci.* **109/110**, 327-330.
11. Antoni F., Fuchs C. and Fogarassy E. (1996) Analytical description of the film thickness distribution obtained by the pulsed laser ablation of a monoatomic target: application to silicon and germanium, *Appl. Surf. Sci.* **96/98**, 50-54.

12. Anisimov S.I., Bäuerle D and Lukyanchuk B.S. (1993) Gas dynamics and film profiles in pulsed laser deposition of materials, *Phys. Rev. B* **48**, 12076-12081.
13. Anisimov S.I., Lukyanchuk B.S. and Luches A. (1995) Dynamics of the three-dimensional expansion in a vapour produced by a laser pulse, *JETP* **81**, 129-131.
14. Kools J.C.S. (1993) Monte Carlo simulations of the transport of laser-ablated atoms in a diluted gas, *J. Appl. Phys.* **74**, 6401-6406.
15. Verardi P., Dinescu M., Craciun F. and Perrone A. (1998) In situ growth of ferroelectric lead zirconate-titanate thin films on Si substrates by pulsed Nd:YAG laser deposition, *Appl. Surf. Sci.* **127/129**, 457-459.
16. Verardi P., Dinescu M. and Craciun F. (1998) Characterization of microstructural and piezoelectric properties of oriented PZT thin films obtained by pulsed laser deposition, *J. Phys. IV France* **8**, Pr-9 121-124.
17. Craciun F., Verardi P., Dinescu M., Dinelli F. and Kolosov O. (1998) Early stages of growth and nanostructure of Pb(Zr,Ti)O₃ thin films observed by atomic force microscopy, *Thin Solid Films* **336**, 281-285.
18. Verardi P., Nastase N., Gherasim C., Ghica C., Dinescu M., Dinu R. and Flueraru C. (1999) Scanning force microscopy and electron microscopy studies of pulsed laser deposited ZnO thin films: application to the bulk acoustic wave (BAW) devices, *J. Crystal Growth* **197**, 523-528.
19. Verardi P., Craciun F., Mirenghi L., Dinescu M. and Sandu V. (1999) An XPS and XRD study of physical and chemical homogeneity of Pb(Zr,Ti)O₃ thin films by pulsed laser deposition, *Appl. Surf. Sci.* **138/139**, 552-556.
20. Craciun F., Verardi P., Dinescu M., Galassi C. and Costa A. (1999) Growth of piezoelectric thin films with fine grain microstructure by high energy pulsed laser deposition, *Sensors and Actuators*, in print.

PIEZOELECTRIC THIN FILMS FOR HIGH FREQUENCY ELECTROACOUSTIC DEVICES

P. VERARDI
CNR - Istituto di Acustica "O.M. Corbino"
Via Fosso del Cavaliere 100, 00133 Rome, Italy

Abstract

Development of SAW (Surface Acoustic Wave) and BAW (Bulk Acoustic Wave) devices operating at microwave frequencies, as well as transducers for micromechanical devices, is strictly connected with the performances of the piezoelectric layer employed for the transduction. The more interesting materials for these applications are ZnO, AlN and PZT. Regardless of the particular material considered, the goal is to obtain a film with physical and electrical characteristics very close to that of the corresponding bulk material. This can be reached by a proper choice of the deposition method and parameters. Common requirements for the deposition methods are the repeatability and the compatibility of the deposition environment with the substrate. Another important requirement is the possibility to deposit sequentially on the same chamber a metallic layer between the substrate and the piezoelectric film.

Between the various technologies proposed for this purpose, the most widely used is the sputtering (DC or RF Magnetron). More recently PLD (Pulsed Laser Deposition) has been successfully introduced and the results are very promising. With both of them, the required deposited compound can be obtained starting from a target of the same composition or better from a target of a pure element reacting with a specific ambient gas before the condensation phase (reactive deposition).

In this study we present the results of our experience on piezoelectric thin film deposition with this two techniques in order to put in evidence and compare their peculiarities.

1. Introduction

High frequency electroacoustics devices are widely applied in electronics circuits for a variety of applications [1]. The propagation velocity of an acoustic wave in solids is about five orders of magnitude less than of those electromagnetic waves. At a frequency of 1 GHz the wavelength of an acoustic wave is few microns. The thickness of the piezoelectric layer of a BAW device corresponds to half wavelength of the launched wave; for the operation at microwave frequencies (above 1 GHz) it means that it should

C. Galassi et al. (eds.), Piezoelectric Materials: Advances in Science, Technology and Applications, 285–292.
© 2000 *Kluwer Academic Publishers. Printed in the Netherlands.*

be less than 1μm tick. Also the generation of SAW on non-piezoelectric substrates needs an active layer that is a fraction of the acoustic wavelength. For these reasons, regardless of the type of the acoustic wave, acousto-electric devices operating at microwave frequencies needs a piezoelectric layer in form of thin film. The development of microwave acousto-electric devices is strictly connected with the evolution of the thin films deposition technologies. Different techniques have been applied for piezoelectric thin films. Reactive vacuum-evaporation method was the first used for the deposition of CdS and ZnO [2]. During the past three decades the best results have been obtained with sputtering in various configurations and modes of operation and it has been the most widely used method for deposition of ZnO [3, 4], AlN [5] and, more recently, PZT. Analysing the wide literature regarding sputtering deposition of ZnO and AlN, RF reactive magnetron sputtering results the most affirmed. In the case of ZnO the target is made of high purity Zn and the atmosphere pure Oxygen or a mixture of Oxygen and Argon with different ratios. The gas pressure is in the range of 10^{-3} mbar and substrate temperature from ambient up to 500 °C while the target-substrate distance is of the order of few centimeters. For AlN the target is made of high purity Al and the reactive gas is a variable mixture of Nitrogen and Argon. Deposited piezoelectric thin films are usually polycrystalline. Any misorientation between various crystallites affects the overall coupling coefficient and lead to the simultaneous generation of longitudinal and shear waves. All the deposition parameters affect the growth of the film especially the orientation and thus the electromechanical coupling coefficient. The definition and the optimization of the set of the deposition parameters is of fundamental importance for the final result. For the AlN with respect to ZnO there is another crucial parameter which affect the stoichiometry of the compound: the base pressure that can be reached by the pumping system before the introduction of the reacting gas. In fact the presence even of small percentage of Oxygen during the deposition can deteriorate the composition. For this it is mandatory a long accurate degassing of the deposition chamber and a residual pressure less than 10^{-6} mbar.

More recently PLD [6], an interesting emerging technique, has been successfully applied to the deposition of piezoelectric thin films [7, 8]. The most evident difference with respect to sputtering is the relatively small deposited area , due to the intrinsic mechanism of material ejection. On the other hand there are some advantages that make PLD competitive with sputtering. As an example complex materials like PZT can be deposited with the same composition of the target at a temperature compatible with Silicon devices technology.

2. Piezoelectric thin films deposition

Piezoelectric thin films, when used as active layers in high frequency BAW transducers need to be deposited not directly on the substrate but over a metallic thin film previously deposited on it. This is referred as the "bottom electrode"; a second metallic layer, the "top electrode" is deposited on the free surface of the piezoelectric film. The requirements for the electrodes are: high electrical conductivity, hardness, good adhesion to the substrate, orientation and lattice constant favourable to oriented growth of the piezoelectric layer. Moreover their thickness should be negligible with respect to acoustic wavelength. Usually electrodes consist on thermally evaporated Au (~1000 Å) over a flash of Cr. Pt instead of Au is preferable for PZT to avoid diffusion of the metal in it. In any case the best results can be obtained if the metallic layers are deposited with the same system used for the piezoelectric one in a sequential way. These considerations can be extended to SAW transducers.

2.1 SPUTTERING DEPOSITION

The sputtering deposition of oriented ZnO [9] and AlN [10] has been performed with a MRC 8620 system equipped with a 6" planar magnetron. In both cases we use reactive deposition starting from a pure Zn or Al target in pure O_2 or a mixture of Ar and N_2 respectively. With a proper choice of deposition parameters we obtain well oriented ZnO and AlN with the c-axis perpendicular to the substrate surface. Optimal deposition conditions are summarized on Table 1. An important factor is the process repeatability. For the AlN this is more critical because of possible strong internal stresses. This problem limits the maximum thickness of the film to 3 μm. Particular care should be taken to the cooling of the substrate after the deposition.

TABLE 1. Set of optimal sputtering conditions for ZnO and AlN.

parameter	ZnO	AlN
target	Zn 99.99 %	Al 99.99 %
reactive gas	O_2 99.99 %	50%N_2 :50%Ar
gas pressure	10^{-3} mbar	4×10^{-3} mbar
substrate temp.	250 °C	350 °C
Substrate-target dist.	8 cm	6 cm
RF power	500 W	500 W
substrate	Silicon	Silicon

2.2 PLD DEPOSITION

The PLD deposition setup is schematically represented in fig. 1. It consists on a stainless steel chamber connected with a pumping system able to pump down up to 10^{-6} mbar. This includes a turbomolecular pump in series with a rotary pump. The beam of the laser is focalized by a simple lens and then feed through a quartz window into the chamber pointing to the target surface with an angle of 45°. The target is rotating at few rpm and simultaneously translated by an external motor in order to sweep the whole target surface and thus avoiding crater formation.

288

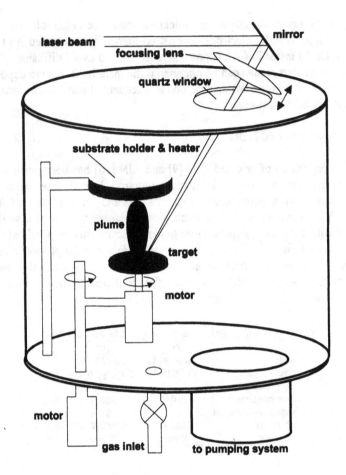

Figure 1. Schematic of the PLD deposition setup

The laser is a Nd-YAG λ=1.06 μm, t_{FWHM} =10 ns, 0.3 J max. energy /pulse, repetition rate=10 Hz. The fluence after the localisation is 25 J/cm^2. The gas necessary to the deposition is admitted into the chamber via a needle valve with a micrometric control. The desired dynamic pressure is reached by an appropriate setting of a throttle valve, in front of the main pump, and the needle valve. The gas pressure during the deposition is sensed by a capacitive gauge. The substrate is held parallel to the target surface and can be heated up to 500 °C. The temperature is monitored by a thermocouple and kept constant within +/- 1 °C by an electronic controller.

2.2.1. *ZnO Deposition*
Deposition of highly oriented ZnO films up to few microns have been obtained by reactive PLD from a pure Zn target in oxygen reactive atmosphere. The optimized set of deposition conditions are reported in Table 2. In Fig. 2 is shown a SEM cross-section of a ZnO film deposited on Si(100). It is clearly visible the columnar structure of the crystallites oriented perpendicularly to the substrate. In order to demonstrate good

quality of the piezoelectric properties, these films have been deposited on Silicon and Sapphire and employed to realize BAW transducers operating in the GHz range [7].

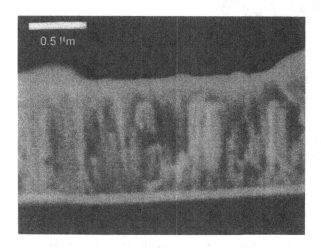

Figure 2. SEM cross-section of ZnO grown on Si(100)

These tests confirm the good quality of the deposited films and encourage further improvements.

2.2.2. *AlN Deposition*
The growth of AlN has been achieved by reactive PLD of a pure Al target in Nitrogen atmosphere. The set of conditions reported in Table 2. is referred to AlN deposited on Si(100) [11]. XRD analysis revealed AlN grown with strong (101) orientation. By changing deposition conditions on different substrates others orientations have been obtained. Realisation of BAW and SAW transducers requires the orientation (002). At present we did not found the conditions which favour this orientation and further studies are in progress on this way.

2.2.3. *PZT Deposition*
Oriented PZT thin films have been obtained by PLD of a commercial PZT4 plate in Oxygen ambient [8]. In Fig. 3 a) is shown a SEM image of the free surface of a PZT film deposited on Si(111), where a relatively compact structure made of submicronic diameter grains can be observed. In Fig. 3 b) is reported an XRD pattern of the same sample. It is clearly evident the peak of the (111) crystalline orientation. The optimal deposition conditions are reported on Table 2. With these conditions all the deposited films were piezoelectric without poling.

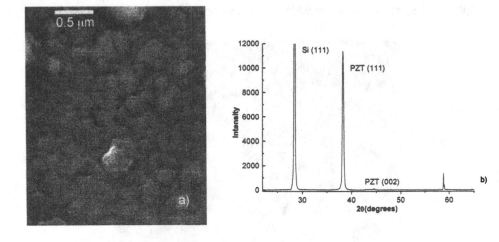

Figure 3. a) SEM image of surface of PZT deposited on Si(111) and b) XRD of the same sample.

TABLE 2. Set of optimal PLD conditions for various piezoelectric films

Parameter	ZnO	AlN	PZT
Target	Zn 99.99 %	Al 99.99 %	PZT 4 plate
Gas mixture	O_2 99.99 %	N_2 99.99 %	O_2 99.99 %
Gas pressure	10^{-1} mbar	5×10^{-2} mbar	150 mbar
Target-substrate dist.	6 cm	4 cm	4 cm
Substrate temp.	250 °C	280 °C	375 °C
Deposition rate	0.2 Å/pulse	0.2 Å/pulse	0.5 Å/pulse
Substrate	Silicon	Silicon	Silicon

3. Piezoelectric thin film characterization

The most important techniques used to characterize deposited films are: optical microscopy, SEM, XRD, XPS, SIMS. They give information on the morphology, structure, composition and orientation. Beside this techniques, for the piezoelectric films the key analysis is the measure of the piezoelectric constants d_{ij} and the electromechanical coupling coefficient k_t. Between the various proposed techniques, we prefer that described in [13] because of its ease use and good estimation of the coefficient d_{33}. This gives a preliminary information on the behaviour of the film as active layer in BAW or SAW transducers.

In Fig. 4 is shown its principle of operation. A narrow electrical pulse is applied to the transducer that converts it on an acoustic wave inside the metallic rod. This mechanical perturbation is applied, through a coupling liquid, perpendicularly to the surface of film under test, which produces an electrical charge displacement at the film surfaces.

The voltage produced by these charges is observed and measured by an oscilloscope. By comparing the voltage produced on the same conditions by the film and by a known piezoelectric plate it is possible to estimate the d_{33} of the film.

For the measurement of the k_t as well as the conversion loss of BAW transducers [3, 12] we use a network analyser HP 8753A with time domain facility in order to simulate the well known pulse-echo technique. For ZnO the measured value of d_{33} was 8 pC/N that is very close to the value for bulk ZnO. From the measurements on a BAW transducer operating around 2 GHz, a k_t=0.25 and conversion loss of 9.5 dB were found.

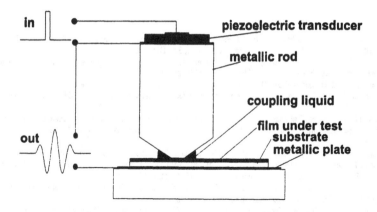

Figure 4. Principle of operation of the device for measuring d_{33}.

For PZT film, at present, only direct measurements of d_{33} have been done. In this case we found a value of 40 pC/N that is about 20% of the PZT bulk constant.

Conclusions

results of the utilization of PLD to the deposition of piezoelectric thin film for high frequency applications have been compared with those obtained with RF sputtering. Although PLD in this field eds further studies, the first results are very promising.

Acknowledgements

he author wishes to thank Dr. Maria Dinescu and Dr. Floriana Craciun for their helpful collaboration. he author gratefully acknowledge the support for this work from NATO-SfP Programme, SfP 97-1934

292

6. References

1. Krishnaswamy, S.V. and McAvoy, B.R. (1993) Thin films in Microwave Acoustics, in Fracombe, M.H. and Vossen J.L. (eds.), *Physics of Thin Films*, , Academic Press Inc., San Diego, **17**, 145-224.
2. Foster, N.F., Coquin, G.A., Rozgonyi, G.A., Vannatta, F.A. (1968) Cadmium sulfide and Zinc Oxide Thin-film Transducers, *IEEE Trans. on Son. and Ultras.*, SU-15/1, 28-41.
3. Foster, N.F. and Rozgonyi, G.A., (1966) Zinc Oxide Film Transducer, *Appl. Phys. Lett.*, **8/9**, 221-223.
4. Hickernell, F.S. (1985) Zinc Oxide Films for Acoustoelectric Device applications, *IEEE Trans. on Son. and Ultras.* SU-32/5, 621-629.
5. Okano, H., Tanaka, N., Takahashi, Y., Tanaka, T., Shibata, K., Nakano, S. (1994) Preparation of aluminum nitride thin films by reactive sputtering and their applications to GHz-band surface acoustic wave devices, *Appl. Phys. Lett.*, **64/2** 166-168.
6. Chrisey, D.B. and Hubler, G.K. (eds.) (1994) *Pulsed Laser Deposition of Thin Films*, John Wiley & Sons, New York
7. Verardi, P., Dinescu, M. (1995) Characteristics of ZnO Thin Film Transducers Deposited by Pulsed Laser Deposition *1995 IEEE International Ultrasonic Symp. Proc. Seattle (WA)*, 1015-1018.
8. Verardi, P., Dinescu, M., Craciun, F., Sandu, V. (1997) Oriented PZT thin films obtained at low substrate temperature by pulsed laser deposition, *Thin Solid Films*, **311**, 171-176.
9. Caliendo, C., D'amico, A., Verardi, P., Verona, E. (1988) Surface Acoustic Wave H$_2$ Sensor on Silicon Substrate *1988 IEEE International Ultrasonics Symposium Proc. ", Chicago (IL)*, 1988, 569.
10. Caliendo, C., Saggio, G., Verardi, P., Verona, E. (1993) Piezoelectric AlN film for SAW devices applications, *1993 IEEE International Ultrasonics Symposium Proc., Baltimore (MD)*, 249.
11. Verardi, P., Dinescu, M., Gerardi, C., Mirenghi, L., Sandu, V. (1997) AlN thin film deposition by laser ablation of Al target in nitrogen reactive atmosphere, *Applied Surface Science*, **109/110**, 371-375.
12. Reeder, T.M. (1967) Microwave Measurement of Thin-Film Transducer Coupling Constant, *Proc of The IEEE, Proc Letters*, June 1967, 1099-1101.
13. Vyun, V.A., Umashev, V.N., Yakovkin, I.B. (1986) *Sov. Phys. PTE* ,6 ,192.

PHOTOELECTRIC AND DIELECTRIC SPECTROSCOPY STUDY OF PZT THIN FILM ELECTRONIC STRUCTURE

V.K. YARMARKIN
A.F. Ioffe Physical-Technical Institute
194021 St.Petersburg, Russia

1. Introduction

Electronic structure of ferroelectric thin films, characterized by electric potential and conductivity distribution through the film thickness, and the presence and the location of trap levels in a band gap influences appreciably the exploiting parameters and reliability of various ferroelectric thin-film devices including capacitors, piezo-, pyro-, electro- and acusto-optic- devices, memories etc. [1].

The existence of non-polarization-dependent steady- photoresponse to UV/visible illumination of metal-PZT-metal structures was established earlier [2, 3]. Different opinions were expressed about the physical nature of this photoresponse. To elucidate it and to ascertain the possibility of obtaining information on PZT thin films electronic structure, a photoelectric and dielectric spectroscopy study of metal-PZT-metal thin-film structures was performed in this work.

2. Experimental procedure

The study was carried out using thin-film capacitor structures prepared on the surface of pre-oxidized Si (100) wafers [4]. The bottom electrode (100-nm Pt on a 10-nm Ti) was sputter-deposited. The sol-gel PZT film thickness was ~200 nm. Ni or Pt top electrode with a diameter of about 0.2-0.5 mm and a thickness of 100 nm were deposited on the film surface by thermal evaporation through a mask. The structures capacitance measured at the frequency of 1 kHz, 40 mV lay in the range 0.25-1.5 nF, while loss tangent equals $(3-5) \cdot 10^{-2}$.

Illumination with non-polarized radiation in the wavelength range of 300-390 nm with an intensity of 0.1-1 mW/cm^2 was carried out using arc mercury lamp with the glass filters. For local optical measurements, a focusing system of MIM-7 microscope giving a light spot with a diameter of about 0.2 mm onto the sample was used. In the range of 400-1200 nm a glow-lamp combined with the glass and the water filters, or with DMR-4 monochromator, was used. The intensity of illumination was varied using mesh filters and monitored by a silicon photo-diode. The electrical resistivity of the structures measured between the top and the bottom electrodes using d.c. voltage of ±1.3 V was ~$5 \cdot 10^{11}$ Ohm in the darkness, and ~$5 \cdot 10^9$ Ohm on illumination (in the 300-

C. Galassi et al. (eds.), Piezoelectric Materials: Advances in Science, Technology and Applications, 293–300.
© 2000 Kluwer Academic Publishers. Printed in the Netherlands.

390 nm range with intensity of 1 mW/cm^2), irrespective of polarity of the applied voltage. The measurements of the short circuit photo current and open circuit photo voltage were carried out by an EM-1electrometer, using 10^9-Ohm and 10^{16}-Ohm load resistors, respectively.

The frequency dependencies of the structures complex capacitance were studied at frequencies from 100 Hz to 100 MHz. Below 1 MHz, we used bridges (TM-351-G, MTsE-12A, and E7-12) which allow direct measurement of the capacitance and loss tangent, while within the 60-400 kHz range the measurements were done with a E9-4 Q-meter, and above 0.5 MHz, with an VM-508 impedance meter. The impedance of the structures $Z(\omega) = |Z| \cdot \exp\{j \cdot \varphi\}$, capacitance $C = Im(Z)/\omega \cdot |Z|^2$ and loss tangent $tg\delta = -Re(Z)/Im(Z)$, were calculated taking into account spurious coupling between elements of the circuit and skin effect in connecting wires [5]. In the above relations, f, $\omega = 2\pi f$ and φ are the frequency, circular frequency and phase of the measuring voltage, respectively.

3. Results and discussion

3.1. PHOTOELECTRIC STUDY OF THE METAL-PZT-METAL STRUCTURES

Fig. 1 shows short circuit photo current I_{sc} and open circuit photo voltage modulus U_{oc} vs. time dependencies measured on Ni-PZT-Pt thin-film structures with 300-390 nm illumination of various intensities. These dependencies can be described by exponential function

$$y(t) \sim [1 - \exp(-t/\tau)] \tag{1}$$

during the rising stage, after the switching on the illumination, and by

$$y(t) \sim \exp(-t/\tau) \tag{2}$$

on the stage of decrease, after the illumination was switched off. The values of τ found by approximation of an experimental curve by dependencies (1), (2) using least square procedure, differ for $I_{sc}(t)$ and $U_{oc}(t)$, and also for their various stages (see Table 1).

Within the experimental error of about 10 %, no difference was observed between the results obtained for the Ni and Pt top electrodes. The polarity of signals in all cases corresponded to the occurrence of negative potential on an illuminated top electrode. Neither the sign, nor the values of I_{sc} and U_{oc} depend on the direction of film remnant polarization. Investigation of photo current I_{ph} and photo voltage U_{ph} dependencies on the electrometer load resistance R_L, with the illumination intensity of 1 mW/cm^2, has shown that I_{ph} values do not depend practically on R_L in the range of $R_L \leq 10^9$ Ohm, and U_{ph} values do not depend on R_L in the range of $R_L \geq 10^{12}$ Ohm; it was a reason of the above-mentioned choice of the R_L values used by measuring of I_{sc} and U_{oc}. Fig. 2a shows the dependencies of the steady state values of I_{sc} and U_{oc} on intensity of illumination J, in the range from 0.014 to 1 mW/cm^2. The dependence $I_{sc}(J)$ is linear, and the change of U_{oc} does not exceed 20% with the change of intensity in ~70 times.

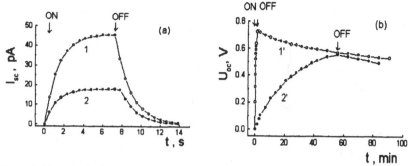

Figure 1. Short circuit photo current (a) and open circuit photo voltage modulus (b) vs. time dependencies for Ni-PZT-Pt thin film structures with 300-390 nm, 1 mW/cm² (1,1') and 0.05 mW/cm² (2,2') illumination.

With transition into the wavelength range of $\lambda \geq 400$ nm, exceeding absorption edge in PZT thin films [6], the described stationary photo response was not observed.

TABLE 1. Time constants τ (in seconds) of $I_{sc}(t)$ and $U_{oc}(t)$ dependencies.

J, mW/cm²	$I_{sc}(t)$		$U_{oc}(t)$	
	rising	decrease	rising	decrease
0.05	2.4	2.6	55	$1.5 \cdot 10^4$
1.0	1.4	1.7	20	$2.0 \cdot 10^4$

In 300-390 nm wavelength range a restoration of the illumination induced U_{oc} values was observed after terminating illumination and subsequently keeping the structures in a short-circuited state in darkness for a relatively short time, from several seconds to several minutes (Fig.2b). It was established that the U_{oc} restoration kinetics can be described by the exponential function (1), time constant τ being increased when the duration of keeping the structures in darkness was increased.

In the 370-1200 nm range the weak non-stationary photo current was observed (Fig. 3). This current, being contrary to that generated at 300-390 nm, corresponds to the positive potential on the top electrode. It falls down to zero during 1-3 min. After terminating the illumination, photo current of opposite direction appears which also falls down to zero. One can see from Fig. 3 that, with displacing of short-wave limit of illumination from 500 up to ~370 nm, the above mentioned stationary photo current, corresponding to negative potential on the top electrode, is superimposed on a non-stationary photo current. The preliminary short-term illumination of the structures by a short-wave radiation ($\lambda = 300$-390 nm) with intensity of about 0.1 mW/cm² for 1 min results, in the range of 500-1200 nm, in an increase of non-stationary photo currents of both directions by ~1.5 times. The non-stationary photo current vanished when the upper limit of short -wave illumination was moved to the range $\lambda \geq 750$ nm (i.e. when the photon energy becomes less than ~1.6 eV), thus indicating that it was not pyro-electric in it's origin. As in this work opaque top electrodes were used and the

296

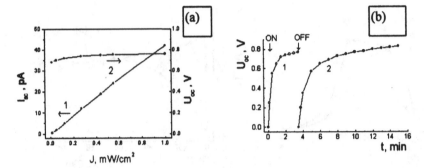

Figure 2. Short circuit photo current (1) and open circuit photo voltage modulus (2) vs. intensity of illumination at 300-390 nm (a), and U_{oc} modulus vs. time dependencies (b) with illumination (1) and after the terminating illumination and subsequent keeping Ni-PZT-Pt structures in darkness in the short-circuited state.

conductivity of the film was small, it is possible to assume that, in the case where the illuminating area considerably exceeds the top electrode area, the photo response develops only in a narrow ring contiguous to the electrode edge. To estimate the width of this ring, the photo current was measured with λ = 300-390 nm as a function of the location of the light spot (0.2 mm in diameter) on the electrode (0.5 mm in diameter). It was observed, that the photo current is proportional to the length of the covered arch of a circle limiting the electrode; the width of the ring where the photo response develops does not exceed 0.01 mm.

The observation of a sharp threshold of the stationary photo response in a wavelength range corresponding to the PZT absorption edge (determined in [6]) allows to conclude that generation of electron-hole pairs occurs in the volume of films. Among the possible origins of the photo-emf in ferroelectrics [7], the following can be excluded in our case: linear and circular photogalvanic effects can be excluded due to their dependencies on the spontaneous polarization, optical rectification does not give rise to a steady-state current or an emf, and the photon drag effect is insignificantly small [7], in comparison with the observed values of U_{oc}. For an estimation of the possible contribution of Dember-emf U_D we make use of the formula [8]

$$U_D = \frac{kT}{e} \cdot \frac{(\mu_n - \mu_p)}{(\mu_n + \mu_p)} \cdot \ln \frac{\sigma_T}{\sigma_B} \tag{3}$$

for the case of bipolar conductivity of semiconductor (here μ_n and μ_p are the mobilities of electrons and holes, σ_T and σ_B are the conductivities of the film in the regions of top and bottom electrodes, respectively). Assuming that σ_T and σ_B values are proportional to intensity of illumination in the corresponding film regions, and taking into account that the reduction of this intensity in 200 nm thick PZT film due to absorption should not exceed 2 orders in magnitude [9], one can see, according to (3), that U_D values cannot exceed 0.12 V. These values are considerably less than the measured U_{oc} values (0.7- 0.9) V, therefore the Dember effect does not contribute significantly.

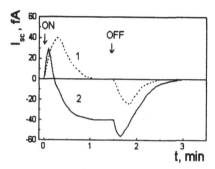

Figure 3. Short circuit photo current vs. time dependencies for Ni-PZT-Pt thin-film structures with illumination in 500-1200 nm (1) and 370-1200 (2) nm wavelength ranges.

It means that the most probable reasons of the observed photoelectric effects in PZT films can be the p-n junction in the film volume, and also the potential barriers at film-electrode interfaces.

The presence of p-n junction in p-type sol-gel PZT films can be caused by formation at their free surface of oxygen-deficient layer having the conductivity of n-type [10]. With illumination of such a p-n junction from the top electrode side, using a photon energy higher than the film's band gap, a negative potential should appear on the top electrode [8], as it was observed in the 300-390 nm wavelength range. Relatively large values of τ obtained (see Table 1) are due to the measuring circuit time constants RC. We believe that this is the reason of the observed exponential character of both $I_{sc}(t)$ and $U_{oc}(t)$ dependencies on signal rising and decrease stages.

The above-mentioned U_{oc} restoration effect in darkness, after keeping the structures in a short-circuited condition for short time, can be explained as a result of the relatively slow (with a time constant larger than the time of U_{oc} restoration) flow of the charge from the p-n junction to the «geometrical» capacity of the structure through the film thickness separating p-n junction from electrodes.

Relatively small non-stationary photo effect observed at 500-1200 nm can be caused by partitioning free carriers released with illumination from the deep trap levels located in PZT band gap ~1.5 eV under the bottom of the conduction band, due to the Shottky barrier electric field near the border of top electrode. It is not a pyro effect, because of the above mentioned spectral threshold near 750 nm. The existence of similar (as in the present case) barrier of 1.83 eV at platinum electrode in the metal-PLT-metal thin-film structures was established in [11]. The above mentioned increase of a photo current in 500-1200 nm wavelength range after preliminary illuminating the structures at 300-390 nm can be explained as a result of increasing trap levels occupancy in the films.

3.2. DIELECTRIC SPECTROSCOPY OF NI-PZT-PT THIN FILM STRUCTURES

The values of C and tan δ over 100 Hz - 100 MHz range are presented in Fig. 4. The pattern of the frequency dependencies, which are characteristic for these structures, is typical for relaxation processes. The experimental values of C and tan δ were used to calculate the complex capacitance of the structures $C^* = C \cdot (1 - j \cdot \tan \delta)$ and to determine

298

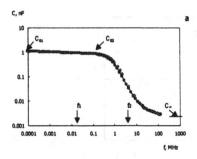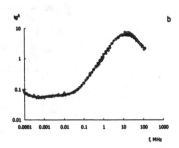

Figure 4. Frequency dependence of capacitance (a) and loss tangent (b) for Ni-PZT-Pt thin film structures.

the real (C') and imaginary (C'') parts of C^*, to be subsequently employed in the data treatment based on the standard Cole-Cole diagrams. The diagram shown in Fig. 5a evidences the existence of at least two relaxation processes in the structures. Bearing this in mind and invoking the d.c. conductivity data, the frequency dependence of C^* was fitted by a function containing two relaxation-type terms and a term taking into account the dielectric losses due to the d.c. conductivity

$$C^* = \varepsilon^x \cdot C_0,\qquad(4)$$

where

$$\varepsilon^x := \varepsilon_\infty + \frac{\varepsilon_{01} - \varepsilon_{02}}{1 + \left(j\omega\tau_1\right)^{1-\alpha_1}} + \frac{\varepsilon_{02} - \varepsilon_\infty}{1 + \left(j\omega\tau_2\right)^{1-\alpha_2}} - \frac{j}{\omega\rho},$$

$C_0 = \varepsilon_0 \cdot A/d$, ε_0 is the dielectric permittivity of vacuum, and A, d and ρ are the area, thickness and effective bulk resistivity of the structure, respectively.

The parameters entering Eq. (4) were found (Table 2) by fitting the calculated curves to experimental points in the Cole-Cole diagram (Fig. 5a) using as the zero approximation for C_∞, C_{01}, C_{02}, $\tau_1 = 1/f_1$, and $\tau_2 = 1/f_2$ the values derived from the frequency dependencies of the capacitance and dielectric losses plotted in Fig. 4. The rms deviation of experimental from calculated data was ~12 %.

The absence of any dispersion of C^* in bulk ceramic PZT samples of the same composition permits to conclude that the features observed in the dielectric spectra can be assigned to interlayer Maxwell-Wagner relaxation originating from the inhomogeneous structure of the films along their thickness. The high-frequency process characterized by a narrow distribution of relaxation times ($\alpha_2 < 0.1$) can be associated with the electrode-film interface, and the low-frequency one, exhibiting a broad relaxation-time spectrum ($\alpha_1 \approx 0.5$), with the presence in the near-electrode film regions of layers with a higher conductivity due to oxygen deficiency [10]. Applying the model of dielectric relaxation in distributed dielectric layers [12] in our case, a distribution function g(t) of the layer's coordinate-dependent time constant t = $\varepsilon_0\varepsilon(x)\rho(x)$ was calculated, using experimental data in Figs. 4 and 5a for the low frequency relaxation process. To estimate the variation of electrical resistivity $\rho(x)$

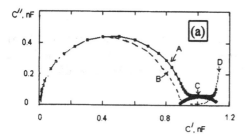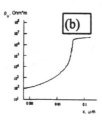

Figure 5. Cole-Cole diagram (a) and distribution of bulk resistivity trough the thickness of the PZT film.
A - experimental data, B, C, D - components of the relaxation process: B - high-frequency process,
C - low-frequency process, D - contribution due to dc conductivity.

through the film thickness, the dielectric permittivity $\varepsilon(x)$ was approximated by a monotonously increasing function with only two of its values fixed, namely, of the surface, $\varepsilon_S = 40$ (for the pyrochlore phase in near-surface regions, as indicated by means of x-ray diffraction and electron-microscopy [13]), and $\varepsilon_B \approx 800$ in the bulk of the film.

TABLE 2. Experimental values of the parameters characterizing the dielectric relaxation in Ni-PZT-Pt structures.

C, nF			τ, s		α		ρ
C_∞	C_{01}	C_{02}	τ_1	τ_2	α_1	α_2	Ohm·cm
0.002	1.15	0.875	$5.56 \cdot 10^{-5}$	$2.17 \cdot 10^{-7}$	0.49	0.05	$5 \cdot 10^8$

The $\rho(x)$ relation presented in Fig. 5b was calculated from the experimentally measured function g(t) by varying the $\varepsilon(x)$ function and taking into account of the above restrictions; the smoothness of $\rho(x)$ function was taken as a good fit-criterion. One can see from Fig. 5b, that in the film thickness region between x = 0.001 and ~0.03 μm, the conductivity is considerably higher than in the $0.03 \le x \le 0.2$ μm region, and is located (as it was discussed in detail in [5]) near the top Ni electrode in the Ni-PZT-Pt structures.

4. Conclusion

A highly inhomogeneous electronic structure of the PZT sol-gel thin ferroelectric films with metal electrodes has been evidenced by the present investigations. This is due to the presence of near-electrode Shottky barriers, p-n junction in the film's volume, relatively high-conductive layer near the top Ni electrode in Ni-PZT-Pt structures and deep trap levels in the PZT band gap.

300

5. Acknowledgement

The author is grateful to Professor V.V. Lemanov, Dr. B.M. Goltsman, Dr. M.M. Kazanin and S.P. Teslenko for their generous help in his work. This work was supported by the Russian Foundation for Basic Research (Grant 98-02-18164).

6. References

1. Scott, J.F. (1998), The physics of ferroelectric ceramic thin films for memory applications, *Ferroelectrics Review* 1, 1-129.
2. Brody, P.S. and Rod, B.J. (1992) Photovoltages in ferroelectric films, *Integrated Ferroelectrics* 2, 1-11.
3. Thakoor, S., Perry, J. and Maserjian, J. (1994) An optical probe for ferroelectric thin film, *Integrated Ferroelectrics* 4, 333-340.
4. Lemanov, V.V., Zaitseva, N.V., Shtel'makh, S.V., Motorny, A.V. and Yarmarkin, V.K. (1995) Structure and properties of sol-gel PZT ferroelectric films, *Ferroelectrics* 170, 231-236.
5. Yarmarkin, V.K. and Teslenko, S.P. (1998) Dielectric relaxation in thin film metal-ferroelectric-metal structures, *Phys.Tverd.Tela* (St. Petersburg) 40, 1915-1918 [*Phys. Solid State* 40, 1738-1741].
6. Sreenivas, K., Sayer, M. and Garrett, P. (1989) Properties of d.c. magnetron-sputtered lead zirconate-titanate thin films, *Thin Solid Films* 172, 251-267.
7. Ruppel, W., Von Baltz, R. and Wurfel, P. (1982) The origin of photo-emf in ferroelectrics and non-ferroelectric materials, *Ferroelectrics* 43, 109-123.
8. Rivkin, S.M. (1963) *Photoelectric phenomena in semiconductors*, GIFML, Moscow.
9. Thakoor, S., and Maserjian, J. (1994) Photoresponse probe of the space charge distribution in ferroelectric lead zirconate-titanate thin film memory capacitors, *J.Vac.Sci.Technol.* A12, 295-299.
10. Mihara, T, Watanabe, H., Yoshimori, H., Paz De Araujo, C.A., Melnick, B. and McMillan, L.D. (1992) Process dependent electrical characteristics and equivalent circuit model of sol-gel based PZT capacitors, *Integrated Ferroelectrics* 1, 269-291.
11. Day, S.K., Lee, J.J. and Alluri, P. (1995) Electrical properties of paraelectric $(Pb_{0.72}La_{0.28})TiO_3$ thin films with high linear dielectric permittivity: Shottki and Ohmic contacts, *Jpn. J. Appl. Phys.* 34, 3142-3152.
12. Kita, Y. (1984) Dielectric relaxation in distributed dielectric layers, *J. Appl. Phys.* 55, 3747-3755.
13. Lemanov, V.V., Mosina, G.N., Sorokin, L.M., Shtel'makh, S.V. and Yarmarkin, V.K. (1996) Surface-layer structure of thin ferroelectric PZT films, *Phys.Tverd.Tela* (St. Petersburg) 38, 3108-3115 [*Phys. Solid State* 38, 1700-1703].

THE INFLUENCE OF INTERFACE ON THE SPONTANEOUS POLARISATION IN PbTiO$_3$ THIN FILMS DEPOSITED ON A SILICON SUBSTRATE

I. BOERASU, L. PINTILIE
National Institute for Materials Physics
Bucharest – Romania
E-mail: boerasu@alpha1.infim.ro

Abstract

Metal/ Ferroelectric/ Semiconductor (MFS) heterostructures have been of interest in the last decade due to possibility of their potential integration in classical silicon technology and because their potential use in memory devices. Research in this field has lead to new generations of sensors and actuators, laser modulators and deflectors, non-volatile random access memories (NRAM's). Unfortunately many at these types of devices remain in the prototype phase because of the impossibility to know and control phenomena which appear at the ferroelectric/ semiconductor interface.

The present paper   presents some results on the investigation of a PbTiO$_3$/ Si heterostructure. A study of the film occurring at the interface and how it influences the ferroelectric properties of the heterostructure can clarify some problems of the reproducibility of the parameters of a MFS heterostructure.

1. Introduction

Ferroelectric thin film received at increasing attention due to their potential use in the next generation at electronic devices, such as ferroelectric memories FRAM [1], pyroelectric detectors [2], dynamic access memories DRAM or static memories SRAM [3], wave guide [4], strong dielectric capacitors [5] and micromotors [6]. In this paper we present a few results of our study of a metal/ ferroelectric/ semiconductor heterostructure.

The PbTiO$_3$ ferroelectric thin film was deposited by a sol-gel technique on a silicon substrate with and without a native SiO$_2$ film. Under investigation is PbTiO$_3$ because it has a large spontaneous polarisation and a high Curie temperature. Wen lead titanate is deposited on a Si substrate with the native SiO$_2$ native film removed we have observed the occurrence of a new film at the ferroelectric-semiconductor interface. It is important to know the influence of the new film on the ferroelectric properties but also it is important to understand the compositional modifications of these materials when

301

C. Galassi et al. (eds.), Piezoelectric Materials: Advances in Science, Technology and Applications, 301–308.
© 2000 *Kluwer Academic Publishers. Printed in the Netherlands.*

they are in form of a film integrated within a M/ F/ S heterostructure. In this study, we have investigated the spontaneous polarisation as a function of annealing temperature.

2. Experimental Procedure

PT ferroelectric thin films can by obtained by physical processes (laser ablation, chatodic pulverisation) [7] or chemical processes: sol-gel and metalorganic vapour deposition MOCVD[8]. We used sol-gel method. The samples were obtained by the sol-gel deposition of a $PbTiO_3$ on p-type 3'' silicon wafers (111) ($\rho = 2 \div 5$ Ωcm) with and without native SiO_2 film removed. The native SiO_2 film was removed from the substrate by chemical etching before the PT thin film deposition. The raw materials for the sol-gel method were the lead acetate and titanium isopropoxide and the solvent was 2-methoxiethanol. The PT film were deposited by spinning at 2 000rpm for 60s, followed by pyrolysis for 2min. in air at 300 °C. To obtained a 0. 8µm final thickness for the amorphous film, the sequence spinning-pyrolisis was repeated 10 times. Finally, the PT thin film was crystallised by conventional thermal annealing in air, for 30min. at temperatures ranging from 575 °C to 650 °C.

Structural and electrical characterisation was performed on the MFS structure. The MFS heterostructures were obtained by deposition of aluminium electrodes of about 2000A thickness on the back silicon surface and semitransparent gold electrodes, of about 600 thickness, on the top PT surface. The gold electrode was shaped intro rectangles of 1. 25 x 1. 5 mm^2 by evaporation through a metallic mask. The geometry of the haterostructureis presented in Figure1.

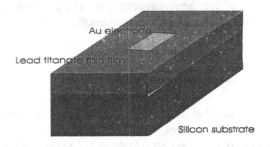

Figure 1. The MFS structure

Electron Dispersive X-ray Spectroscopy (EDXS), X-Ray Diffraction (XRD), Scanning Electron Microscopy (SEM), Atomic Force Microscopy (AFM), and electrical measurements (capacitance-voltage characteristics) made the study of the ferroelectric-semiconductor heterostructure.

3. Results and Discussions

3.1 STRUCTURAL ANALYSIS

3.1.1 Electron Dispersive X-ray Spectroscopy analysis

The first structural studies on the deposited PT thin film have been the EDXS investigation using a Jeol 6300A scanning electron microscope. Energy Dispersive X-Ray Spectroscopy revealed a film with a good crystaline structure and single phase ($PbTiO_3$) having a Pb/ Ti ratio about 0. 9 which lead to the conclusion of a good stoichiometry.

3.1.2   X-Ray Diffraction analysis

The XRD study was done using a DRON-02 diffractometer with a $CuK\alpha$ radiation (λ = 1. 5432nm). The analysis of the diffraction patterns revealed a perovskite structure and the relatively large peaks indicate small crystallite dimension (Figure 2).

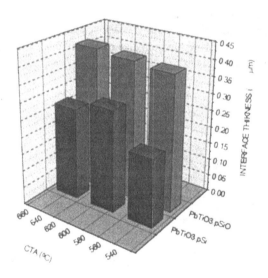

Figure 2. The XRD patterns for MFS heterostructure

3.1.3   Scanning Electron Microscopy analysis

The SEM image (Figure 3) was performed using a Jeol 6300A scanning electron microscope and shows that the final thickness of the film is about 0. 8μm. It also reveals for PT deposited on Si substrate with native SiO_2 film removed, the occurrence of an interface film between the silicon substrate and the lead titanate thin film. The SEM pictures show that the thickness of the interface film increases with increasing the PT annealing temperature.

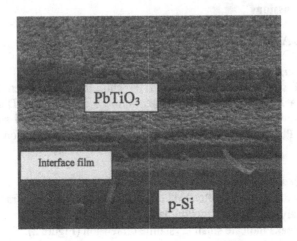

Figure 3. The SEM image for PbTiO₃ / p-Si annealing at 650 °C/ 30 min.

3.1.4 Atomic Force Microscopy analysis

The AFM was performed using a Jeol 6300A scanning electron microscope and analysis had revealed a crystallite medium size of about 50 ÷ 75 nm, depending on the annealing temperature (figure 4).

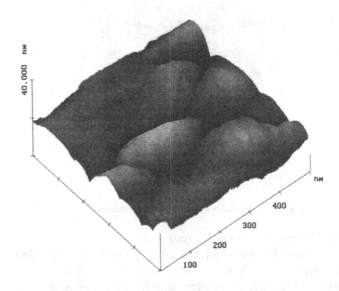

Figure 4. AFM for PT/ p-Si, 650 °C/ 30 min.

3.1.5 Capacitance-voltage characteristics

The capacitance-voltage characteristics of the PT/ Si heterostructures were determined using a Hewlett Packard 4194A Impedance/ Gain Analyser (Figure 5).

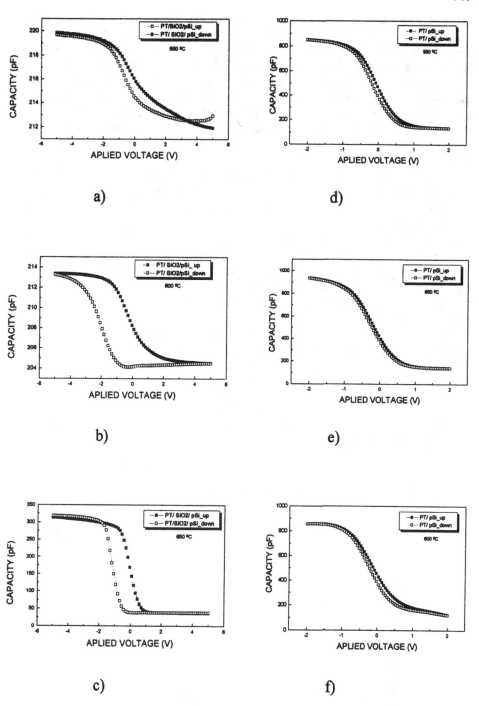

Figure 5) The C - V characteristics for
PbTiO₃ / SiO₂ / pSi annealed at a) 550 °C; b) 600 °C; c) 650 °C;
respectively for PbTiO₃ / pSi annealed at d) 550 °C; e) 600 °C; f) 650 °C.
The measurement frequency was 100 kHz.

The C-V characteristics reveal a clockwise hysteresis loop. This hysteresis may be connected with the transformation of the initial amorphous oxide thin film into a ferroelectric phase after annealing. The direction of the initial amorphous oxide thins film into a ferroelectric phase after annealing. The direction of the loop is an evidence for the existence of spontaneous polarisation in the PT thin film. For the p-type Si, the clockwise loop is due to the existence of the spontaneous polarisation in PbTiO₃ while a loop in the other sense is due to the electric conduction in oxide [9].

An important quantity in the C-V characteristics is the capacitance of the MFS structure in accumulation. This capacitance can be related to the capacitance of the oxide films.

From Figure 5 a), b) and c), the Cox measured in the accumulation region is the total series capacitance of the oxide films: PbTiO₃ and SiO₂. The relative dielectric constants of PbTiO₃ and SiO₂ layers are 80 and 3. 8. Because the SiO₂ native films are very thin, we can approximate $C_{ox} = C_{pt}$.

In case of PT deposited directly on p-Si the capacitance in accumulation is given by the series capacitance of the PT film and of the insulating layer occurring at the PT/ p-Si interface during annealing. This layer could have a high dielectric constant and alter the capacitance of the ferroelectric film. However, it cannot be explained at this moment why the capacitance of the MFS structure in accumulation is larger to PT deposited on p-Si than to of PT deposited on SiO₂/ Si.

The C-V curves show a hysteresis loop after annealing. The primary effect at the MFS structure[10] is the change of the threshold voltage V_T depending on the sweeping direction (up or down). Using the difference between the V_T measured from the two C-V curves we can determine the memory widow [11] :

$$\Delta V = V_T^{UP} - V_T^{DOWN}$$

The dependence of the memory window on the annealing temperature is presented in Figure 6 for the two structures.

Figure 6. The memory window for PT/ SiO₂/ p-Si and for PT/ p-Si

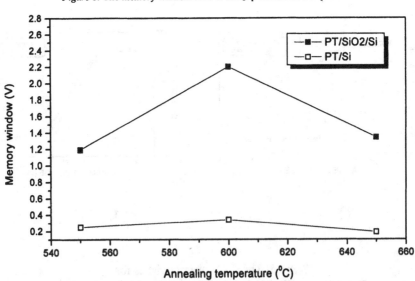

The memory window is larger in the case of PT films deposited on SiO_2/ Si. It has a maximum for the sample annealed at 600 °C. This could be related to the crystallite dimension. At low annealing temperature (550 °C) the crystallites are small and some pinning effects can alter the polarisation switching. At high annealing temperatures the crystalinity is enhanced (see X-ray) and the crystallites are larger. This can lead to a lower density is film. The effect is a decrease of the polarisation because the applied voltage is divided between the crystallites and the pores of the film. In both cases lower polarisation leads to a lower memory window.

In case of PT films deposited directly on p-Si the memory window is very small as a consequence of the insulating layer that occurs at the PT/ p-Si interface (as we see in the SEM analysis). The dielectric constant of this layer could be smaller than of the PT film, so that a large fraction on the applied voltage falls on the interface layer. An other effect of this layer could be the occurrence of an ionic polarisation that cannot be reversed by the applied voltage. In both cases the effect is a reduction of the ferroelectric polarisation that leads to a smaller memory window.

4. Conclusions

Using the sol-gel method, PbTiO3 thin films were deposited onto p-Si substrate with and without native SiO_2 films. X-ray diffraction revealed that the pyrolized films are amorphous but after annealing, the $PbTiO_3$ thin films crystallise only in the desired phase. The microstructure characteristics (AFM, SEM, EDXS) show for all PT films a relative dense structure with a uniform grain size, a good stoichiometry and the occurrence of an interface film between the PT film and the p-Si substrate when the PT is deposited on the p-Si substrate with native SiO_2 film removed. The MFS structures were electrically characterised by capacitance-voltage measurements. These measurements confirm hysteresis behaviour and the memory window. The PT/ SiO_2/ p-Si has good ferroelectric properties in comparison with PT/ p-Si.

The alteration of the ferroelectric properties in case of the PT/ p-Si can be explained by the occurrence of an interface layer between the PT film and the p-Si substrate. The most probable mechanism for the development from the layer is the bi-directional diffusion of ions and a possible chemical reaction that leads to the development of a third phase during annealing.

5. Acknowledgements

The authors gratefully acknowledge the competent assistance of Dr. Marin Alexe, Max Planck Institute fur Mikrostrukturphysik, Halle/ Salle Germany, who performed the microstructure characterisation (AFM, SEM, EDXS)

6. References

1. T. Sumi et al., Integr. Ferroelectr. 6, 1 (1995)
2. R. W. Whatmore, P.C. Osbond, and N.M. Shorrocks, Ferroelectrics 76, 351 (1987)

308

3. L.D. MaMillan, M. Huffman, T.L. Roberts, and M.C. Scott, Integr. Ferroelectr. 4, 319 (1994)
4. S.R.J. Brueck and R.A.Myers, Mater. Res. Soc. Symp. Proc. 341, 243 (1994)
5. R.E. Newnham and G.R. Ruschau, J. Am. Ceram. Soc. 74, 463 (1991)
6. D.Q. Xiao, J.G. Zhu and Z.H. Qian, Ferroelectrics 151,27 (1991)
7. S. Amoruso et al., J. Appl. Phys. 78, 494 (1995)
8. P.K. Larsen et al., J. Appl. Phys. 76, 2405 (1995)
9. P. Gaucher, J.Hector, and J.C. Kurfiss, NATO ASI Ser., Ser. E 284, 147 (1994)
10. K. Aizawa and H. Ishiwara, Jpn. J. Appl. Phy., 33, 5178 (1994)
11. P. Richmann, " MOS Fiel Effect Transistors and Circuits", Jhon Wiley & Sons, N.Y., (1973)

COMPACT PIEZOELECTRIC ULTRASONIC MOTORS

KENJI UCHINO AND BURHANETTIN KOC
International Center for Actuators and Transducers, Materials Research Laboratory,
The Pennsylvania State University
University Park, PA 16802, USA

Abstract

This paper reviews recent developments of compact ultrasonic motors using piezoelectric resonant vibrations. Following the historical background, ultrasonic motors using the standing and travelling waves are introduced. Driving principles and motor characteristics are explained in comparison with the conventional electromagnetic motors.

1.   Introduction

In office equipment such as printers and floppy disk drives, market research indicates that tiny motors smaller than 1 cm^3 would be in large demand over the next ten years. However, using the conventional electromagnetic motor structure, it is rather difficult to produce a motor with sufficient energy efficiency. Piezoelectric ultrasonic motors, whose efficiency is insensitive to size, are superior in the mm-size motor area.

In general, piezoelectric and electrostrictive actuators are classified into two categories, based on the type of driving voltage applied to the device and the nature of the strain induced by the voltage: (1) rigid displacement devices for which the strain is induced unidirectionally along an applied dc field (servo displacement transducers and pulse drive motors), and (2) resonating displacement devices for which the alternating strain is excited by an ac field at the mechanical resonance frequency (ultrasonic motors). The AC resonant displacement is not directly proportional to the applied voltage, but is, instead, dependent on adjustment of the drive frequency. Although the positioning accuracy is not as high as that of the rigid displacement devices, very high speed motion due to the high frequency is an attractive feature of the ultrasonic motors.

The materials requirements for these classes of devices are somewhat different, and certain compounds will be better suited for particular applications. The servo-displacement transducer suffers most from strain hysteresis and, therefore, a PMN electrostrictor is preferred for this application. The pulse-drive motor requires a low permittivity material aiming at quick response with a limited power supply rather than a

309

C. Galassi et al. (eds.), Piezoelectric Materials: Advances in Science, Technology and Applications, 309–320.

small hysteresis, so that soft PZT piezoelectrics are preferred to the high-permittivity PMN for this application. On the contrary, the ultrasonic motor requires a very hard piezoelectric with a high mechanical quality factor Q_m, in order to minimize heat generation and maximize displacement. Note that the resonance displacement is equal to $\alpha \cdot dEL$, where d is a piezoelectric constant, E, applied electric field, L, sample length and $\alpha \cdot$ is an amplification factor proportional to the mechanical Q.

This paper deals with ultrasonic motors using resonant vibrations, putting a particular focus on miniaturized motors. Following the historical background, various ultrasonic motors are introduced. Driving principles and motor characteristics are explained in comparison with the conventional electromagnetic motors.

2. Classification of ultrasonic motors

2.1 HISTORICAL BACKGROUND

Electromagnetic motors were invented more than a hundred years ago. While these motors still dominate the industry, a drastic improvement cannot be expected except through new discoveries in magnetic or superconducting materials. Regarding conventional electromagnetic motors, tiny motors smaller than 1cm long are rather difficult to produce with sufficient energy efficiency. Therefore, a new class of motors using high power ultrasonic energy, ultrasonic motor, is gaining wide spread attention. Ultrasonic motors made with piezoceramics whose efficiency is insensitive to size are superior in the mini-motor area. Figure 1 shows the basic construction of an ultrasonic motor, which consists of a high-frequency power supply, a vibrator and a slider. Further, the vibrator is composed of a piezoelectric driving component and an elastic vibratory part, and the slider is composed of an elastic moving part and a friction coat.

Though there had been some earlier attempts, the practical ultrasonic motor was proposed firstly by H. V. Barth of IBM in 1973 [1]. A rotor was pressed against two horns placed at different locations. By exciting one of the horns, the rotor was driven in one direction, and by exciting the other horn, the rotation direction was reversed. Various mechanisms based on virtually the same principle were proposed by V. V. Lavrinenko [2] and P. E. Vasiliev [3] in the former USSR. Because of difficulty in maintaining a constant vibration amplitude with temperature rise, wear and tear, the motors were not of much practical use at that time. In 1980's, with increasing chip pattern density, the semiconductor industry began to request much more precise and sophisticated positioners which do not generate magnetic field noise. This urgent request has accelerated the developments in ultrasonic motors. Another advantage of ultrasonic motors over the conventional electromagnetic motors with expensive copper coils, is the improved availability of piezoelectric ceramics at reasonable cost. Japanese manufacturers are producing piezoelectric buzzers around 30 - 40 cent price range at the moment.

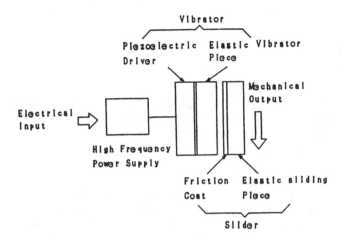

Figure 1. Fundamental construction of ultrasonic motors.

Let us summarize the merits and demerits of the ultrasonic motors:

Merits
1. Low speed and high torque   -- Direct drive
2. Quick response, wide velocity range, hard brake and no backlash
                                  -- Excellent controllability
                                  -- Fine position resolution
3. High power / weight ratio and high efficiency
4. Quiet drive
5. Compact size and light weight
6. Simple structure and easy production process
7. Negligible effect from external magnetic or radioactive fields, and also no generation of these fields

Demerits
8. Necessity for a high frequency power supply
9. Less durability due to frictional drive
10. Drooping torque vs. speed characteristics

2.2   CLASSIFICATION AND PRINCIPLES OF ULTRASONIC MOTORS

From a customer's point of view, there are rotary and linear type motors. If we categorize them from the vibrator shape, there are rod type, shaped, ring (square) and cylinder types. Two categories are being investigated for ultrasonic motors from a vibration characteristic viewpoint: a standing-wave type and a travelling-wave type. The standing wave is expressed by

$$u_S(x,t) = A \cos kx \cdot \cos \omega t, \qquad (1)$$

while the propagating wave is expressed as

$$u_p(x,t) = A \cos (kx - \omega t). \tag{2}$$

Using a trigonometric relation, Eq. (2) can be transformed as

$$u_p(x,t) = A \cos kx \cdot \cos \omega t + A \cos (kx-\pi/2)\cdot\cos (\omega t-\pi/2). \tag{3}$$

This leads to an important result, i. e. a propagating wave can be generated by superimposing two standing waves whose phases differ by 90 degree to each other both in time and in space. This principle is necessary to generate a propagating wave on a limited volume/size substance, because only standing waves can be excited stably in a finite size.

The standing-wave type is sometimes referred to as a vibratory-coupler type or a "woodpecker" type, where a vibratory piece is connected to a piezoelectric driver and the tip portion generates flat-elliptical movement. Figure 2 shows a simple model proposed by T. Sashida [4]. A vibratory piece is connected to a piezoelectric driver and the tip portion generates flat-elliptical movement. Attached to a rotor or a slider, the vibratory piece provides intermittent rotational torque or thrust. The standing-wave type has, in general, high efficiency, but lack of control in both clockwise and counterclockwise directions is a problem.

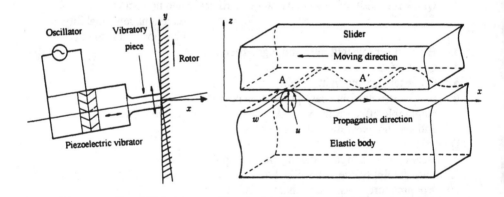

Figure 2. Vibratory coupler type motor.   *Figure 3.* Principle of the propagating wave type motor.

By comparison, the propagating-wave type (a surface-wave or "surfing" type) combines two standing waves with a 90 degree phase difference both in time and in space. The principle is shown in Fig. 3. A surface particle of the elastic body draws an elliptical locus due to the coupling of longitudinal and transverse waves. This type requires, in general, two vibration sources to generate one propagating wave, leading to low efficiency (not more than 50 %), but it is controllable in both the rotational directions.

3. Conventional motor designs

3.1 SASHIDA MOTOR

Figure 4 shows the famous Sashida motor [5]. By means of the travelling elastic wave induced by a thin piezoelectric ring, a ring-type slider in contact with the "rippled" surface of the elastic body bonded onto the piezoelectric is driven in both directions by exchanging the sine and cosine voltage inputs. Another advantage is its thin design, which makes it suitable for installation in cameras as an automatic focusing device. Eighty percent of the exchange lenses in Canon's "EOS" camera series have already been replaced by the ultrasonic motor mechanism. Most of the studies on ultrasonic motors done in the US and Japan have been modifications of Sashida's type.

The PZT piezoelectric ring is divided into 16 positively and negatively poled regions and two asymmetric electrode gap regions so as to generate a 9th mode propagating wave at 44 kHz. A proto-type was composed of a brass ring of 60 mm in outer diameter, 45 mm in inner diameter and 2.5 mm in thickness, bonded onto a PZT ceramic ring of 0.5 mm in thickness with divided electrodes on the back-side. The rotor was made of polymer coated with hard rubber or polyurethane. Figure 5 shows Sashida's motor characteristics.

Canon utilized the "surfing" motor for a camera automatic focusing mechanism, installing the ring motor compactly in the lens frame. It is noteworthy that the stator elastic ring has many teeth, which can magnify the transverse elliptical displacement and improve the speed. The lens position can be shifted back and forth through a screw mechanism. The advantages of this motor over the conventional electromagnetic motor are:

Silent drive due to the ultrasonic frequency drive and no gear mechanism (i. e. more suitable to video cameras with microphones).
Thin motor design and no speed reduction mechanism such as gears, leading to space saving.
Energy saving.
A general problem encountered in these travelling wave type motors is the support of the stator. In the case of a standing wave motor, the nodal points or lines are generally supported; this causes minimum effects on the resonance vibration. To the contrary, a travelling wave does not have such steady nodal points or lines. Thus, special considerations are necessary. In Fig. 4, the stator is basically fixed very gently along the axial direction through felt so as not to suppress the bending vibration. It is important to note that the stop pins which latch onto the stator teeth only provide high rigidity against the rotation.

3.2   INCHWORM DEVICES

Although the motion principle is different, inchworm devices move fast in apparent similarity to ultrasonic motors. The inchworm is driven by a rectangular wave below the resonance frequency, and moves intermittently and discretely. Sophisticated linear

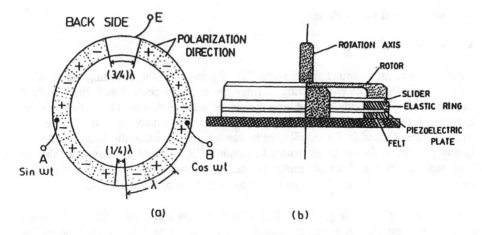

Figure 4. Stator structure of Sashida's motor.

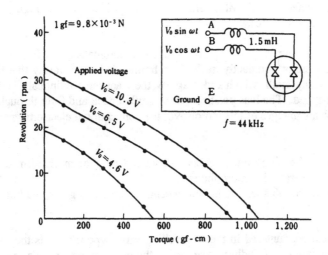

Figure 5. Motor characteristics of Sashida's motor.

walking machines have been developed by two German companies. Philips fabricated a linear drive inchworm using two d_{33} (longitudinal mode) and two d_{31} (transverse mode) multilayer actuators (Fig. 6) [6]. Very precise positioning of less than 1 nm was reported. The problems with this type of device are: (1) audible noise, and (2) heat generation, when driven at high frequency. Physik Instrumente manufactured a two-leg inchworm [7].

A pair of inchworm units consisting of two multilayer actuators, are coupled with 90° phase difference in time so as to produce a smooth motion instead of a discrete step motion.

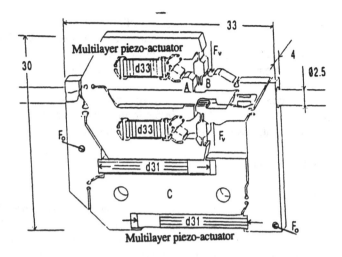

Figure 6. Philips' inchworm.

4. Compact motor designs

4.1 TRAVELING WAVE TYPES

Using basically the same principle as Sashida's, Seiko Instruments miniaturized the ultrasonic motor to as tiny as 10 mm in diameter [8]. Figure 7 shows the construcion of this small motor with 10 mm diameter and 4.5 mm thickness. A driving voltage of 3 V and a current 60 mA provides 6000 rev/min (no-load) with torque of 0.1 mN·m. Seiko installed this tiny motor into a wrist watch as a silent alarm. Rotating an imbalanced mass provides enough hand shake to a human without generating audible noise. AlliedSignal developed ultrasonic motors similar to Shinsei's, which would be utilized as mechanical switches for launching missiles [9].

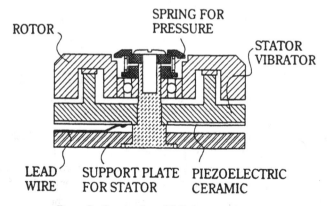

Figure 7. Construction of Seiko's motor.

316

A significant problem in miniaturizing this sort of travelling wave motor can be found in the ceramic manufacturing process; without providing a sufficient buffer gap between the adjacent electrodes, the electrical poling process (upward and downward) easily initiates the crack on the electrode gap due to the residual stress concentration. This may restrict the further miniaturization of the traveling wave type motors. To the contrary, standing wave type motors, the structure of which is less complicated, are more suitable for miniaturization as we discuss in the following. They require only one uniformly poled piezo-element, less electric lead wires and one power supply.

4.2   STANDING WAVE TYPES

4.2.1   Linear motors

Uchino et al. invented a -shaped linear motor [10]. This linear motor is equipped with a multilayer piezoelectric actuator and fork-shaped metallic legs as shown in Fig. 8. Since there is a slight difference in the mechanical resonance frequency between the two legs, the phase difference between the bending vibrations of both legs can be controlled by changing the drive frequency. The walking slider moves in a way similar to a horse using its fore and hind legs when trotting. A test motor 20 x 20 x 5 mm^3 in dimension exhibited a maximum speed of 30 cm/s and a maximum thrust of 0.9 kgf with a maximum efficiency of 20%, when driven at 98 kHz at 6V (actual power = 0.7 W). This motor has been employed in a precision X-Y stage.

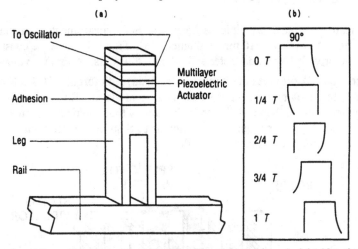

Figure 8. π-shaped linear ultrasonic motor. (a) construction and (b) walking principle. Note the 90 degree phase difference like human walk.

We further miniaturized this "Shepherd" shape by 1/10 into a "Dachshund" shape by reducing the leg length (Fig. 9) [11]. According to this miniaturization, we utilized a unimorph type drive mechanism in conjunction with a coupling mode between 1st longitudinal and 4th bending modes. Under 100 V_{p-p} applied (0.6 W), this linear motor exhibited the maximum speed of 160 mm/sec and the thrust of 0.25 - 1.00 N.

One ceramic multilayer component actuator was proposed by Mitsui Chemical [12]. Figure 10 shows the electrode pattern. Only by the external connection, a combined vibration of the longitudinal L_1 and bending B_2 modes could be excited.

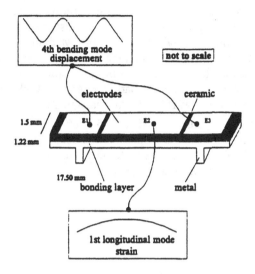

Figure 9. Miniature ultrasonic linear motor.

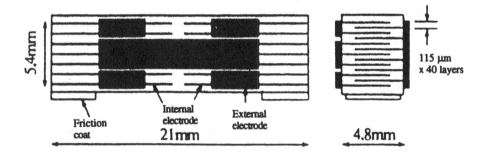

Figure 10. Multilayer ceramic simple linear motor (Mitsui Chemical).

4.2.2 Rotary motors

Hitachi Maxel significantly improved the torque and efficiency by using a torsional coupler, and by the increasing pressing force with a bolt [13]. The torsional coupler looks like an old fashioned TV channel knob, consisting of two legs which transform longitudinal vibration generated by the Langevin vibrator to a bending mode of the knob disk, and a vibratory extruder. Notice that this extruder is aligned with a certain cant angle to the legs, which transforms the bending to a torsion vibration. This transverse moment coupled with the bending up-down motion leads to an elliptical

318

rotation on the tip portion. A motor with 30mm x 60mm in size and 20 - 30⁰ in cant angle between a leg and a vibratory piece provided the torque as high as 1.3 N·m and the efficiency of 80%. However, this type provides only unidirectional rotation.

The Penn State University has developed a compact ultrasonic rotory motor as tiny as 3 mm in diameter. As shown in Fig. 11, the stator consists basically of a piezoelectric ring and two concave/convex metal endcaps with "windmill" shaped slots bonded together, so as to generate a coupled vibration of up-down and torsional type [14]. Since the component number and the fabrication process were minimized, the fabrication price would be decreased remarkably, and it would be adaptive to the disposable usage. When driven at 160 kHz, the maximum revolution 2000rpm and the maximum torque 0.8mN·m were obtained for a 5 mmØ motor. Notice that even the drive of the motor is intermittent, the output rotation becomes very smooth because of the inertia of the rotor. Figure 12 shows motor characteristics plotted as a function of motor size for modified "windmill" motors [15].

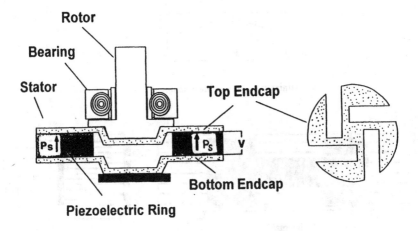

Figure 11. "Windmill" motor with a disk-shaped torsional coupler.

5. Integrated motor designs

We will introduce an ultrasonic motor fabricated on a Si substrate, jointly developed by MIT and Penn State [16]. After coating a PZT thin film on a Si membrane (2.2mm x 2.2mm), an 8-pole stator (1.2mm and 2.0mm inner and outer. diameters) was patterned on the surface electrode. The 8 segmented electrode pads were driven in a four phase sequence repeated twice. A small contact glass lens was used as a rotor. This simple structure provided already 10^3 time higher torque than the conventional electrostactic MEMS motors. Another intriguing surface acoustic wave motor has been proposed by Kurosawa and Higuchi [17]. Rayleigh waves were excited in two crossed directions on a 127.8⁰-rotation Y-LiNbO₃ plate with two pairs of interdigital electrode patterns. A

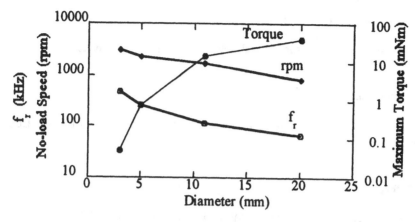

Figure 12. Radial mode resonance frequency, no-load speed and starting torque vs. diameter of the stator. Speed and torque were measured at 15.7 V.

A slider was composed of three balls as legs. The driving vibration amplitude and the wave velocity of the Rayleigh waves were adjusted to 6.1 nm and 22 cm/sec for both x and y directions.

It is important to note that even though the up-down vibrational amplitude is much smaller (< 1/10) than the surface roughness of the LiNbO$_3$, the slider was transferred smoothly. The mechanism has not been clarified yet, it might be due to the locally enhanced friction force through a ball-point contact.

6.   Summary

Ultrasonic motors are characterized by "low speed and high torque," which are contrasted with "high speed and low torque" of the conventional electromagnetic motors. Thus, the ultrasonic motors do not require gear mechanisms, leading to very quiet operation and space saving.   Negligible effects from external magnetic or radioactive fields, and no generation of these fields are suitable for the application to electron beam lithography etc. relevant to the semiconductor technology. Moreover, high power / weight ratio, high efficiency, compact size and light weight are very promising for the future micro actuators adopted to catheter or tele-surgery.

For the further applications of the ultrasonic motors, systematic investigations on the following issues will be required:

(1)   development of low loss & high vibration velocity piezo-ceramics,
(2)   piezo-actuator component designs with high resistance to fracture and good heat dissipation,
(3)   ultrasonic motor designs;
    a. motor types (standing-wave type, traveling-wave type, hybrid type, integrated type),
    b. simple displacement magnification mechanisms of vibratory piece

320

　　　　　　(horn, hinge-lever),
　　　　　c. frictional contact part,
(4)　　　　inexpensive and efficient high frequency/high power supplies.

7. Acknowledgement

This work was partially supported by the Office of Naval Research through Contract No. N00014-96-1-1173.

8. References

1.　Barth H. V., (1973) Ultrasonic Motors, *IBM Technical Disclosure Bull.* **16**, 2263.
2.　Lavrinenko, V.V., Vishnevski S. S. and Kartashev I. K. (1976). Piezoelectric Motors, *Izvestiya Vysshikh Uchebnykh Zavedenii, Radioelektronica* **13**, 57.
3.　Vasiliev, P. E. et al. (1979) UK Patent Application GB 2020857 A.
4.　Sashida, T. (1982) Ultrasonic Motors *Oyo Butsuri* 51, 713.
5.　Sashida, T. (1983) Ultrasonic Motors, *Mech. Automation of Jpn.*, **15** (2), 31.
6.　Coster, M.P.A (1994) Walking Piezo Motor *Proc.4th Int'l Conf. on New Actuators*, Germany 2.6, p.144,
7.　Gloess, R. (1994) A High Resolution Piezo Walk Drive *Proc. 4th Int'l Conf. on New Actuators*, Germany P26, p.190,
8.　Kasuga, M. T. Satoh, N. Tsukada, T. Yamazaki, F. Ogawa, M. Suzuki, I. Horikoshi and T. Itoh, (1991) Compact Ultrasonic Motor *J. Soc. Precision Eng.*, 57, 63
9.　Cummings J. and Stutts, D. (1994) Design for Manufacturability of Ceramic Components *Am. Ceram. Soc. Trans.*, 147.
10.　Uchino, K. Kato K. and Tohda, M. (1988) Ultrasonic Linear Motor Using a Multilayered Piezoelectric Actuator, *Ferroelectrics* 87, 331.
11.　Bein, T., Breitbach E. J. and Uchino, K. (1997) Linear Ultrasonic Motor Using the First Longitudinal and the Fourth Bending Mode, *Smart Mater. Struct.* 6, 619.
12.　Saigo, H. (1994) 15th Symp. Ultrasonic Electronics (USE 94), No. PB-46, p.253
13.　Kumada, A. (1985) Kumada Motors *Jpn. J. Appl. Phys.*, 24, Suppl. 24-2, 739.
14.　Koc, B. Dogan, A. Xu, Y. Newnham R. E. and Uchino, K. (1998) An Ultrasonic Motor Using a Metal-Ceramic Composite Actuator Generating Torsional Displacement, *Jpn. J. Appl. Phys.* 37, 5659.
15.　Koc, B, Xu Y. and Uchino K., (1998) Composite Ultrasonic Motors *Proc. IEEE Int'l Ultrasonic Symp.*, Sendai, Japan [in press].
16.　Flyn, A. M., Tavrow, L. S., Bart, S. F., Brooks, R.A.. Ehrlich,, D.J.,Udayakumar, K.R and Cross, L. E. (1992).*J. Microelectro-mechanical Systems*, 1, 44
17.　Takahashi, M, Kurosawa, M., and Higuchi, T. (1994).*Proc. 6th Symp. Electro-Magnetic Dynamics '94*, No. 940-26 II, D718, 349.

For further reading:
18.　Uchino, K. (1998) Piezoelectric ultrasonic motors: overview" *Smart Mater. Struct.* 7, 273.
19.　Uchino, K. (1996) *Piezoelectric Actuators and Ultrasonic Motors*, Kluwer Academic Publishers, MA.

PIEZOELECTRIC ACTUATORS USING ULTRASONIC WAVES

F. BASTIEN AND J.F. MANCEAU
LPMO-CNRS, 32 avenue de l'Observatoire - 25044 Besançon cedex
FRANCE
Email: fbastien@univ-fcomte.fr

Abstract

Among the different types of piezoelectric actuators, we focus on actuators where movement is obtained by friction on a surface subjected to an elliptic displacement of at least one point.

The elliptic motion is obtained by either a combination of two vibration modes or by using a travelling wave in an annular disk. In fact, between the two possibilities, different intermediate solutions are possible. We present a new class of actuators called quasi-travelling waves actuators.

In a second part, we discuss the limiting performance factors and the relationship with material science. After a general discussion, we emphasise the problem of the reduction of the size of the motor. Several examples of these new types of motors will be given including small size motors, with dimensions in order of magnitude of a few millimetres, where the use of bulk ceramic is not easy. Direct deposition of a ceramic layer would be more efficient, but usually, the thickness of the layer (obtained by sol-gel technique or cathodic sputtering) is too small (few micrometers).The development of ceramic deposition by screen-printing to provide thicker layers would be a large improvement in the piezoelectric material domain.

1. Introduction

Piezoelectric actuators already have some industrial applications. Actuators using direct expansion of piezoelectric material are perhaps the most usual but in that case the displacement is very small even if the force is large. In order to get unlimited movement, vibrating piezoelectric motors have been developed using different principles. Classification of these vibrating piezoelectric motors is difficult, nevertheless the general idea is always based on a vibration with a component perpendicular to the friction surface and a component tangential to the friction surface.

The classification is often based on the choice of the vibrating modes. These can be two identical modes, two different modes, or even only one mode. (See chap 2 of Ueha's book [1]). The type of mode can be chosen among flexural, longitudinal, radial,

C. Galassi et al. (eds.), Piezoelectric Materials: Advances in Science, Technology and Applications, 321–333.
© 2000 *Kluwer Academic Publishers. Printed in the Netherlands.*

torsional...In fact, this classification is not easy to use because the difference between classes is not always very clear and some motors are difficult to classify.

In all cases, we have to choose a method to generate the vibration. Usually the excitation is due to the vibration of one or several piezoelectric ceramics. We have also to choose a friction layer between the "vibrating stator" and the "rotor". Among different types of motors [1,2], the most usual is the travelling wave motor [3]. We will present the advantages of this type of motors and its generalisation using new structures.

Many types of ultrasonic motors have been developed during the last fifteen years in several countries. Their abilities are now well known (high torque at low speed, high holding torque, self-holding torque, compactness and small thickness, non-sensitivity to electromagnetism...) and they are already used for some industrial applications.

However, there is a growing need for low-cost integrated micro-systems [4] and motors have to be included in such devices. Therefore, the fabrication of micromotor is an important domain of research with important problems to be solved.

For example, the use of classical machining techniques becomes difficult when the size of the structures is reduced to some square millimetres. Some motors have been made in this manner [5] but their cost is high, due to the precision of the pieces and the assembly time required. Furthermore, they are not compatible with batch fabrication processes.

This has led us to consider new technologies like silicon bulk micro-machining. This implies new constraints such as the difficulty of obtaining circular structures, as such shapes are used for many types of motors. To solve this, we have developed new principles of ultrasonic actuators to produce linear or circular motions in rectangular or square structures [6, 7], since these shapes are the easiest to obtain with silicon micro-machining.

When designing small size motors, the use of bulk piezoelectric ceramics becomes difficult. To solve this we have used thin deposited piezoelectric layers. Several kinds of materials and techniques are available to obtain such layers and some tests have already been performed; the key-factor being to obtain a sufficient thickness in order to optimise the excitation of the structures.

In this paper, we focus on a new type of circular motor in clamped square structures. First, we review their principles and the validation we have performed. Then, for the piezoelectric travelling wave motor, we will discuss the limiting factors and how they change related to the scaling factor.

2. The ultrasonic travelling wave motor

2.1 PRINCIPLE

By generating a flexural travelling wave in a thin structure, the points on the surface move elliptically. If a moveable object ("rotor") is pressed on a plate ("stator") in the area where a travelling wave is present, a movement is generated. The theory of friction between stator and rotor is rather complex, but experiment shows that the system works even with small vibration amplitude relative to the machining precision of rotor and stator.

2.2   THE CLASSICAL STRUCTURE: ANNULAR PLATE MOTOR

The necessity to produce a travelling wave limits the possibilities of structures. Up to now travelling wave motors use an annular structure. In order to generate a travelling wave in such a structure, we have to combine the nth flexural vibration mode with the same mode geometrically shifted a quarter of a wavelength, both modes being excited at the same frequency with a phase shift of $\pi/2$ (Figure 1). We then obtain a travelling wave propagating along a circular trajectory with n points of contact with the rotor.

In fact, this case is a special one and a quasi-travelling wave can be produced in many different structures. The simplest cases correspond to rectangular or square membranes.

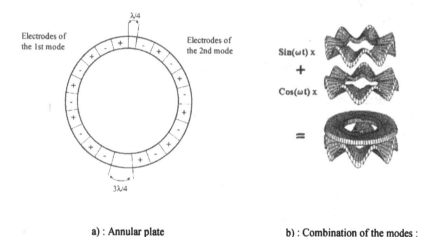

a) : Annular plate   b) : Combination of the modes :

Figure 1 : Excitation of the modes in an annular plate.

3.   The ultrasonic quasi-travelling wave motor

3.1 PRINCIPLE

Let us consider a structure having two vibration modes with the same eigen frequency. The first mode is excited with an amplitude proportional to $\sin(\omega t)$ where ω is the angular frequency and the second mode with an amplitude proportional to $\cos(\omega t)$. If you find a curve along which the elongation for the first mode is $\Psi_1 = A \sin(ks)$ where k is a constant and s the curvilinear abscissa of the curve and $\Psi_2 = A \cos(ks)$ for the second mode, it is easy to generate a travelling wave. The first mode excited with

$V_1 = A_1 \sin(\omega t)$ and the second mode with $V_2 = A_1 \cos(\omega t)$ give $\Psi_1 = A_1 \sin(\omega t) \sin(ks)$ and $\Psi_2 = A_1 \cos(\omega t) \cos(ks)$ respectively and then the resulting elongation along the curve is $\Psi = \Psi_1 + \Psi_2 = A_1 \cos(\omega t - ks)$.

Obviously, it is not possible to find a curve where the previous condition is completely fulfilled. It is often possible to find a curve where $\Psi = A(s) \cos(\omega t - k(s) s)$, with $A(s)$ and $k(s)$ are almost constant with curvilinear abscissas. Consequently, quasi-travelling wave motors can be realised using different structures.

We note that the travelling wave does not have to be necessarily produced along the whole curve. If along some parts of the curve the so-called wave is present, these corresponding areas will be chosen as contact areas. The condition for the structure is the presence of two vibration modes with the same eigen frequency or almost the same frequency. Square membranes have this property and for the rectangular membrane, a good choice of the ratio width over length also gives the property.

Follows the choice of the structure, the trajectory corresponding to the quasi-travelling wave can be found by intuition for simple cases but a more general method has been developed.

3.2 MODELLING

We have developed a modelling method for the case where a thin plate is used as a stator.

1-First, we determine the geometric parameters corresponding to a case with two modes having the same eigenfrequency ω. We find the neutral fibre displacement for these two modes with the same eigenfrequency

2-Both modes are excited by a sinusoidal signal, the first one as $\sin(\omega t)$ the second as $\cos(\omega t)$

3-We find the contact area between the stator and the movable object (rotor)

4-Inside this "contact area", we find areas where the condition concerning the phase is fulfilled. The phase must change almost linearly with the position along the curve. That is to say, we have to find a curve in the contact area along which the phase gradient is constant or almost constant.

5-In the area of the previous curve, we compute the horizontal speed at the time that corresponds to the maximum vertical displacement. Let us remark that this time is generally not the same for all the points. On the final figure, we show a field of speed with a length of the arrow proportional to the particular horizontal speed (see fig 3 and fig 4). From this figure, it is easy to see the existence of a curve along which the horizontal speed is tangential to a given direction. If the rotor has an effective contact in the neighbourhood of this curve, we have a movement. This method has been described in detail in [8]. We will give here some examples of the results.

3.3 NEW STRUCTURES: AN ULTRASONIC MOTOR WITH A SQUARE STATOR

If we consider a clamped square plate, the easiest way to get a travelling wave is to use (1.2) and (2.1) modes that are the same modes but are shifted of a quarter of a wavelength. Combining these two modes as explained for the annular plate produces a travelling wave propagating along curve (C) (Figure 2). If we press a rotor against a selected part of this structure, we can get a rotating movement.

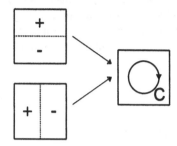

Figure 2: Combination of the (2.1) and (1.2) modes

Another example is given for (2.3) and (3.2) modes in a square clamped plate (Figure 3(a)). Along the dotted circle, we can get a circular movement (Figure 3(b)). Obviously as in the previous case, the direction of the movement can be inverted by changing the phase shift between the temporal excitation of the two modes.

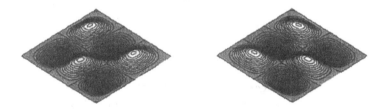

a) : Amplitude of the (2.3) and (3.2) modes

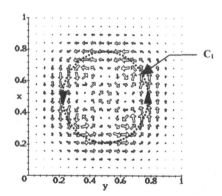

b) : Horizontal particle speed at the maximum displacement for the combination of the modes

Figure 3 : Production of a quasi-travelling wave in a clamped square structure

326

Both previous cases are given as basic examples and many other types of combinations are possible.

For a non-circular structure like a rectangular plate, obtaining a travelling wave is not so obvious. Nevertheless, it is possible to find a particular geometry allowing us to have two modes with nearly the same frequency. For example, it is possible to use modes 1-3 and 2-1 in a rectangular plate with the right ratio between length and width (see Figure 4)

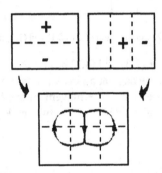

Figure 4 : Combination of mode 2.1 and 1.3 in a rectangular plate.

It is also possible to produce some linear movement using other combination. As a theoretical example, let us examine the combination of 2-2 and 3-0 modes in a rectangular all-sides-free plate. One of the main problems is to hold this type of structure. In that case, neither the amplitude nor the phase follows precisely the criteria of constant amplitude and phase gradient. Despite this fact, the "quasi-travelling wave" is sufficient to produce movement, for example along, the y=1 line (see Figure 5)

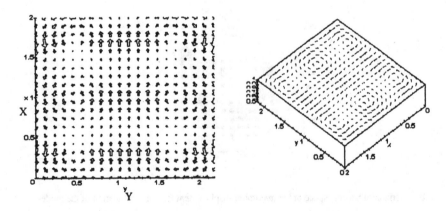

Figure 5 : Horizontal particle speed at the contact point.

4. Some prototypes

4.1 EXPERIMENTAL VALIDATION

In order to validate the principle we built a 20x20mm square plate prototype (in beryllium bronze), which was designed to obtain the rotation of a 14mm diameter rotor. However, when using the combination of modes (1,2) and (2,1), we obtained poor performance. In order to improve our results we kept the same principle but used another combination of modes of higher rank (3.2) and (2.3), which allowed the generation of movement through three contact points along the contact curve. We then obtained encouraging performances for such a motor [9] including clockwise and counter-clockwise movement, no-load rotating speeds of 250 rpm and a maximal mechanical power corresponding to a torque of 1 mN.m at about 100 rpm (See figure 6). A prototype of a linear motor has been built. We obtained a linear speed of about 2 cm/s. This actuator can support a pressing force of about 6 N. The pulling force is about 0.1 N. This system is more efficient for charge displacement than for pulling.

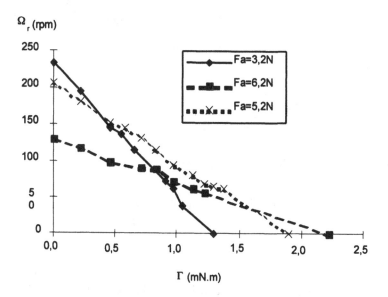

Figure 6 : Speed versus torque characteristic (F_a= pressing force)

4.2 CLASSICALLY MACHINED MICROMOTOR

Our first goal has been to generate the movement of a 5 mm diameter rotor using 10*10 mm square structures. At this scale, we are at the boundary between two different approaches [10] : classically machined structures with bulk ceramics assembled on them or silicon micro-machining with thin deposited piezoelectric layers. Both techniques have been investigated.

The structure used here is a clamped square plate. In this case, the easiest way is to use the (1.2) and (2.1) modes, but with this combination, the performance is rather low.

Therefore, we used a combination of modes of a higher ranks namely (3.2) and (2.3) modes which allowed us to generate a movement through the three contact points described previously. This motor was also made of beryllium bronze. The design is directly inspired from the 20x20 mm prototype, but with a 75 % decrease of the surface and about 40 % of the global thickness (Figure 7). Excitation of the modes was achieved with a 10x10x0.2 mm piezoelectric plate with deposited electrodes. Initial results show the rotation of a 5mm-diameter rotor positioned as shown in Figure 8.

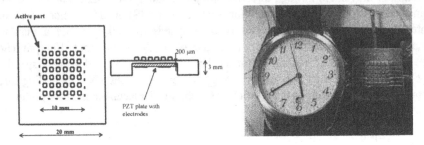

Figure 7 : Sketch and Top view of the 10x10 mm beryllium bronze

Figure 8 : Rotation of a 5 mm diameter rotor

We were able to measure speeds over 300 rpm (clockwise and counter-clockwise) with voltages under 40V using the combination of (2.1) and (1.2) modes, which provided much better results than with the bigger prototype. The (2.3) and (3.2) modes have also been tried successfully. The torque has been estimated at about 10 μNm.

4.3   SILICON MICRO-MACHINED STRUCTURES, USING DEPOSITED PIEZOELECTRIC LAYERS

Considering the mechanical properties of silicon and the sizes of a few millimetre, if we want to keep the resonant frequencies of our structures in the practical range of ultrasonics, we have to use membranes with a thickness between 10 and 50 μm. Bulk-piezoelectric plates are about 200 μm thick which are large compared with the silicon thickness. The thickness of piezoelectric thin layers obtained by cathodic sputtering is limited to some micrometers. PZT obtained by sol gel method is also below one micrometer. Nevertheless, we have obtained the motion of small particles with a sol gel PZT layer (thickness 0.64 μm). It is necessary to have a higher thickness of PZT to obtain larger amplitude in order to move heavier object. Therefore for micro-motors, development of thick deposited layers (in the 100 μm range) is very important.

5. Limiting factors

5.1 INTRODUCTION TO THE LIMITING FACTORS

The different factors that limit the performances of ultrasonic motors depends upon the basic principle and its technical data. We must have in mind the final use of the micromotor (or micro-actuator) taking into account for example the performance, dimensions, size and price. Using micromotors is not so obvious today because the whole system has to be achieved including micromotors and electronic control. For example, in microsurgery, where chemical or biochemical analysers have to manipulate small quantities of fluid, a new microsystem has to be designed. In all cases, the determination of the limiting factors is useful. Let us first remember the limiting factor for electrostatic and electromagnetic motors.

For electrostatic motors, the main limitation is due to the breakdown voltage between electrodes. In the case of electromagnetic motor, the limitation is the thermal dissipation of the windings. As a function of the scale factor, the limitation can change when the aim is to downsize the motors. The advantage of the electromagnetic motor over the electrostatic motor is clear for centimetre-range diameters. This advantage becomes less obvious for millimetre sizes and of course for sub-millimetre sizes. For this size, the electrostatic motor presents some convenient aspects. Nevertheless, micromachining using micro-electronic techniques has difficulty to make structures with a low friction coefficient, and it is difficult to obtain a shaft to transmit the movement (and torque). This difficulty limits the electrostatic motor but not the ultrasonic motor because it uses friction.

5.2 LIMITING FACTORS OF THE PIEZOELECTRIC MOTOR

In case of the piezoelectric travelling wave motor, the torque is mainly a function of:
- The maximum pressing force of the rotor upon the stator. This force depends on the aptitude of the ceramic to maintain a maximised amplitude of vibration
- The friction layer capability to transmit the mechanical power. The properties of the friction layer are a compromise between friction and slip. Indeed, no points of the surface of the stator have the same speed. It seems that often the limiting factor is not the friction layer if we do not consider the time of use.
- The stickiness of the ceramic on the stator is a critical factor. Increase of temperature due to losses in the ceramic can also be a limitation.

The particle speed limit of the material can play a limiting role, mainly for large amplitude vibrations, but this factor is generally not the ultimate one. For example, a 1 μm amplitude vibration at a frequency of 50 kHz gives a particle speed of about 0.3 m.s^{-1}. This speed is below the elastic limit of many materials.

For the rotation velocity limiting factors are:

- The amplitude of the vibration but also the h/λ ratio where h is the distance between the neutral line of flexion of the stator and the contact surface, and λ is the wavelength. It is interesting to increase h but this also increases the rigidity of the stator. Making teeth on the stator, which increases h but not the rigidity, can palliate this. Nevertheless, the height/width ratio of teeth cannot be too large in order to avoid spurious modes.
- The maximum speed is limited by the properties of the ceramic, which has to generate large amplitude of vibration in a rather thick structure.

To conclude, the limiting factor is not completely determined for all types and all sizes of ultrasonic motors but ceramics improvement is an important challenge for ultrasonic motors.

6. The scaling factor of the piezoelectric motors

In order to evaluate the ability of the motor to be miniaturised. let us now try to determine the dependency of the parameters on the dimensions. We suppose that for a given geometry, we reduce all sizes proportionally. Keeping the same material and structure, we note [L^n] when a parameter varies with the dimensions to the power of n.

Let us assess the flexure of the structure such as a beam or a membrane when a piezoelectric ceramic is glued on it. As a first approximation, at the maximum electric field (breakdown field), the ceramic produces a force proportional to the perpendicular section surface so we can establish that force varies with [L^2]. The flexion moment varies with [L^3]. A beam has rigidity proportional to its inertial moment, which corresponds to [L^4]. Therefore, the maximum vertical displacement of the beam noted Δ varies with [L].

In order to produce a movement, the rotor to the stator exerts a pressing force and then by the way of the friction layer with its friction coefficient μ, it produces a force μF.

The maximum force corresponds to the one exerted to eliminate the displacement Δ.

For a given mode of vibration of the structure, the maximum force depends on the product of the displacement and the characteristic length: Δ.[L]. The pressing force varies then as [L^2] and it is the same law for the force F_T. The torque Γ is then proportional to the product F_T by the radius. We can say that the torque Γ of the motor varies as [L^3].

This rule is experimentally verified as shown in Figure 9 where different experimental results obtained in the literature are reported. The relationship between the maximum torque and the diameter, or a representative dimension φ, can be expressed proportional to φ$^{3.2}$, which is very similar with the previous result.

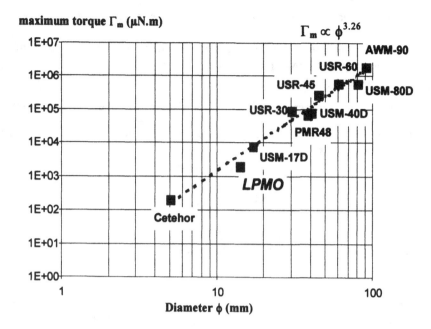

Figure 9: Maximum torque as a function of diameter for different travelling waves motor according to the literature [1,2]

(USM= Matsushita motors; USR= Shinsei motors, PMR=Physik Instrument motors; AWM= AEG motors; CETEHOR=Centre Technique de l'industrie Horlogère, Besançon, France; LPMO: 20*20 mm motor)

This rule can be used if the diameter is not too small. Effectively, there are two other limiting factors. First, the decrease of the amplitude of the vibration leads the motor to be increasingly sensitive to fabrication defects, because the precision of the surface does not increase with the size reduction. For example, a 10 cm-diameter stator having an amplitude of vibration of 1 μm yields to an amplitude of 10 nm when the diameter decreases to 1 mm. This is comparable to the surface roughness. Another difficulty is the augmentation of the frequency when the diameter gets smaller. This augmentation produces more losses in the ceramic.

Let us now evaluate the increase of the frequency. The structure can be basically approximated to a mass m and a rigidity k. The frequency changes as $\sqrt{k/m}$ and the rigidity as $\dfrac{EI}{l^3}$ with E the Young modulus, I the moment of inertia of the structure and l the length. Because I varies as $[L^4]$, k as $[L]$, the mass as $[L^3]$, then $\sqrt{k/m}$ varies as $[L^{-1}]$. This means that if the frequency is 15 kHz for a diameter of 10 cm, it reaches 1,5 MHz in the case of a diameter of 1 mm.

Let us give some order of magnitude for the torque. If a motor would have a torque of 1 Nm for a diameter of 10 cm, we evaluate 250 μNm at 5 mm and 16 μNm at 2 mm. For a micromotor with a diameter of some millimetres, we can say that a torque of about 100 μNm seems to be a reasonable objective. As a comparison, a watch motor should have a torque of 15μNm in order to generate the movement of the needle at a speed of 1 rpm.

Let us examine the variation of the maximal rotation speed. The evaluation of the speed is obtained by the following method. For a distortion of the neutral line expressed by y=A.sin(kx).cos(ωt) (where k=2π/λ and λ is the wavelength of the distortion). The maximal horizontal-displacement d is proportional to $h\dfrac{dy}{dx}$, where h is the distance between the neutral line and the friction surface (about half the thickness of the stator). So d can be expressed as:

$$d = A.h.k.\cos(kx).\cos(\omega t)$$

Then, the horizontal speed of the particles is:

$$v = A.h.k.\omega.\cos(kx).\sin(\omega t)$$

The maximal speed is:

$$v_{max} = A.h.k.\ \omega = A2\pi h/\lambda.\ \omega$$

The maximal rotational speed Ω of the rotor is (R: radius of the stator):

$$\Omega = \frac{v_{max}}{R} A \frac{2\pi h}{\lambda} \frac{1}{R} \omega$$

Because A ∝ [L], ω ∝ [L⁻¹] and h/λ ∝[L⁰], and if we consider keeping the same rank of mode then Ω ∝ [L⁻¹].The rotation speed increases when the dimensions decrease. Consequently, the maximal power $P_{max}=\frac{1}{2}\Gamma_{max}.\Omega_{max}$ is proportional to [L²] and the maximal power per volume unit is proportional to [L⁻¹] which is favourable to the small motor.

In fact, as mentioned previously, several factors limit this advantage. When the dimensions decrease, the rank of the modes is usually lower, then h/λ tends to vary as [L] and the ratio of the power to volume does not change with the dimensions.

To summarise, miniaturisation is not only limited by technical difficulties but also by the increase of losses with frequency and by defects encountered in machining (particularly surface flatness) which are increasingly perceptible.

Among the technical difficulties when reducing the size, assembling the ceramic on the stator becomes increasingly critical. It would be preferable to use other methods such as film deposition of PZT, especially with a diameter lower than 1 cm.

Unfortunately, methods such as sol-gel or cathodic pulverisation do not provide a thickness of more than 10μm. For a thickness about 100μm, required generally for a few millimetre size motors, the most suitable method seems to be screen-printing. In that case, the whole structure that includes stator and ceramic must be sintered at a very high temperature, which is not always compatible with the material used. Some studies are currently being undertaken in order to reduce the sintering temperature of thick piezoelectric films.

7. Discussion and conclusion

Good material properties are essential to set up a travelling wave piezoelectric motor and even more for its miniaturisation.

Among the subjects currently under or interesting to study, are:
* The friction layer
* Piezoelectric ceramic with low losses and high coupling factor
* Achieving a better adhesion of ceramic on metal. For micromotors, achieving a piezoelectric layer about 100µm at a low sintering temperature.
* To replace the teeth, it would be interesting to produce hard materials in compression but soft concerning flexion. Composite materials can be used with a "soft" component, but it is difficult to find a soft material with low mechanical losses.

In conclusion, small motors (a diameter between 5 and 2 mm) seem to be the best choice for the near future. It seems possible to fabricate very soon a micromotor with a diameter of 5 mm, a thickness of 3 mm, a maximum torque of 500 µN.m and a maximum rotation speed of about 200 rpm.

8. References

[1] Ueha S., Tomikawa Y., (1993) *Ultrasonic motors - Theory and applications*, Oxford Science Publications. Oxford

[2] Wallaschek J., (1995) Piezoelectric ultrasonic motors, *International Journal of Intelligent Material Systems and Structures*, **6**, 1, 71-83.

[3] Ueha S., (1989) Present state of the art of ultrasonic motors, *Japanese Journal of Applied Physics*,**28-1**, 3-6.

[4] Fujita H., (1996) Future of actuators and microsystems, *Sensors and Actuators A*, **56**, . 105-111.

[5] Duffait R., Manceau J.F., Biwersi S., (1998) Miniaturisation des moteurs piézoélectriques, 1er Colloque MAGELEC, Toulouse, 28-29 Avril 1998.

[6] Manceau J.F., Bastien F., (1996) Linear motor using a quasi-travelling wave in a rectangular plate, *Ultrasonics*, **34**,. 257-260.

[7] Biwersi S., Manceau J.F., Bastien F., *(1996)* Ultrasonic motors using quasi-travelling waves in different types of plates, Proc. 3rd France-Japan and 1st Europe-Asia Congress on Mecatronics, Besançon, 1-3 Octobre 1996, pp 741-745.

[8] Manceau J.F., Biwersi S., Bastien F., (1998) On the generation and identification of travelling waves in non-circular structures - Application to innovative piezoelectric motors, *Smart Materials and Structures*, **7 (3)**, 337-344.

[9] Biwersi S., Manceau J.F., Bastien F., (1997) Rotating ultrasonic motor using vibration modes in a clamped square plate, 2nd World Congress on Ultrasonics, Yokohama, 24-27 Août 1997, pp. 432-433.

[10] Biwersi S., Manceau J.F., Bastien F., (1997) Linear and circular ultrasonic actuators: miniaturisation and new concepts, Eurosensors XI, Warsaw, 21-24 Sept.1997, Proceedings pp. 775-778.

MEMS APPLICATION OF PIEZOELECTRIC MATERIALS

D. HAUDEN
Laboratoire de Physique et Métrologie des Oscillateurs – UPR 3203
Institut des Microtechniques de Franche-Comté – FR 0067
CNRS – Université de Franche-Comté
32, avenue de l'Observatoire – 25044 Besançon Cedex France

Abstract

Among the different principles to act on microstructures to fabricate microactuators, or to transduce electromechanical phenomenon to detect physical quantities or chemical or biological species, the piezoelectricity is one of the most popular in microsystems.

This paper will present piezoelectric materials already used as active electromechanical components which combined with other materials lead to new microelectromechanical devices.

Micromachined quartz crystal enables to fabricate new sensors for inertial measurements or microbalances for gas and liquid analysis. Thin solid films of piezoelectric sputtered on silicon are fabricated for integrated microsensors.

Ferroelectric materials in bulk or thin films are also combined with metals, glass or silicon to conceive and fabricate new microactuators and micromotors.

Material properties, technologies used, design methods and fabrication processes will be presented and illustrated with several examples.

1. Introduction

After a brief introduction on MEMS and Microsystems definitions and a state of the art on the different principles of actuation, materials used will be presented as substrates or active materials.

Among the different materials, piezoelectric substrates and layers have a particular interest for the MEMS applications.

Different examples illustrate the physical, technological and design problems to use piezomaterials in miniaturized components and systems as sensors, actuators, motors or more complicated systems.

A specific approach for hybrid or integrated with electronics will be also exposed to underline the connection, assembly and packaging, difficulties in MEMS applications.

C. Galassi et al. (eds.), Piezoelectric Materials: Advances in Science, Technology and Applications, 335–346
© 2000 *Kluwer Academic Publishers. Printed in the Netherlands.*

2. MEMS and Microsystems : state of the art and basic examples

Since about 20 years, MEMS applications are developed based on the concept to integrate together different properties like mechanics and electronics or optics, mechanics and electronics in the same miniaturized components.

This way was open with the development of integrated (or hybrid) sensors fabricated either from the microelectronics C-MOS process and using piezoresistors or capacitances as transducers : pressure sensors [1], accelerometers, etc, or from Surface Acoustic Waves Sensors (SAW) on piezosubstrates or silicon/piezo thin film composites when the technology is similar to the microelectronic processes. A chemical sensor based on complete integrated SAW sensor system with the electronics of oscillators and the frequency mixers which provides a frequency shifts proportional to the detected gas mass adsorbed on one of both SAW delay lines made on a ZnO/Silicon composite membrane is commercialized in the Netherlands.

Microsensors for inertial measurements (acceleration, rotation rate), for pressure, temperature, magnetic field, for liquid and gas detectors, are the world wide market in MEMS applications. More difficult is the <u>actuator</u> miniaturization to provide new functions combining electronics and movement (kinematics), because this specificity of micromechanics is not closely compatible with the silicon processes in electronics.

The mechanical displacement, the torque, the mechanical power, the friction, the surface and volume degradations have a major importance in micromechanics and only a weak influence in electronics. Then the combination of both needs a new physical approach for material, active forces, architecture. For instance, the ratio surface/volume being more important in miniaturized components, then electrostatic effect or surface viscosity or surface tension are not anymore neglectable, but at the opposite they could applied for acting actuators. Moreover, because parts of the microsystem must move with several degrees of freedom, they must have a free part and clamping points.

Consequently, they need the development of an extra technology process to realize this function. This leads to introduce the sacrificial layer concept [2] in the C-MOS technology with an etching compatibility between the different materials (figure 1).

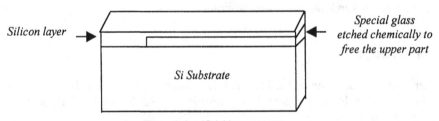

Figure 1. Sacrificial layer concept
(Glass is selectively etched with an acid and silicon in KOH solution)

If we look at this very sample example, we can say that the integrated actuator concept introduces the third dimensions in the MEMS applications, third dimension which is the base of mechanical applications. So, microactuators are based on three new principles :

- 3-Dimension architectures

- driving forces provided by physical effects or active materials
- movement control which needs microsensors

The most compatible example with microelectronics is based on polysilicon surface machined structures acting with electrostatic forces and controlled with capacitive measurements. The figure 2 shows a micromechanical example fabricated at Sandia labs which is an oblique mirror where the slope is controlled with a comb drive system (electrostatic forces) though a planetary reduces (microwheels).

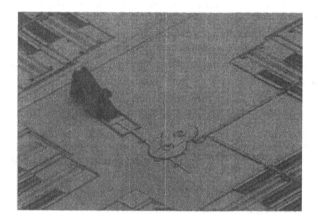

Figure 2. Adaptable micromirror adjustable with an electrostatic force (Sandia Labs).

Only a thermally active bimorph is also IC compatible [3].
Every other modes of actuation need special materials as Shape Memory Alloys (SMA) [4], Magnetic or Giant Magnetostrictive (GMA) [5], or piezoelectric or electrostrictive materials. The shape and the material composition, different from silicon components and lead to complicated process for integration with electronics. Preferentially, the microactuator is connected with its electronics in an hybrid circuit.
The table I summarize the materials used in MEMS and Microsystems and the acting effects applied with them.

TABLE 1. Main material used in MEMS and Microsystems

Materials	Substrate	Actuation	Physical effects and force
Silicon and compounds (SiO_2, SiPoly, Si_3N_4)	X	X	Electrostatic forces Piezoresistive detection
Glass	X	Sacrificial layer with Silicon	Electrostatic forces Capacitives detection
Plastics + Air or liquid or gel (PVDF)	X X	X X	Electrostatic Fluidic Piezoelectric
Quartz (Monocrystal)	X	X	Piezoelectric Electrostatic
Ceramics	X (Al_2O_3)	X	Piezoelectric Electrostrictive
Metals SMA Magnetic GMA	X	Bimorph X X X	Thermal Thermal Inductive Inductive

This table shows that piezoelectric materials play a important role in MEMS.

3. Bulk, thin films or thick layers ?

The piezoelectric materials are intrinsically interesting for actuation either using the static piezo behavior like extension or bending deformation, or resonant modes of vibrations, then the displacement is amplified because the Q-factor of the material.

But, because the applications of piezoelectricity for MEMS require different specifications for the architecture, the environment, the control and command signals, etc. the piezomaterials must be chosen specially for each microcomponents in order to match better the requirements. So, we have a double choice : the material and the technology available to design the microcomponent with this material. The material could be structured from bulk product, or thin film definition or thick layer chemically grooved or deposited.

Two kinds of piezoelectric materials can be distinguished : monocrystals which are only available in bulk and polycrystals which could processed in bulk either films, sometimes in thick layers.

Monocrystals are quartz crystals, $LiNbO_3$, $LiTaO_3$, $GaPO_4$, monocrystals are ceramics (all kinds), ZnO, AlN and also polymer films (PVDF).

Monocrystals are grooved either with an hydrothermal technique (quartz, $GaPO_4$) or with a Chokralsky method. In both cases, only bulk materials are available. Then thin plates or wafers are fabricated by sawing, lapping and polishing with a thickness greater than 120 μm.

Polycrystals are fabricated in bulk plates or thin films (thickness in less than few microns) with a sol-gel and sintering technique for ceramics or with a thin film deposition under vacuum as sputtering method (PZT, ZnO, AlN), or by laser deposition (PZT). The deposition of the thin film can be done on different substrates : glass, silicon, metals.

For MEMS, silicon substrate is the most important to integrate the electronics within the MEMS. The table II presents different possibilities of thin films deposited on silicon and silicon carbide [6].

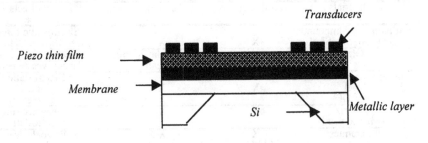

(SH and Lamb waves excitation devices scheme)

TABLE 2. Piezoelectric thin films deposition Silicon or SiC substrates (F. Bastien et al)

Materials combinations	1	2	3	4	5	6
Transducers (1000 A)	Al	Pt	Pt	Pt	Pt	Al
Piezo. Thin films (1 to 2 μm)	AlN	ZnO	PZT	ZnO	PZT	AlN
Metallic layer (ground)	Al	Ti/Pt	Ti/Pt	Ti/Pt	Ti/Pt	Al
Membrane (1 to 25 μm)	SiC	SiC	SiC	Si	Si	Si

The main technical problem is to obtain a thick layer of PZT ceramic where the thickness is greater of 15 micrometers (15 to 60 μm). Up today, no deposition method is available to reach good layer mechanical and piezoelectrical properties in this dimension. Moreover any assembly technique (smart cut technique) is reproducible to obtain a PZT/Silicon structure where the PZT is equivalent to the silicon membrane (few tens of μm). Nevertheless, it is a great challenge for piezoelectric motors actuators and microsystems to develop a piezo thick layer technique combined with the silicon process.

4. Piezoelectric micromachined sensors

As shown for electrostatic MEMS, the sensor applications using piezoelectricity is most developed compared with microactuators and micromotors. Almost, every quantities were tested for measurements techniques based on piezoelectric effects : physical quantities (forces, mass, acceleration, rotation, pressure electric field and magnetic field), chemical detection or measurements by adding or not a sensitive and selective coating, biology species with a selective membrane. In this two last cases of microbalances, the mass variations are transduced into electric values either with bulk resonators (BAW) or Surface Acoustic Waves lines or resonators (SAW). Also, piezoelectric transducers are the excitator/sensor for acoustical imaging in the medical echographic systems or in the NDE system.

The examples described below are taken in the different application where micromachining is partly used to fabricate the complete sensors.

4.1. MICROMACHINED QUARTZ SENSORS

They are based either on BAW resonator properties or SAW plates for microbalances or pressure sensors.

The BAW resonators/sensors are fabricated on a silicon way as the tuning fork resonators : chemical etching and flexural or torsional modes for the detection [7].

The microfabrication of this kind of structures is summarized in this way :

-   Starting with polished wafers of quartz crystal (Z-cut, with a thickness of 127 μm)

- Double sides coated with Cr/Au films in vacuum chamber (thermal evaporation)
- Paterned of the double sides photo resist through 2 photolithographic masks in a double side mask aligner (UV line)
- Chemical etched of the gold non protected with the illuminated resist
- Chemical etched of the quartz crystal non protected with the gold layer

The etching is composed with 50% HF, 40% NH$_4$F and water. The etching is done at 85°C.

Figure 3 gives the edges of etched quartz cantilevers along Y-axis (Z-cut) and sides and the end of the cantilever are vertically defined. That is not the case for other crystallographic cuts and for instance for + X-end and – X-end.

Figure 3. Y-axis cantilever etched from a Z-cut

The second step is the electrodes paterning performed by a second photolithographic step and a gold chemical etching. The figure 4 presents the batch processed force sensors micromachined in Z-cut quartz crystal plates.

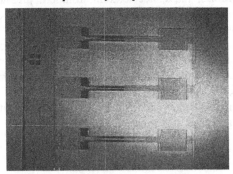

Figure 4. Micromachined quartz crystal force sensors

4.2. SH AND LAMB WAVE SENSORS

The second class of examples is based on SH or Lamb waves on quartz or composite piezo/silicon sensors for gas or liquid detections. Generally, microbalances are designed from a classical BAW resonator. We will focus here on SAW sensors where the thickness of the plate or membrane is in the order of the wave length value. That is mean that only guided acoustic modes can propagate in it or can be trapped along the plate structure in a resonant configuration. Two examples are presented here. The first one [8] is made on an AT quartz plate of 127 μm trapping a resonant transverse wave (figure 5). This device is tested as a microbalance used to control the Al deposition in a sputtering equipment. All resonant modes are frequency shifted when a layer is deposited.

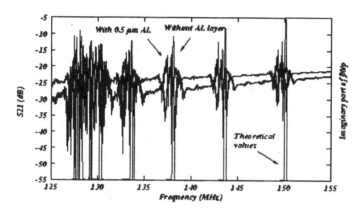

Figure 5. Experimental influence of Al loading on the electrical transmission spectrum
of transverse wave resonator on AT-cut quartz thin plate.
Comparison with theoretical value of wave number βo

The second example shows the influence of the liquid on lamb A_o mode of a sensor structure in composite AlN/Al/Si. The purpose is to detect liquid mass variations in the case of chemical or biology species detection by trapping them on the surface. The figure 6 presents the results when the device is in air (continuous line) and in water (broken line). The so mode vibrates at 9.12 MHz and the frequency shift given is equal to 3.5 MHz.

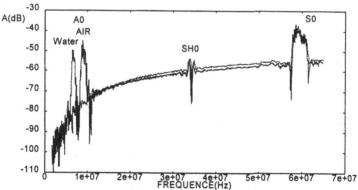

Figure 6. Transfer function in air (continuous line) and in water (broken line) *(th$_{AlN}$ = 3 μm, th$_{Si}$ = 9μm)*

4.3. ULTRASONIC TRANSDUCERS FOR ACOUSTICAL IMAGING

The last example, I would like comment is a micromachined transducer for acoustical imaging [9] studied for a miniaturized endoscopic echographic head. It is designed to have a focused point on the axis, and as shown on the figure 7, the scanning is provided with an ultrasonic micromotor.

Micro-motor   Ultrasound transducer

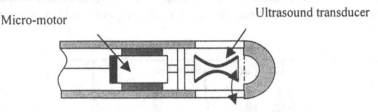

Figure 7. Scheme of a miniaturized endoscopic system

The ultrasonic transducer presented on the figure 8 is composed with 5 annular 60 μm thick rings in PZT micromachined with an high resolution ultrasonic grinding method.

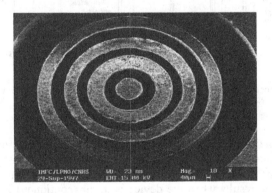

Figure 8. Test model of a PZT annular transducer machined in a 1 mm plate
(aspect ratio close to 5)

The modeling of the transducer is made with a new simulation softwave using the MODULEF-FEM developed with coupled modes, recalibration option and liquid coupling loads [10].

5. Piezoelectric microactuators and ultrasonic motors

Several actuation principles can be used in piezoelectric microactuators or micromotors. The most well-known is the inch worm based system where the movement is generated with a periodic sequence of extension –compression pulse following by a clamp step. Other names are given for motors (hybrid stepping for instance). The rotor or the slider is carried by friction with the surface of the stator or of the end of the effector.

This principle could consider general in ultrasonic linear microactuator and micromotors.

The figure 5 shows a linear micromotor based on the Rayleigh wave properties in LiNbO₃. The Rayleigh wave has one elliptic polarization in the saggital plane, then the slider is propagating along the acoustic beam [11]. Figure 9a is the principle of the micromotor and figure 5b is the schematic of the linear surface wave micromotor.

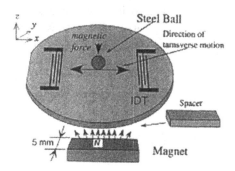

Figure 9a. Set up for experiments for the optimum preload conditions

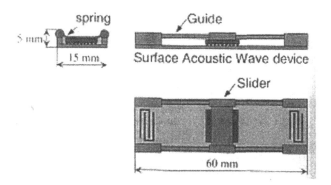

Figure 9b. Future design of a surface acoustic wave linear motor

Maximum speed measured is 0.7 m/s for a maximum acceleration of 900m/s². The output force is 0.5 mN when the driving voltage is in order of 120 V at 10 MHz, and pre-load is 310 MPa.

The figure 10 presents a rotation ultrasonics motor based on a travelling wave propagations at the stator surface generated from two standing waves in quadrature. This is a general principle for such kind of motor where the rotor waves by friction on the travelling wave. The diameter is 5 mm and the technology in micromechanics similar to those used in watch industry [12].

Figure 10. Miniaturized ultrasonic motor

To miniaturize more the motors and to involve into batch process microfabrication compatible with the silicon process, it is necessary to adjust the thickness of the piezoelectric drivers versus the stator thickness. Works were done by J.F. Manceau *et al.* Using a thin layer of PZT sputtered up to 3.5 μm thick, ultrasonics silicon based micromotors wave fabricated in a composite structure. The steps of the process are summarized in the flow chart of the process presented on the figure 11.

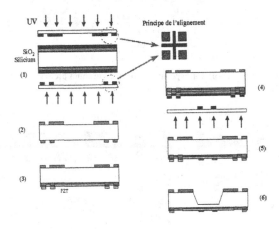

Figure 11. Flow chart of the microfabrication of the piezoelectric/silicon stator of an ultrasonic micromotor

1- Patterning of the alignment cross and the etching windows
2- development
3- Ti/Pt layer and PZT sputtered layer
4/5- lift off patterning of electrodes
6- Silicon chemical etching to get the statoric membrane

These process is performed on 3" wafers for 2 types of micromotors : 10 × 10 mm membranes leading to 5 mm of diameter ultrasonic motors and 3.5 × 3.5 mm to fabricated 2 mm of diameter micromotor.

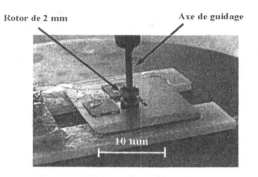

Figure 12. 2 mm ultrasonic micromotor fabricated with a silicon membrane and a PZT thin film

The figure 12 shows the 2 mm ultrasonic micromotor assembled with its ratio and ready to run [12].

6. Conclusion and perspective

The general problem of the piezoelectric materials used in microsensors and microactuators was presented to underline the potentiality and the actual limits to increase this mode of detection or actuation in MEMS and microsystems applications. The main limit for piezoelectric microsensor is the temperature influence in resonant and sensors. Moreover it is also limited by one extra drift the intrinsic aging of the devices which is difficult to avoid or modelized.

The microactuators need large movements in extension and compression. The multilayer piezoelectric stacks could be one of the solution ; combining extension and flexure could also be efficient [13]. Moreover, the technical compatibility of the piezotechnique and the silicon or glass batch process must be improved to lead better devices and cheaper systems. But it needs to developed low temperature deposition techniques or new method to bound piezoelectric and silicon wafers.

Furthermore, the piezo materials are in a good competition for dedicated MEMS applications today and tomorrow.

References

[1]   Wenk B., Ruethmuller W., Wagner B., Bereler W., (1991) Piezoresistive silicon low pressure sensor, *Eurosensors V Conference*, Editor A. d'Amico, ELSEVIER, Roma.

[2]   Tai Y.C., Fau L.S., (1989) Muller R.S., IC-processed micromotors :design, technology and testing, *Proc. of the IEEE Micro Electro Mechanical Systems Workshop*, Salt Lake City, Uthah,. 1-6.

[3]   Ruethmuller W., Bereler (1988) W., Thermally excited silicon microactuators, *IEEE. Trans. on Electron. Devices*, **35** (6), 758-763.

[4]   Bertch A., Zissi S., Calin M., Ballandras S., Bourjault A., Hauden D., André J.C., (1996) Conception and realization of miniaturized actuators fabricated by Microstereophotolithography and actuated by shape memory alloys, *Proc. of Mechatronics Symposium '96*, Besançon, 631-634

[5]   Fuhuda T., Hosohai H., Ohyama H., Hashimoto H., Arai F., (1991), Giant magnetostrictive alloy (GMA) applications to micromobile robot as a microactuator without power supply cable, *Proc. Of the IEEE Micro Electro Mechanical Systems Workshop*, Nara, 210-2.15.

[6]   Laurent Th. (1997), Génération, propagation et détection d'ondes de Lamb. Modélisation et applications aux membranes des couches minces piézoélectriques sur silicium, French PhD Thesis, Besançon.

[7]   Truchot Ph. (1995), Définition et modélisations analytiques et numériques de structures vibrantes en quartz en vue de la réalisation de microcapteurs à sortie fréquentielle; French PhD Thesis, Besançon.

[8]   Ballandras S., Briot J.B., Martin G., (1997) Theoretical and experimental study of gravimetric sensitivity of transverse waves on thin quartz plates, *Proc. of the 11th European Forum in Time and Frequency*, Neuchâtel, 163-166.

[9]   Bourriot N., Ballandras S., Steichen W., Maitre P., Piwakowski B., (1997) Design and fabrication of a miniaturized annular array, *IEEE Ultrasonics Symp. Proc.* Toronto, 1713-1716.

[10]   Piranda B., (1998) Etude du rayonnement et des couplages parasites sur antennes acoustiques par une méthode mixte éléments finis/éléments de frontière, French PhD Thesis, Besançon.

[11]   Kuribayashi Kurosawa M., Tahakashi M., Higuahi T., (1998) Elastic contact conditions to optimize function drive of surface acoustic wave motor, *IEEE. Trans. on UFFC* **45**(5), 1229-1237.

[12]   Manceau J.F., Bastien F. and Duffait R., (1994) Ultrasonics micromotors stator evaluation, *IEEE Ultras. Symp. Proc.*, Cannes, 539-543.

[13]   Glumac D. E. and Robbins W.P., (1995) A planar unimorph-based actuator with large vertical displacement capability. Part I : Experiment, *IEEE Trans. on Ultrasonics, Ferroelectrics, and Frequency Control*, **45**(5), 1145-1150.

FABRICATION AND MODELING OF MINIATURIZED PIEZOELECTRIC TRANSDUCERS FOR IMAGING APPLICATIONS

S. BALLANDRAS, B. PIRANDA, W. STEICHEN*, P.MAÎTRE, N. BOURRIOT, P. GUINE
*Laboratoire de Physique et Métrologie des Oscillateurs du C.N.R.S.
associé à l'Université de Franche-Comté
32 Avenue de l'Observatoire, 25044 Besançcon Cedex
*THOMSON MICROSONICS - TMX
399 Route des Crètes, B.P. 256
06904 SOPHIA ANTIPOLIS*

Abstract

Development of new technologies for microsystems design and fabrication pushes toward the new concept of micro-electro-mechanical systems. In this approach, the implementation of miniaturized ultrasound transducers for biomedical application represents a very important issue from both scientific and market points of view. The proposed work consists in applying ultrasound micromachining processes to manufacture very small PZT-based transducers capable to be integrated in microcatheter for detection or imaging purpose. It is shown that such fabrication approach is compatible with power full modeling tools based on finite elements analysis.

1. Introduction

Recent advances in Micro-Electro-Mechanical Systems (MEMS) have created a strong interest in the design and fabrication of small piezoelectric transducers for various applications in the sensor and actuator domain [1]. Much of work has been devoted to the manufacture of small piezo-ceramics devices exhibiting high aspect ratios [2] for the development of miniaturized acoustic probes dedicated to endoscopic imaging. In this area, new fabrication techniques such as the LIGA process [3] or silicon micromachining [4] appear necessary to overcome intrinsic limitations of classical sawing technologies. These approaches dramatically reduce the dimensions of acoustic imaging probes and implement new structures or patterns of piezoelectric transducers to improve their typical characteristics. It is now possible to combine such miniaturized probes with micro-electro-mechanical devices such as micromotors and front end electronic control modules to provide fully integrated

C. Galassi et al. (eds.), Piezoelectric Materials: Advances in Science, Technology and Applications, 347–355.

endoscopes. Consequently, the fabrication of smart catheters of diameter smaller than 1.5 mm can now be considered as a realistic issue.

However, the reliable design of piezoelectric transducers remains rather difficult to achieve because of the complexity of the structures and principles on which they are based. For example, the characterization of acoustic properties of materials such as polymers widely employed in the fabrication of the transducers is rarely satisfactory for precise modeling of the acoustic behavior of the probe. Taking into account the effective structure of the transducer also requires precise control of assembly processes to obtain the theoretically predicted characteristics of a given device.

Finally, the effective working conditions of the device (radiation in water, in viscoelastic solids or in layered media) are generally not considered when designing the probe because they require advanced theoretical models based on finite element and boundary element methods [6].

Nevertheless, many authors have shown interest in numerical calculation tools for modeling piezoelectric transducers [7, 8]. Considering these different aspects of the problem, our research group has initiated the development of a design environment for the fabrication of miniaturized piezoelectric transducers used in acoustic imaging.

It is based on finite element and boundary element methods implemented in a set of programs starting from the MODULEF package [9] for 2D and 3D problems including graphic procedures to animate the studied structures. These tools have been validated for a large number of structures, including the radiation conditions in fluids [10, 11]. At the same time, a model updating procedure has been developed [12] to solve the problem of the precise characterization of material constants, allowing us to deduce very efficiently not cuts elastic and dielectric losses, but also shear elastic constants for which experimental measurements remain difficult and not accurate.

The present paper demonstrates the application of the above design package for the implementation of a miniaturized annular array transducer devoted to endoscopic acoustic imaging. The first section reviews the finite element calculation. From a technological point of view, different solutions have been considered to identify a simple approach well-adapted to the fabrication of complicated 3D transducers. The second part of the paper deals with ultrasonic micromachining techniques to design and build high aspect ratio 3D pattern transducers. Test structures have been manufactured first to demonstrate the technological limits of the approach. Finally, a 2mm diameter annular array composed of 4 rings has been fabricated. The last section of the article shows the comparison between experiments and finite element predictions, emphasizing the good agreement between theory and experiments. As a conclusion, improvements of the present work are directed toward the fabrication of a 7 channel annular transducer in a 1.5 mm. diameter catheter.

2.   Finite element calculations

In this section, brief description of the finite element method for piezoelectric transducers is presented. More precise developments and explanations about this subject are given in [6-14] providing a comprehensive outlook of the topic. The philosophy of this calculational technique consists in minimizing the variations of the Lagrangian established for piezoelectric materials. For this purpose, the Lagrangian is discretized to simulate finite dimension structures, and the elastic, piezoelectric and dielectric properties of the corresponding domain are approximated using polynomial developments. This operation corresponds to a division of the transducer into elementary volumes of finite dimensions (see fig. 1) for which derivations and integrations of the equilibrium equations are performed.

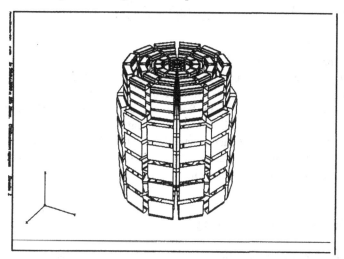

Figure 1. Typical example of a mesh used to simulate annular piezoelectric transducers

Finally, integrals of the discretized Lagrangian are reduced to a system of linear algebraic equations which is solved to find the nodal values of the unknown of the problem, namely the mechanical displacements u_i and the associated scalar electric potential ϕ. The minimization of the Lagrangian is assumed for any variation δu_i and $\delta \phi$ of the unknowns, yielding the following relation for a monochromatic electrical excitation (in $e^{j\omega t}$) :

$$
\begin{bmatrix} K_{uu} - \omega^2 M & K_{u\phi} \\ K_{\phi u} & K_{\phi\phi} \end{bmatrix} \begin{Bmatrix} u_i \\ \phi \end{Bmatrix} = \begin{Bmatrix} f \\ Q \end{Bmatrix} \quad (1)
$$

where K_{uu} and M represent respectively the stiffness and mass matrices of the purely elastic part of the problem related via the angular frequency ω, $K_{u\phi}$ is the piezoelectric matrix and $K_{\phi\phi}$ the dielectric matrix. The right hand side of eq. (1) is composed of the mechanical and electrical loads (respectively f and Q forming the load vector), which are left free in the case of vibration in air. Since the matrix of eq. (1) is not definite positive and may exhibit singularities, a resolution   algorithm based on the Crout factorization is implemented, providing u_i and ϕ at each node of the mesh and for each ω. A problem involving many different materials is also reduced to the general form of eq.(1) without any loss of generality. The admittance of the transducer is obtained by multiplying the matrix of eq. (1) by the solution vector and by performing the sum of all the electrical charges on the active electrode (cf. fig. 1). For an electrical excitation of magnitude 1 Volt, the admittance Y is given by :

$$Y(\omega) = j\omega \times Q_{tot}(\omega) = j\omega \sum_n Q_n(\omega) \text{ (2)}$$

where n is the index denoting the nodes of the active electrode. This calculational principle is valid for both 2D and 3D models. It is emphasized that the matrices K of eq. (1) are generally complex to take into account elastic and dielectric losses of the materials. The case of radiation in fluids has been implemented using Green's function and boundary integrals inserted in the finite element equations as boundary conditions. More details about such calculations will be found in [6, 10, 11]. For the present development, only the operation in air has been simulated and compared to experiment (see section 4 of this paper).

3.   Ultrasound micromachining

As noted in the introduction, different microfabrication techniques are available now for building piezoelectric transducers. For example, the LIGA process can be advantageously used for the production of transducer arrays based on 1-3 composite structures [2]. Although this technique appears one of the most promising for high density arrays, it seems at the present time rather difficult to create multilayers devices and it requires very good control of the PZT material when firing and polarizing the final transducer. For the present work, another technique has been investigated based on the direct machining of a piezoceramic layer.

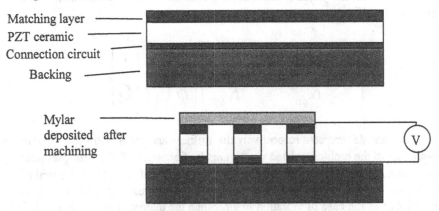

Figure 2. Scheme of a multilayer structure machined in one pass using ultrasound techniques

The basic idea is to build the final device in the multilayer structure of fig. 2, composed of a backing, a connection circuit, a PZT plate and a conductive matching layer, in one pass. A mylar foil is deposited *in fine* to ground the different rings of the array, the excitation being applied via the connection circuit.

Ultrasound micromachining has been widely used for the fabrication of quartz resonators requiring high precision and quality control during the fabrication process.

It is well adapted to the machining of piezoelectric materials since a minimum amount of heat is produced during the process, which theoretically reduces the problems of depolarization encountered with classical sawing techniques.

Furthermore, this technique can be applied for range of materials of different nature (single crystals, ceramics, polymers) and it is consequently well suited for etching multilayer structures.

The principle of ultrasound micromachining shown in fig. 3 consists in reproducing a form patterned on what is called a " sonotrode ". This operation is performed in a liquid containing abrasive grains of boron carbide (in our application) which are mechanically activated by the motion of the sonotrode. A piezoelectric transducer is used to excite the sonotrode at its resonance frequency (20 kHz) for which the amplitude of the motion at its end is a maximum. A radial and longitudinal precision of respectively 5 µm and 10 µm are achievable with such a process. The resolution is directly governed by the tip of the sonotrode (typically 50 µm in this case).

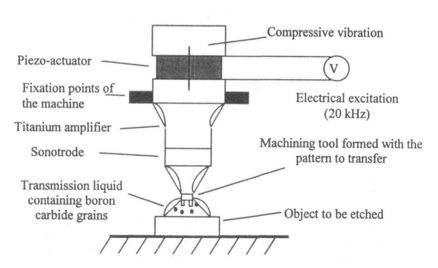

Figure 3. General principle of ultrasound micromachining

Different tests have been performed to check the validity of this approach for machining PZT layers. First, the evolution of the coupling coefficient versus width over thickness ratio has been measured and compared to results using classical sawing techniques. A 235 µm thick PZT plate glued on a backing was used. It was shown [13] that similar electrical characteristics can be measured for structures built

352

using both techniques. Aspect ratios close to 10 (see fig. 4) were obtained with a coupling coefficient of 0.65 (max. 0.7 for aspect ratios close to 3). Another test consisting on the fabrication of 100 and 50 μm width ridges in a 4 layer plate (mylar + 2matching layers + PZT glued on a backing) has been performed successfully, demonstrating the efficiency of the approach.

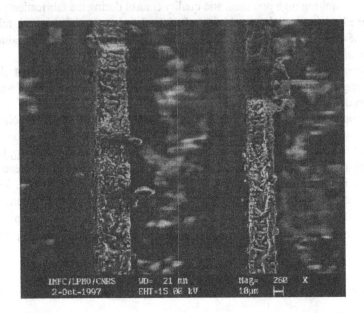

Figure 4. SEM photograph of high aspect ratio PZT beams (35 and 27 μm width) built in a 235 μm thick plate

4.   Miniaturized annular array

Considering the results shown in the previous section, one can investigate the manufacture of 7 ring annular arrays of a diameter 1.3 mm this was designed initially to provide efficient acoustic focusing by an electronically controlled phase shift the excitation of the different rings [14]. However, due to the low adherence of the PZT on the backing which becomes critical element widths smaller than 50 μm, larger scale devices have been fabricated and tested first. A 0.5 mm thick PZT plate (Ferroperm) has been glued to a backing (thickness 2 mm.) using an electrically conductive epoxy based layer (Protavic CM326 LVA+LVB). This was also used to build the matching layer (100 μm thick), deposited by a serigraphy process. 4 elements of 300 μm width have been machined in this plate, allowing us to obtain a 4 channel array within a diameter of 2 mm. Figure 5 shows a SEM view of the final device, demonstrates out the quality of ultrasound machining for complicated miniaturized patterns such as annular arrays.

The electrical response of the different rings has been measured using a network analyzer in reflection mode allowing us to compare the effective acoustic behavior of the array with finite element predictions. Figure 6 shows the comparison between theory and experiments for the first inner ring of the array. The good agreement between both curves indicates that the principal modes of the structure are correctly represented even by using interpolation polynomials of degree 1. The frequency shift between the 2 curves are due to a lack of precision in the characterization of the PZT provided by Ferroperm. The first peak is more attenuated than what was predicted, due to the difficulty correctly estimating acoustic losses in the backing or in the epoxy layers. Nevertheless, it is shown that the general behavior of the device is faithfully simulated by simple finite element calculations in the spectral domain, allowing us to precisely design and optimize any 3D transducer structure.

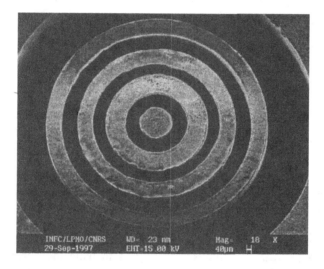

Figure 5. SEM view of the miniaturized annular array (ext. diameter 2mm, thickness 600 μm)

5.   Conclusion

An ultrasound micromachining process has been implemented and tested for the fabrication of piezoelectric transducers using PZT and polymers in one pass. This approach represents a very flexible solution for complicated 2D or 3D structure to be machined in ceramics like PZT. It allows a significant reduction of complexity for the fabrication of multilayer devices since it is able to machine different materials without loss of precision due to alignment after etching operations. Moreover, large aspect ratios (up to 10) can be achieved for very small element width (down to 27 μm). The ability of the process to machine complicated 3D structures has been demonstrated by manufacturing a 2 mm diameter 4 channel annular array.

Experimental measurements of the electrical admittance of this device have been compared to theoretical calculations using our home made 3D finite element package,

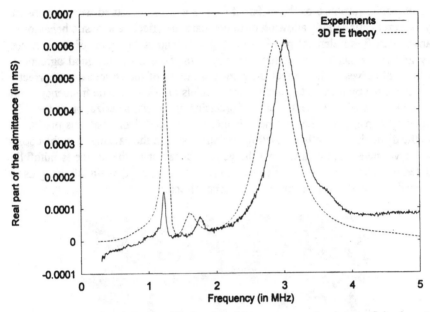

Figure 6. Electrical admittance of the first ring of the array : comparison between finite element predictions and experimental data

pointing out the agreement between the different results. It is then possible to efficiently use finite element calculations for an optimized design procedure even if some material properties are not precisely characterized.

Some difficulties must be overcome to fabricate 7 channel annular arrays in less than 1.5 mm diameter. The main problem to be solved consists in the improvement of the adherence between the different layers used to build the transducer. In this approach, the use of conductive epoxy-based polymers would be avoided at the backing level and replaced by more efficient gluing. This operation may be even performed after machining the PZT and the matching layer with the coherence of the structure being ensured by the mylar or any additional layer removed after gluing the PZT on the backing. Direct contact of the rings across the backing is currently being studied

6. References

1. Trimmer, W.S., (1997) *Micromechanics and MEMS : Classic and Seminal Papers to 1990*, IEEE Press, Piscataway
2. Hirata, Y.,Okuyama, H., Ogino, S., Numazawa, T., Takada, H. (1995) Piezoelectric composites for micro-ultrasonic transducers realized by deep etch X-ray lithography, *Proceedings of the IEEE MEMS'Workshop*, 191-196
3. Becker, E.W., Ehrfeld, W., Hagmann, P., Maner, A., Münchmeyer, D., (1986) Fabrication of microstructures with high aspect ratios and great structural heights by synchrotron radiation

lithography, galvanoforming, and plastic moulding (LIGA process), *Microelectronic Engineering*, 4, 35-56

4. Kaminsky, G., (1985) Micromachining of Silicon mechanical structures, *Journal of Vacuum Science Technology*, B3, 4, 1015-1024

5. Knapen, P.M., Meyer, B.J., de Vries, P.F., (1995) Micromotor and guide wire, in particular for guiding catheters, with such a micromotor, *Int. Patent n° WO 95/32 539*

6. Lerch, R., Landes, H., Friedrich, W., Hebel, R., Hoess,A., Kaarman, H.T., (1992) Modeling of acoustic antennas with a combined finite element/boundary element method, *Proceedings of the IEEE Ultrasonics Symposium*, 581-584

7. Allik, H. and Hugues, T.J.R., (1970) Finite element method for piezoelectric vibration, *Int. J. Num. Meth. Eng.*, 2, 151-157

8. Steichen, W., Vanderbork, G., Lagier, Y., (1988) Determination of the power limits of a high frequency transducer using the finite element method, *Power Sonics and Ultrasonics Trans. Dev.*, 160-174

9. Bernadou, M., George, P.L., Hassim, A., Joly, P., Laug, P., Muller, B., Perronet, A., Saltel, E., Steer, D., Vanderbork,G., Vidrascu, M., (1988) MODULEF : Une bibliothèque modulaire d'éléments finis, I.N.R.I.A. Ed., Rocquencourt

10. Piranda, B., (1998) Etude du rayonnement et des couplages parasites sur antennes acoustiques par une methode mixte éléments finis/éléments de frontière, Thèse de Doctorat en Sciences pour l'Ingénieur, Université de Franche-Comté, Besançon

11. Piranda, B., Steichen, W., Ballandras, S., (1998) Comparison between different finite element/boundary formulations for modeling acoustic radiation in fluids, *Proceedings of the IEEE Ultrasonics Symposium*, 1073-1076

12. Piranda, B., Steichen, W., Ballandras, S., (1998) Model updating applied to ultrasound piezoelectric transducers, *Proceedings of the IEEE Ultrasonics Symposium*, 1057-1060

13. Bourriot, N., Ballandras, S., Steichen, W., Maître, P., Piwakowski, B., Gelly, J.-F., Piranda, B., Müksch, R., (1997) Design and fabrication of a miniaturized annular array, *Proceedings of the IEEE Ultrasonics Symposium*, 1713-1716

14. Bourriot, N., Ballandras, S., Steichen, W., Piwakowski, B., Sbai, K., Gelly, J.-F., Piranda, B., (1997), Design of a miniaturized annular transducer for intravascular acoustic imaging, Microsim'97, Lausanne

FLEXTENSIONAL COMPOSITE TRANSDUCERS: DESIGNING, FABRICATION AND APPLICATION

Aydin Dogan, Kenji Uchino*, Robert E. Newnham *
Anadolu University, Ceramic Engineering Department
26470 Eskisehir TURKEY; * The Pennsylvania State University,
Materials Research Laboratory, University Park- PA 16802

Abstract

The design, fabrication and application of flextensional composite transducers are summarized in this study. The moonie and cymbal type flextensional transducer consist of a ceramic driving element sandwiched between two truncated conical metal or plastic endcaps. These transducers can be used as both sensors and actuators. Piezoelectric, electrostrictive and antiferroelectric-ferroelectric switching types of ceramic can be used as the driving element in either single layer or multilayer form. In the cymbal and moonie transducer designs, the flexibility and durability of metals and the driving power of the ceramic element are combined.

Poled ferroelectric ceramics (Curie groups ∞m) possess three independent piezoelectric coefficients: d_{31}, d_{33}, and d_{15}. Each of the piezoelectric coefficients can be used as the driving element of a composite cymbal or moonie transducer. Cymbal transducers can also be fabricated from ring shaped ceramics sandwiched between truncated metal endcaps. Polarization and electric field directions of the samples can be altered systematically to make use of the three different piezoelectric coefficients as driving elements. The cymbal is a versatile performer. Desired actuation and sensing performance can be tailored by engineering the flexibility of the endcaps or changing the cavity dimensions beneath the endcaps. Possible design changes can be also investigated with FEM by using computational tools. Cymbal and moonie transducers can be used as hydrophones, acceleration sensors, positioners, and in many other applications.

1. Introduction

Accentric materials show piezoelectricity, which is defined as the ability of developing an electrical charge proportional to a mechanical stress. The piezoelectric response may be either direct or converse. For the direct piezoelectric response, the electrical displacement is proportional to the stress applied to the material. In the converse effect, material strain is proportional to the field applied across the material. This quality of piezoelectric materials has led to their use in transducers, which convert electrical energy to mechanical energy, and vice versa. These electromechanical transducers have

C. Galassi et al. (eds.), Piezoelectric Materials: Advances in Science, Technology and Applications, 357–374.
© 2000 *Kluwer Academic Publishers. Printed in the Netherlands.*

found applications where they are used in either active, e.g. ultrasonic probes, or passive modes, e.g. hydrophones.

A number of monolithic materials exhibit piezoelectric behavior. Ceramics, polymers and as well as their composites belong to this materials group. Conflicting goals in optimizing physical and electromechanical properties of transducers have led the researchers to look at the composite materials.

Superior properties have been obtained with piezocomposites consisting of an active piezoelectric ceramic in an inactive polymer. The microstructural arrangement of component phases in the composite, sometimes referred to as connectivity, is a critical parameter for the electromechanical performance of the composite. For a composite containing two phases, there are sixteen different connectivity patterns. Over the past two decades, researchers have investigated several methods to process piezocomposites and improve their properties. The 0-3, 3-1, and 2-2 connectivities have been worked on extensively. There are several excellent review papers on connectivity patterns and the processing techniques used to form piezocomposites [1], [2]. Various composite have different applications. Competition is still going on to determine the composite with the best electromechanical performance.

Second generation piezocomposites include the moonie and cymbal with 2-(0)-(2) connectivity. Ceramic-metal composites generally have a simple design with metal faceplates, shells or caps that couple the ceramic to the surrounding medium. The metal component transfers the incident stress to the ceramic or the displacement to the medium. Flextensional transducers are good examples of ceramic-metal composites. In flextensional transducers, the flexural vibration of the metal shell causes an extensional vibration of the piezoelectric ceramic. The moonie and cymbal transducers possessing 2-(0)-2 connectivity are miniaturized versions of flextensionals. This paper describes the moonie and cymbal type composite transducers.

2. Principle

The moonie and cymbal transducers consist of an electro-active ceramic driving element sandwiched between two metal endcaps with shallow cavities on their inner surface. Designs of the moonie and cymbal transducers are illustrated graphically in Figure 1 and Figure 2. In the case of the moonie, the cavities are in the shape of a half moon, whereas the cymbal has a truncated cone-shaped cavity. The presence of these cavities allows the metal caps to serve as mechanical transformers for converting and amplifying a portion of the incident axial-direction stress into tangential and radial stresses of opposite sign. Thus, the d_{33} and d_{31} contributions of the PZT now add together (rather then subtracting) in the effective d value of a device, such as a hydrophone or accelerometer. Regarding the converse effect, the radial motion of the ceramic driving element is transferred and amplified by the metal endcaps in axial direction. Hence, moonie and cymbal can be used as both sensors and actuators.

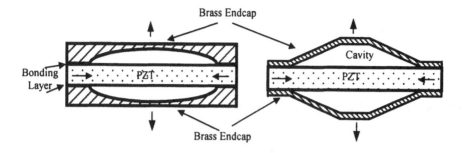

Figure 1. Schematic of Moonie Transducer.   Figure 2. Schematic Cymbal Transducer.

2.1 FIRST APPROXIMATION FOR THE CHARGE CALCULATION ON THE CYMBAL AND MOONIE TRANSDUCER

Let us assume that we have a flextensional transducer which has cone shaped endcaps The charge created under an applied stress on the cymbal transducer at uniaxial direction perpendicular to the metal endcaps can be estimated as follows.

The force is applied to the cone on the summit and transferred to the piezoelectric ceramic. The essential parameters for the calculation are marked on Figure 3.

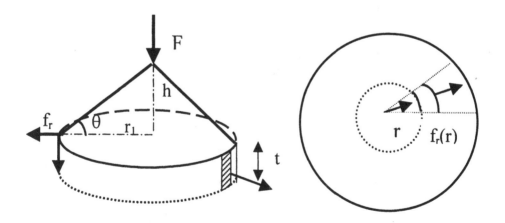

Figure 3. Front and top view of flextensional tranduser cone shaped endcap
Where: h, cavity depth; r_1, cavity radius; t, thickness of ceramic,
θ angle between cone and piezoelectric ceramic.

360

$$\frac{h}{r_1} = \tan \theta \qquad (1)$$

The total force on the entire circumference is

$$2\pi r_1 f_r = \frac{F}{2 \tan \theta} \qquad (2)$$

and the force per unit area at r= r₁ is

$$\frac{f_r}{t} = \frac{F}{4\pi r_1 t \tan \theta} \qquad (3)$$

where t is the thickness of the ceramic disk.

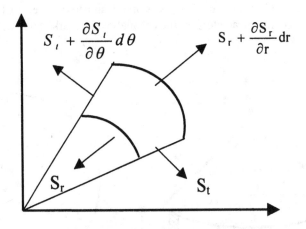

Figure 4. Stress distrubition on a finite segment of ceramic element.

When a force balance is written and solved for plane stress case, the following equation is obtained.

$$S_r = S_t = \frac{F}{4\pi r_1 t \tan \theta} \qquad (4)$$

The charge generation is calculated from,

$$\int_0^{r_1} \underbrace{d_{31}(S_r + S_t)}_{P(r)} 2\pi r dr = d_{31} \frac{r_1}{2t} \frac{F}{\tan\theta} \qquad (5)$$

when two endcaps are used:

$$Q = d_{31} \frac{r_1}{t} \frac{F}{\tan\theta} \qquad (6)$$

$$C = \epsilon \frac{\pi r_1^{\,2}}{t} \qquad (7)$$

$$V = \frac{Q}{C} = d_{31} \frac{1}{\epsilon \pi r_1} \frac{F}{\tan\theta} \qquad (8)$$

where is the Q charge; C, Capacitance; V, voltage, and ϵ the dielectric permittivity.

For example: For a flextensional transducer with conical endcap, which has 9.0 mm cavity diameter and 0.2 mm cavity depth and consisting of 1.0-mm thick PZT-5A ceramic disc with 153 pC/N d_{31} coefficient, the calculated effective charge is around 17,500 pC/N

The effective piezoelectric coefficients of moonie and cymbal transducers, with a 12.7-mm outer diameter and 9.0-mm cavity diameter and 0.2-mm cavity depth consisting of PZT-5A ceramic, are shown in Figure 5. In moonie and cymbal structure d_{31} and d_{33} coefficients work together (in general in piezoceramics they work always against each other due to poisson effect). This is part of the reason of high effective charge coefficient. The moonie shows highly position dependent behavior for the effective piezoelectric coefficient. The cavity beneath the endcap of the moonie actuator plays a crucial role on the characteristics of the moonie [3]. The effective piezoelectric coefficient increases with increasing cavity diameter and cavity depth. The effective piezoelectric coefficient of the moonie transducer decreases rapidly with increasing the endcap thickness. The piezoelectric charge coefficient of moonie transducers also show position dependent behavior similar to that of the displacement.

Placing a groove 9.0-mm in diameter, 0.2-mm in depth, and 1.0-mm in width in the

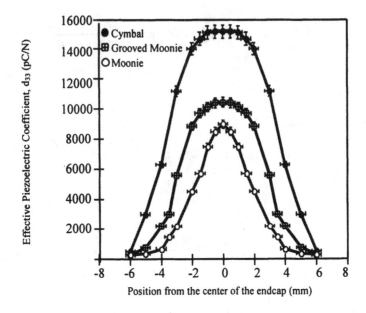

Figure 5. Effective piezoelectric charge coefficient of same size various design of flextensional transducers.

brass endcaps increased the effective piezoelectric coefficient almost 20%. The piezoelectric charge coefficient is simply the ratio of the applied stress and the induced charge. Thus, we can say that the groove on the endcaps increases the stress transformation ratio and therefore the generated charge. The moonies with grooved endcaps also showed less position-dependent behavior. For the 2-mm diameter section at the center of the samples, the effective piezoelectric coefficient is about 11,000 pC/N.

With the cymbal endcaps, the piezoelectric coefficients increased almost 60%. For a cymbal 12.7-mm in diameter and 1.7-mm in total thickness, an effective piezoelectric coefficient of more than 15,000 pC/N was measured over the 4-mm diameter center section of the cymbal transducer. We have concluded that the thick metal region near the edge of the moonie metal endcaps is a passive region, which does not assist stress transfer, and acts to decrease the total efficiency. Cymbal endcaps transfer the stress more efficiently and improve the energy transfer markedly. The calculated result for conical endcap and the measured result for cymbal with truncated conical endcaps are in very good agreement.

3. Fabrication Method

When fabricating the composite moonie and cymbal transducers, piezoelectric, electrostrictive, or antiferro- to- ferroelectric ceramics can be used as the driving element in either single layer or multilayer form. Shallow cavities beneath the endcaps of the moonie transducers are machined into the inner surface of each of the end caps. Brass, phosphor bronze, other flexible metals and their alloys, and even acrylic can be used as endcap materials.

For cymbals the truncated conical endcaps are punched using a specially designed punch to fabricate the transducers rapidly at minimal cost. Shaping and cutting operations are carried out simultaneously during processing. The final product requires only surface treatment for good bonding. Moreover, endcaps can be easily fabricated from metal sheets by punching. With this fabrication routes it is possible to fabricate identical endcaps with minimal labor.

The metal moonie and cymbal endcaps are bonded to the electro-active ceramic disks around the circumference using soldering alloy or epoxies such as Eccobond epoxy resin from Emerson & Cuming, taking special care not to fill the cavity. The thickness of the epoxy bonding layer must be very thin, approximately 20 μm. Using silver epoxy as a gluing agent, composite actuators may also be stacked together at the center of the endcaps to achieve still higher displacements. Moonie and cymbal transducers can be easily fabricated 3-50 mm in diameter and 1-3 mm in total thickness. Currently most of the studies are done on the 12.7-mm diameter 1.5-mm thick samples.

4. Design Optimization With the Support of Finite Element Analysis

Finite element programs ATILA, MARK, and ANSYS were used for the design and development stages of the moonie and cymbal transducers. The cymbal actuator is a second-generation moonie-type composite developed using FEA analysis in collaboration with experiment. Finite Element Analysis has identified high stress concentration in the metal endcaps just above the edge of the ceramic metal bonding layer near the edge of cavity [4].

The stress concentration on the brass endcap just above the bonding layer reduces the effective force transfer from the PZT to the cap. It is possible to eliminate part of the stress concentration by removing a portion of the endcap just above the bonding region where the maximum stress concentration is observed. An enhancement in properties has been observed by introducing a ring shaped groove on the exterior surface of the endcaps [5]. By moving the groove toward the edge of the actuator, the displacement increases. The highest displacement was achieved when the groove was above the edge of bonding layer. It is found that the deeper and wider the groove, the higher the displacement [6].

In reality, placing a ring shaped groove on the endcap does not eliminate the stress but further concentrates it into a very narrow region. Stress concentrations at the groove edges are a potential source of fatigue and may eventually produce failure under long-term usage. Moreover, additional labor is required to machine the groove into the endcaps.

The cymbal transducer with truncated endcaps has been designed to remove much of the stress concentration and to produce higher and more reproducible displacements.

364

Although this new design looks similar to the earlier moonie design, it has a different displacement mechanism. Displacement is primarily a flexural motion of the endcap for the original moonie design, but for the new design, the displacement is created by the combination of flexural and rotational motions. Figure 6 shows the displacement values of the different endcap designs with a fixed cavity depth (0.20-mm) and diameter (9.0 mm). A linear Voltage Differential Transformer (LVDT) was used for displacement analysis of the transducers. A variable electric field up to 1kV/mm with triangular waveform at 0.1 Hz is applied to the sample. A moonie actuator with 0.30-mm thick brass endcaps provides a 22-μm displacement. Using the ring-shaped groove design, the displacement was increased to 32 μm with a groove 9.0 mm in diameter, 0.2 mm in depth, and 1.0 mm in width machined into the brass endcaps of the same actuator. A cymbal actuator with uniformly thick punched endcaps exhibits around 40-μm displacements, about twice the moonie displacement, and about 50 times larger than uncapped PZT.

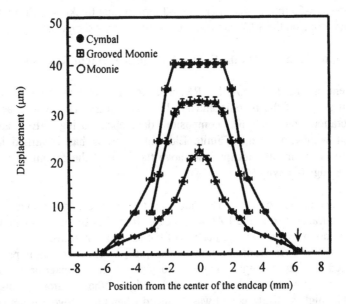

Figure 6. Displacement characteristics of moonie and cymbal flextensional transducers.

4.1 EFFECT OF SIZE AND SHAPE

The dimensions of the cavity beneath the endcaps play a crucial role in the cymbal performance. Engineering the flexibility of the endcaps or changing cavity dimensions can tailor the desired actuation. The displacement, which is the result of converse piezoelectric effect, has a quadratic relationship with the cavity diameter. Figure 7 shows the calculated and experimental results of displacement cavity relation of cymbal transducer with constant endcap thickness (0.25mm) and PZT thickness (1.0 mm). Figure 8 shows the optimal cavity depth, which is around 0.25 mm for a cymbal transducer with constant cavity diameter (9.0 mm) and PZT thickness (1.0 mm).

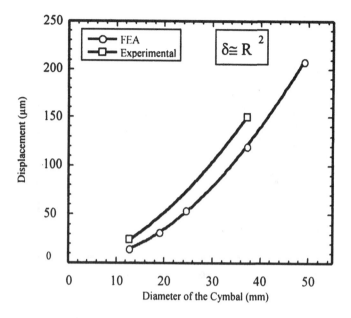

Figure 7. Displacement cavity relation of cymbal transducer with constant endcap thickness (0.25mm) and PZT thickness (1.0 mm).

4.2 EFFECT OF ENDCAP MATERIAL

The flexibility and modulus of elasticity of the endcap material are important parameters defining the important actuator properties: displacement, response speed and generative force, of moonie and cymbal actuators and transducers. Figure 9 shows the effects of the Young's modulus of the metal endcaps and the hardness of the PZT ceramics on the displacement of composite cymbal actuators. Increasing the Young's modulus of the metal endcaps reduces the displacement of the cymbal actuator. This reduction is almost linear and the displacement of the highest Young's modulus metal endcaps is approximately 55% lower than that achieved using the most compliant metal endcaps. Similar behavior is observed for all PZT ceramic types. The linear decrease of the displacement with the increase if the Young's modulus of the metal confirms the spring-like nature of the endcaps.

Figure 10 shows the net displacement of the cymbal actuators made with different metal endcaps. The net displacement is the displacement produced by the actuator when it is electrically driven. If the actuator is loaded, a free deflection is produced and at each load, the application of an electric field produced a net displacement. The free deflection of the composite is related to the spring characteristics of the composite. Metals with low modulus of elasticity show higher displacement and higher free deflection. On the contrary, the maximum load, defined as the load for which 90% or the initial net displacement is maintained, increased with the modulus of the elasticity of the metal endcaps.

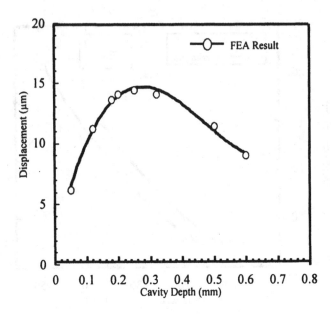

Figure 8. The optimal cavity depth for a cymbal transducer with constant cavity diameter (9.0 mm) and PZT thickness (1.0 mm).

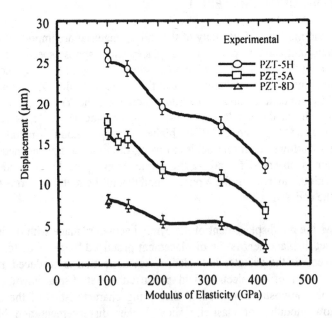

Figure 9. Effect of endcap modulus of elasticity on the displacement performance.

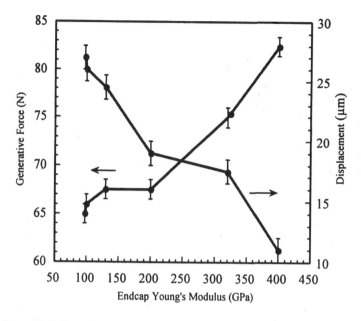

Figure 10. Trade- off between generative force and displacement for the cymbal.

4.3 EFFECT OF ELECTRO-ACTIVE CERAMIC DRIVING ELEMENT

PZT ceramics, PMNPT relaxor ferroelectrics and PNZST type antiferro-ferroelectric ceramic elements can be used in single layer or in multilayer form. Typical displacement hysteresis graphs of 12.7-mm diameter 1-mm thick samples made of all three kinds of electro-active ceramics materials are shown in the Figure 11. Displacement hysteresis for the cymbal transducers, 12.7 mm diameter 1.0 mm PZT thickness, and 9.0 mm cavity diameter 0.2 mm cavity depth, fabricated with all three kind of electro-active ceramic element are shown in Figure 12. Cymbal actuators with soft PZT ceramics exhibit linear displacement with a rather large hysteresis, which is the indication of losses. The cymbal actuator with the PMN-PT type of ceramic driving element shows larger displacement with lower losses. However, it exhibits a nonlinear displacement consistent with the relaxor characteristics of the PMN-PT ceramics. Linear displacement characteristics can be achieved by using charge driving electrical circuitry. PNZST is an antiferroelectric-ferroelectric type phase transition material. Unique property of this material is the volumetric expansion under applied electric field. With cymbal endcap design, this volumetric expansion is converted to a negative axial displacement.

4.4 VARIOUS DRIVING MODES TO ACTIVATE THE CYMBAL TRANSDUCER

Poled ferroelectric ceramics (Curie groups ∞m) possess three independent piezoelectric coefficients: d_{31}, d_{33}, and d_{15}. By careful designing, each of the piezoelectric coefficients can be activated as the driving mode of a composite cymbal or moonie transducer.

368

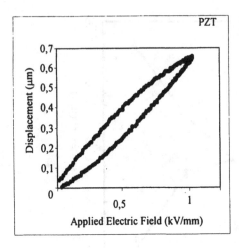

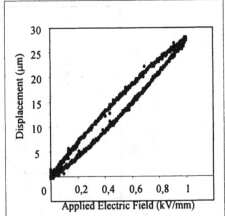

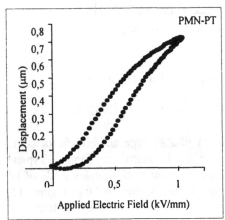

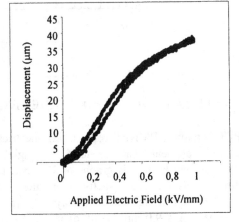

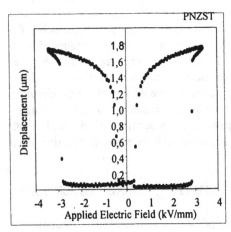

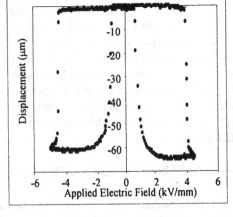

Figure 11. Hysteresis of various electro-active ceramics.

Figure 12. Hysteresis of cymbals with various electro-active ceramics.

Cymbal transducers can be fabricated from ring shaped ceramics sandwiched between truncated metal endcaps. Polarization and electric field directions of the samples can be altered systematically to make use of the three different piezoelectric coefficients as driving power. Figure 13 shows four combinations between electric field and internal polarization of the ceramic driving element. In case one, the radial mode d_{31} of the piezoelectric ceramic element is used to activate the endcaps. In case two, the longitudinal coefficient d_{33} is used to activate the metal endcaps. The shear mode d_{15} of the piezoceramic possesses the highest value piezoelectric coefficient value. Cases 3 and 4 are designed to activate the ceramic in a shear mode. Figure 14 shows the displacement versus inner ring diameter of the piezoelectric driving element relation of the cymbal transducer. From the graph, it may be seen that the d_{31} mode, which exhibits the highest displacement, is still the best mode to drive the cymbal transducer. Although d_{15} is the highest piezoelectric coefficient, it is not effective for driving cymbal transducer in the present design. It has to be kept in mind that the shear mode is a rather soft mode. A rigid solid disc has to be placed between the rings. In general increasing the size of the inner diameter decreases the displacement value, because of decreasing volume of the ceramic body and volumetric efficiency of transducer. From the manufacturing point of view, case 1 is the best, because it is easy to fabricate and easy to apply electric field without developing a short circuit. The other three cases are more difficult to fabricate and requires rather careful electric wiring.

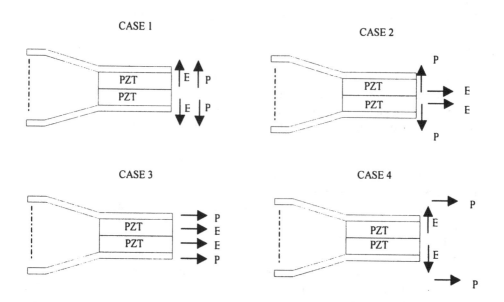

Figure 13 Various driving modes of the cymbal transducer with a ceramic ring.

370

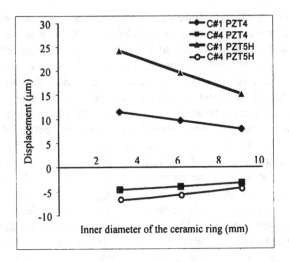

Figure 14. FEA results for displacement ceramic inner diameter relation.

5. Application of the moonie and cymbal transducers

Moonie and cymbal transducers have great potential for both sensor and actuator applications [7], [8]. They can also be utilized as the switching element in valve designs. There is a volume change inside the moonie and cymbal transducers during cycling. This volume change can be utilized in minipump applications.

5.1 ACTUATORS

Flextensional moonie and cymbal actuators with their moderate generative force and displacement values fill the gap between multilayer and bimorph actuators. Each solid-state actuator design has attractive features that can be exploited for certain applications. The advantages of the moonie and cymbal actuators are the easy tailoring of the desired actuator properties by altering the cavity size and endcap dimensions. Easy fabrication is another advantage. Several features of the various solid-state actuator designs are listed in Table I. It is rather difficult to compare the different actuators because of differences in geometry and various operating conditions for specific applications. To make a fair comparison, similar dimensions for each actuator were selected, and the measurement conditions are those specified in Table I. The rainbow actuator also partially covers the gap between multilayer and bimorph actuators [9]. For that type of actuator, a reduction step during processing of the ceramic element at high temperature results in a semiconducting layer and stress-bias. Although it shows flexural motion, the rainbow can be categorized as a monomorph or a unimorph type of actuator. The effective coupling factor of rainbow is theoretically smaller than the moonie and cymbal. High applied electric field, position-dependent displacement and cost are the main disadvantages of the rainbow actuator in comparison with the cymbal. In the moonie and cymbal design, multilayer piezoelectric ceramics can be used as the driving element to reduce the applied voltage. The moonie and cymbal actuator can be used as a micropositioner for applications requiring small size with relatively quick response. OMRON Corporation has already succeeded in using the multilayer moonie actuator for an optical scanner [10]. Other applications for the cymbal and moonie include sensing and vibration suppression elements in the automotive and aerospace industry, switching element in valve design, micropositioners requiring small size with relatively quick response for precise positioning device in CD-ROM and magneto-optic memory storage driver, mini-pumps, relays, and switches, printer hammers, and linear and rotary ultrasonic motors

Table I. Comparison of the Solid State Ceramic Actuator Designs.

Features	Multilayer	Bimorph	Rainbow	Cymbal	Moonie
Dimensions	5x5x12. (LxWxT) mm^3	12.7x10x0. 6 (LxWxT) mm^3	F 12.7 mm T= 0.5 mm	F 12.7 mm T= 1.7 mm	F 12.7 mm T= 1.7 mm
Drive Voltage (V)	100	100	450	100	100
Displacement (µm)	10	35	20	40	20
Contact surface (mm^2)	25	1	1	3	1
Generative Force (N)	900	0.5-1	1-3	15-60	3
Position dependent of displacement	None	Maximum at the tip	Maximum at the center	Maximum at the center but more diffuse	Maximum at the center
Stability under loading	Very high	very low	low	high	low
Fastest Response Time (µsec)	1- 5	100	100	5-50	5-50
Fabrication method	Tape casting and cofiring at 1200 °C	Bonding ceramic element with metal shim	Reducing ceramic element at 950 °C	Bonding ceramic element with metal endcaps	Bonding ceramic element with metal endcaps
Fabrication Cost	high	low	medium	low	medium

5.2 HYDROPHONE APPLICATION

Because of their very high piezoelectric charge coefficients, moonie and cymbal transducers can be used as hydrophones, accelerometers and air acoustic transducers. The advantages of the cymbal-type hydrophone are very large d_h (hydrostatic charge) and g_h (hydrostatic voltage) coefficients along with lightweight and inexpensive fabrication. Cymbal also has excellent potential for use as a shallow water projector.

Cymbal has a Q less than 10 when water loaded. The moderate TVR exhibited by a single element device can be greatly enhanced by incorporating them into a close packed array [11], [12]. Hydrophone figures of merit ($d_h.g_h$) of some of the widely used composites and single element transducers are compared in the plot in Figure 15. Due to the size dependence of some transducers, the figure of merit is calculated for a 1-cm^2 transducer for a valid comparison. Cymbal exhibits the highest figure of merit among all composites. Figure 16 shows the pressure dependence of the effective d_h and g_h coefficient of identical transducers with different cap materials and 0.25 mm cavity depth. These data clearly show that caps made of stiffer metals are capable of withstanding higher pressures without degradation in performance. Stiffer caps are not as efficient in transferring stress to the piezoceramic, which is why the effective d_h coefficient drops for cymbals with stiffer.

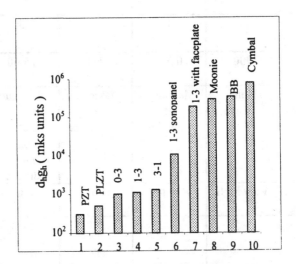

Figure 15. Hydrophone figures of merit ($d_h.g_h$) of some of the widely used composites and single element transducers.

5.3. ULTRASONIC MOTOR APPLICATION

An ultrasonic motor was derived from moonie and cymbal design [13]. The motor is named the windmill because of the appearance of the slitted endcaps. Detailed information concerning this motor can be obtained from the article of K. Uchino and B. Koc "Compact piezoelectric ultrasonic motors" in this proceeding.

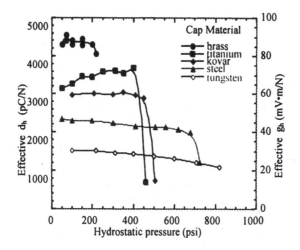

Figure 16. The pressure dependence of the effective d_h and g_h coefficients for same size cymbal transducers with different cap materials.

5.4 ACCELEROMETER APPLICATION

Cymbal transducer has been investigated for accelerometer applications [14]. A high effective piezoelectric charge coefficient (d_{33}) of the cymbal transducer was observed around 15000 pC/N, which is much higher than that of piezoelectric ceramic, around 550 pC/N. With this feature, the cymbal transducer is a good candidate for highly sensitive accelerometer applications. Figure 17 shows the Log sensitivity frequency relation of cymbal accelerometers with various endcaps in comparison with PZT itself. Cymbal accelerometers have more than two orders of magnitude higher sensitivity than PZT ceramics at low frequencies.

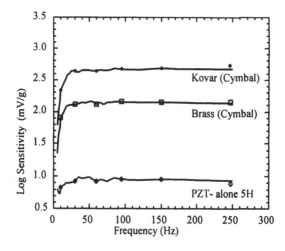

Figure 17. Acceleration sensitivity of cymbal in comparison with PZT itself.

REFERENCES

1. Newnham R.E., Skinner D.P. Cross L.E., "Connectivity and Piezoelectric-Pyroelectric Composites," *Mater Res Bull* 1978, 13, 525
2. Gururaja, T.R., A. Safari, R.E. Newnham and L.E. Cross, "Piezoelectric Ceramic-Polymer Composites for Transducer Applications," *Electronic Ceramics*, Edited by L.M. Levinson. Marcell Dekker. Inc., New York, 92 (1987)
3. A. Dogan, S. Yoshikawa, K. Uchino, R.E. Newnham, "The Effect of Geometry on the Characteristics of the Moonie Transducer and Reliability Issue", *IEEE Ultrasonic Symposium Proceedings* , Vol. II. pp. 935-939, 1994
4. Q. C. Xu, S. Yoshikawa, J. R. Belsick and R. E. Newnham, "Piezoelectric composites with high sensitivity and capacitance for use at high pressure," *IEEE Trans. on UFFC* Vol. 38, pp. 634-639, 1991.
5. Y. Sugawara, "Development of metal-ceramic composite piezoelectric actuators and their applications," M.S. Thesis, Sophia University, Tokyo Japan, 1991.
6. K. Onitsuka, A. Dogan, Q.C. Xu, S. Yoshikawa, R.E. Newnham," Design Optimization for Metal-Ceramic Composite Ceramic Composite Actuator, "*Ferroelectrics* Vol. 156, pp. 37-42, 1994.
7. J. F. Tressler, "Smart ceramic-metal composites for active vibration control," *M.S. Thesis*, Ceramic Science, The Pennsylvania State University, 1993.
8. K. Onitsuka, A. Dogan, J.F. Tressler, Q.C. Xu, S. Yoshikawa, R.E. Newnham, " Metal-Ceramic Composite Transducer, The Moonie", *J. Int. Mat. Sys. & Struct.*, Vol. 6, pp. 447-455, 1995.
9. H. Goto, K. Imanaka, "Super compact dual axis optical scanning unit applying a torsional spring resonator driven by a piezoelectric actuator," *Proc. of SPIE*, Vol. 1544, pp. 272-281, 1991
10. G. Haertling, "Rainbow ceramics, a new type of ultra high displacement actuator," *Bull. of Am. Ceramic Soc.*, Vol. 73, No 1, pp. 93-96, January 1994
11. J.F. Tressler, A. Dogan, J.F. Fernandez, J.T. Fielding, K. Uchino, R. E. Newnham "Capped Ceramic Hydrophone" *IEEE-UFFC Ultrasonic Symposium Proceeding*, Seattle 1995.
12. J.F. Tressler, R. E. Newnham, W.J. Hughes "Capped Ceramic underwater sound projector: The 'cymbal' transducer " *J.Acoust. Soc. Am. 105 (2), pp 591-600 Febuary 1999*
13. B. Koc, A. Dogan, Y. Xu, R.E.Newnham,and K. Uchino " An Ultrasonic Motor Using a Metal-Ceramic Composite Actuator Generating Torsional Displacement," *Jpn. J. Appl. Phys. Vol.* 37 (1998) pp 5659-5662
14. B. Koc, A. Dogan J.F. Fernandez, R.E. Newnham, and K. Uchino "Accelerometer application of the moonie and cymbal transducers" *Jpn. J. Appl. Phys. Vol.* 35 (1996) pp 4547-4549.

A DISK TYPE PIEZOELECTRIC TRANSFORMER WITH CRESCENT SHAPE INPUT ELECTRODES

BURHANETTIN KOC AND KENJI UCHINO
International Center for Actuators and Transducers
Materials Research Laboratory
The Pennsylvania State University, University Park PA 16802

Abstract

For most PZT based piezoelectric materials the shear mode coupling constant is twice as large as the transverse mode coupling. From this motivation, a new circular piezoelectric transformer was designed. Non-concentric input electrodes and non-uniform output poling lead to the usage of shear mode piezoelectric effect at (3,0) radial mode resonance frequency. A prototype three layered transformer with 25.4-mm in diameter and 1.5 mm in thickness, operating at 153 kHz, was fabricated and its characteristics were measured. Characteristics, such as efficiency, step-up ratio and temperature rise of the proposed transformer are presented.

1. Introduction

One of the bulkiest components in information processing equipment (such as note-book-type personal computers) is the power supply, specifically the electromagnetic transformer used in power supply. Losses such as skin effect, thin wire loss and core loss of the electromagnetic transformer increase rapidly as the size of the transformer is reduced. Therefore, it is difficult to realize miniature low profile electromagnetic transformers with high efficiency.

High efficiency, small size, no electromagnetic noise are some of the attractive features of piezoelectric transformers making them more suitable for miniaturized power inverter elements such as lighting up the cold cathode fluorescent lamp (CCFL) behind a color liquid crystal displays (LCD) or generating high voltage for air-cleaners.

The original design to transform an input ac voltage to step up or step down using converse and direct piezoelectric properties of ceramic materials was proposed by Rosen [1]. The principle of this type of transformer is to excite a piezoelectric element (Figure 1) at its mechanical resonance frequency. Applying an electrical input to one part of the piezoelectric element generates a mechanical vibration and then this mechanical vibration is converted into electrical voltage from the other part of the piezoelectric plate.

375

C. Galassi et al. (eds.), Piezoelectric Materials: Advances in Science, Technology and Applications, 375–382.
© 2000 *Kluwer Academic Publishers. Printed in the Netherlands.*

The initial Rosen type transformer had major reliability problem, which is the easy mechanical breakdown at the center position due to the coincidence of the residual stress concentration and the vibration nodal point. In addition to improved mechanically tough ceramic materials, by redesigning the electrode configuration and exciting a third longitudinal resonance mode of the rectangular plate, the piezoelectric transformer shown in Figure 2a was commercialized by NEC in 1994 for miniaturized back-light inverter to light up the cold cathode fluorescent lamp (CCFL) behind a color liquid crystal displays (LCD) [2].

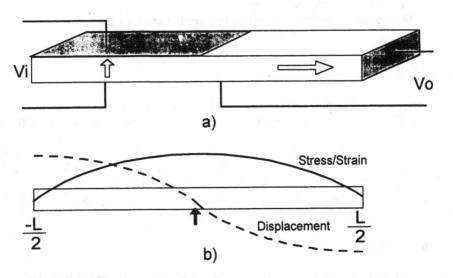

Figure 1. a) Rosen type piezoelectric transformer, b) Stress/strain and displacement distribution at first longitudinal resonance mode.

Another third mode transformer shown in Figure 2b was commercialized, recently, by Mitsui-Sekka to use it also for miniaturized power inverter element [3]. In addition to no electromagnetic noise generation, small size and high efficiency are some of the advantages of the piezoelectric transformers over the electromagnetic types particularly for computer applications.

The third mode transformers have some advantages over original Rosen type first mode transformer, such as follows:
i) for the third mode transformer, stress concentration on the bar is separated into three different points which makes the transformer mechanically stronger. The electric field and stress concentration are not at the same point on the piezoelectric bar, while for the first mode transformer, the maximum stress and nodal point are at the center of the piezoelectric bar that makes the transformer mechanically weaker. This is the major problem of the first mode transformer. Moreover, in order to get higher efficiency the transformer has to be clamped only at a nodal point. Because the third mode transformers have three nodal points on the bar, they can be clamped at two points

easily without affecting the vibration modes. Therefore, third mode transformers can be packaged more safely.

ii)   the third mode transformers are more efficient (96%) than the first mode transformers (90 %).

iii) impedance matching of the third mode transformer with CCFL is better than the first mode transformer.

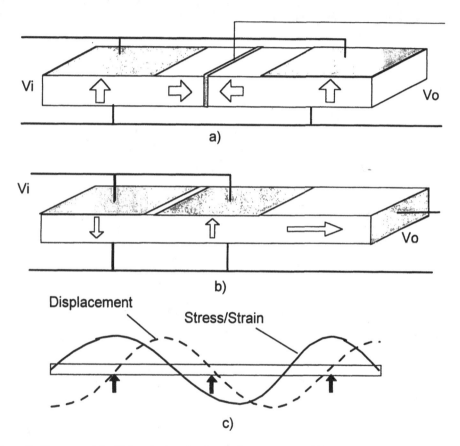

Figure 2.   Structures of the third mode piezoelectric transformers, proposed by (a) NEC, (b) Mitsui-Sekka, (c) stress/strain and displacement distribution.

For most PZT based piezoelectric materials, the shear mode coupling constant is twice as large as the transverse mode coupling.   From this motivation, a new disk type piezoelectric transformer was designed and it will be introduced in the following section.

2. Disk Type Piezoelectric Transformer

All the rectangular type transformers explained in the previous section are using the transverse mode coupling constant (k_{31}) of the piezoelectric materials. For most piezoelectric materials, however, the shear mode coupling constant (k_{15}) is twice as large as the transverse mode coupling constant (k_{31}). From this motivation, we designed a new circular piezoelectric transformer, which uses shear (k_{15}) or planar (k_p) mode coupling constants of the piezoelectric effect (Fig 3). The transformer is operating at the third radial resonance frequency of the piezoelectric disk.

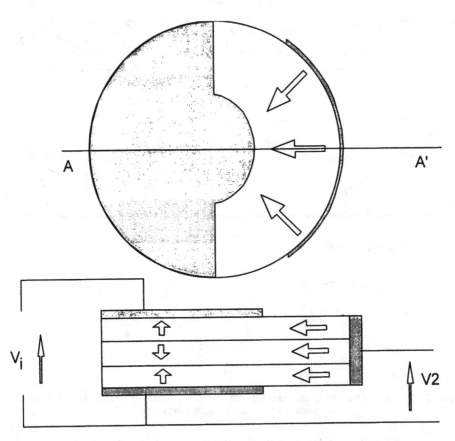

Figure 3. Newly designed circular type piezoelectric transformer structure.

The behavior of the transformer was modeled using ATILA finite element software. The fundamental and third harmonic mode shapes and potential fields are shown in Figures 4 and 5. When the transformer is driven at the first radial resonance frequency it uses planar mode coupling constant effectively. If the transformer is driven at the third harmonic radial frequency, it uses shear mode coupling coefficient. This can be seen from the mode shape and potential field shown in Figure 5.

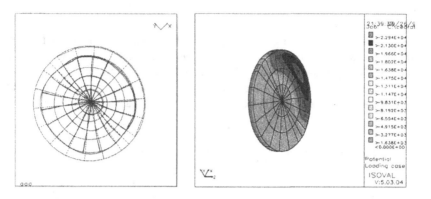

Figure 4. First mode potential field and displacement shape of the circular transformer (67 kHz FEM calculations using ATILA©).

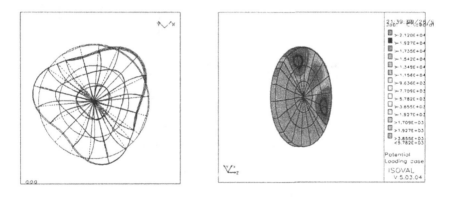

Figure 5. Third mode potential field and displacement shape of the circular transformer (152 kHz FEM calculations using ATILA©).

3. Experimental

The transformer structure shown in Figure 4 was fabricated using commercially available piezoelectric ceramic material (APC International, USA). Hard piezoelectric ceramic disks 25.4 mm in diameter and 0.5 mm in thickness were sputtered with

380

platinum electrodes according to the proposed electrode configuration. The transformer was subjected to a polarization treatment by applying a DC voltage of 3.0 kV/mm at 150 °C across the input and output electrical terminals. Three identical single layer transformers were then stacked using adhesive epoxy as the last step of the fabrication process.

In order to clarify the operating frequency of the transformer, the input (output terminal open-circuited) and output admittance spectra were measured and the results are shown in Figure 6. The possible operating frequencies of the transformer are first three radial mode resonance frequencies of which both input and output terminals can excite. The best performance was obtained when the transformer was driven at the third mode resonance frequency.

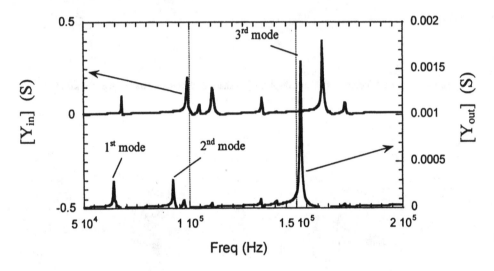

Figure 6. Input and output admittance spectrum.

The transformer low field voltage step-up ratio and efficiency were measured for different resistive loads and the results are shown in Figure 7. The output power on the resistive loads was 1.5 Watts and the transformer was driven around 153 kHz. When measuring the efficiency, the operating frequency of the transformer was tuned so that the maximum efficiency could be obtained for different resistive loads. The efficiency for a resistive load ranging from 100 to 1200 kΩ is found to be around 90 %. For the same range of load resistance, the voltage step up ratio was increased from 70 to 320. These primary results clearly indicate that the proposed transformer can realize a step up ratio and power high enough to light up the cold cathode fluorescent lamp (CCFL) for back-lights in color liquid crystal displays (LCD) when it is operated at the third mode resonance frequency.

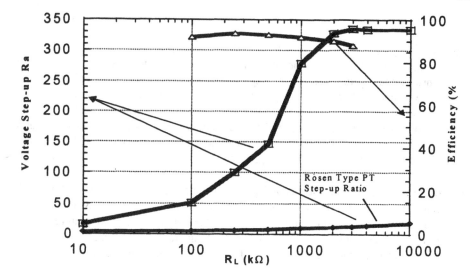

Figure 7. Step-up ratio and efficiency as a function of load resistance.

Figure 8 shows the output/input voltage step-up ratio and temperature rise of the piezoelectric transformer with variable output power for a constant resistive load of 110 kΩ. Temperature rise for a 110 kΩ resistive load is not significant up to 10 Watts of output power. When the output power is further increased, the temperature of the transformer was observed to increase by more than 80 °C from room temperature. The transformer step-up ratio, however, decreased gradually from 100 to 70 for the same load condition.

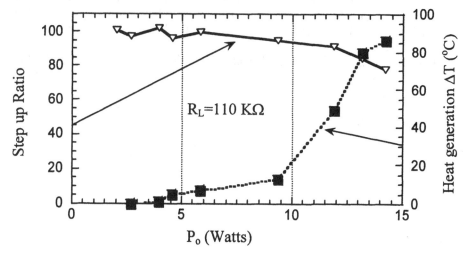

Figure 8. Step-up Ratio and Temperature rise as a function of output power on a resistive load.

4. Conclusions

A new disk type piezoelectric transformer design was proposed using three layered commercially available hard piezoelectric disks. Non-concentric input electrodes and non-uniform output poling lead to the usage of shear mode piezoelectric effect at (3,0) radial mode resonance frequency. A prototype three layered transformer of 25.4-mm in diameter and 1.5 mm in thickness, operating at 153 kHz, was fabricated and its characteristics were measured. The transformer can generate enough step-up ratio and power to light up CCFL for back-light of color liquid crystal displays (LCD) when it is operated at the third mode resonance frequency.

Heat generation and step up ratio as a function of output power was also investigated. Temperature rise for a 110 kΩ resistive load is not significant up to 10 Watts range of output power. When the output power was further increased the temperature of the transformer increased more than 80 °C from room temperature. The transformer step-up ratio however decreased gradually from 100 to 70 for the same load condition.

Acknowledgements

This work was partially supported by the Turkish Higher Educational Council and the Office of Naval Research, USA.

5. References

1. Rosen, C.A. (1959), Electromechanical Application, in H.W. Katz (eds), *Solid State Magnetic and Dielectric Devices*, New York: Wiley, pp. 171-229.
2. Kawashima, S., Ohnishi, O., Hakamata, H., Tagami, S., Fukuoka, A., Inoue, T. and Hirose, S. (1994) Third Order Longitudional Mode Piezoelectric Ceramic Transformer and its Application to High-Voltage Inverter, *IEEE Int. Ultrasonic Symp,* Cannes, 525-530.
3. Kanayama, K., Maruko, N. and Saigoh, H. (1998) Development of the Multilayer Alternately Poled Piezoelectric Transformers, *Jpn. J. Appl. Phys.* Vol. 37, pp.2891-2895, 1998.
4. Tsuchiya, H. and Fukami, T. (1986) Design and Principles for Multilayer Piezoelectric Transformers, *Ferroelectrics*, **68**, 225-234.

DESIGN AND PERFORMANCE OF A LINEAR PIEZOELECTRIC WEDGEWORM ACTUATOR

GARY H. KOOPMANN, GEORGE A. LESIEUTRE, JEREMY
FRANK, AND WEICHENG CHEN
Center for Acoustics and Vibration
157 Hammond Building,
Penn State University, University Park, PA 16802

Abstract

A new concept in linear piezoelectric actuators is developed for applications in adaptive, conformable structures for flow control. Motivated by a desire for high actuation force (>1kN) and simplified drive signals, the design takes advantage of self-locking wedges to lock the clamping elements. The concept relies heavily on knowledge and manipulation of the friction coefficients between several surfaces, so the choice of coatings and lubricants are a major part of the investigation. Since the wedges are self-locking in one direction, the actuation force is limited only by the size of the piezoceramic and the strength of the actuator structure. The device contains a single piezoceramic stack (8x8x42 mm, PZT 5H), so the drive signals and amplifiers are drastically simplified from previous designs. A prototype of the concept is developed and experimentally tested. At a drive frequency of 200 Hz, the free velocity is 8 mm/s with a travel of 25 mm. An actuation force of 250 N is achieved with the prototype. The wedge concept also reduces the amount of precision necessary in machining and assembling the device.

1. Introduction

The possible applications of a high force, high displacement linear actuator are numerous, and the development of such a device has received considerable attention in recent years, as evidenced by the patents disclosed.[1-16] The clamping mechanism in most of the designs consists of a piezoceramic element clamping directly (more or less) onto a moving part. Since the displacements of the piezoceramics are usually on the order of 10-20 microns, this means that the machining accuracy must be within a few microns to ensure a high clamping force (Figure 1a). Such tight tolerances can be very expensive, and any inaccuracy quickly leads to a loss of actuation force. The new

383

C. Galassi et al. (eds.), Piezoelectric Materials: Advances in Science, Technology and Applications, 383–390.
© 2000 *Kluwer Academic Publishers. Printed in the Netherlands.*

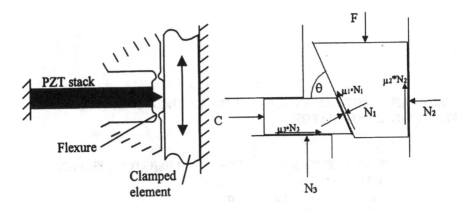

Figure 1 a) Direct clamping method b) Passive wedge clamping method

wedge concept employs a passive mechanism for clamping, relaxing the need for low surface tolerances and eliminating the piezoelectric clamping stacks (Figure 1b). For certain values of the parameters shown the wedge will lock, preventing the upper block from sliding down.

2. Design of the Wedgeworm Actuator

The motion of a wedgeworm actuator is quite similar to that of previously designed inchworm-type linear actuators[17-23.] (Figure 2). A central pusher element consisting of two halves is separated by the piezoelectric stack and held together by flexures cut from the structure. The flexures are used to apply a pre-compressive force to the stack, necessary to keep the stack from going into tension. As seen in Figure 1b), the left wedges are constrained to move horizontally and are in contact with the pusher surface such that the surfaces are self-locking and the pusher cannot slide down. For this schematic, the right wedges are not in contact with the pusher. The right and left wedges are rigidly attached and pulled left or right such that only one set or the other is in contact with the pusher at any time. Here, the left wedges are pulled to the right by a spring force, keeping them in constant contact with the pusher. Motion of the device is as follows: 1) The piezoceramic stack is de-energized and the pusher is locked in place by the left wedges. 2) The stack is energized with voltage. The bottom of the pusher is locked by the clamp and cannot move down, but the top is able to extend up. As the pusher moves up, the top left wedge moves to the right so that contact is maintained. 3) The stack is de-energized. Now, the top is locked in the extended position, so the flexures pull the bottom of the pusher up, completing one step. This time, the bottom left wedge moves to the right to maintain contact with the pusher. The process is repeated at a high driving frequency (~200 Hz) to give a high actuation rate and smooth motion.

In order to use the wedge concept as an actuator clamping mechanism, it is necessary to understand the physics that make the wedges self-locking. The equilibrium equations

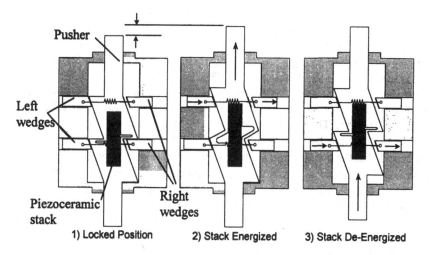

Figure 2 Operational mechanism of the wedgeworm concept

for a static system consisting of two wedges can be simplified to a single equation relating two applied forces on the wedges, C and F (Figure 3). If the ratio is negative the wedges are self-locking. If C/F is positive, then a downward force F must be supported by a horizontal force C, so the wedges are not self-locking. In Figure 3, the self-locking equation is a function of the coefficient of friction at three contacting surfaces and the wedge angle. As a possibility for unlocking the wedge for two-way motion, a change in the coefficient of friction at the wedge interface is assumed. This could be accomplished by extending a teflon (low friction) pad from the wedge surface with a piezoceramic stack. For the actuator to have the maximum amount of travel, the wedge angle θ should be as high as possible. The problem is subject to two constraints, however. First, the self-locking equation must be negative (locked) using the original friction parameters. With the teflon pad extended, the equation should be positive (not locked), which gives the second constraint on the parameters. A non-linear simulated annealing optimization routine is used to maximize the wedge angle subject to the self-locking and non-self-locking constraints. A summary of the optimization problem is shown in Figure 4. The objective function is to maximize

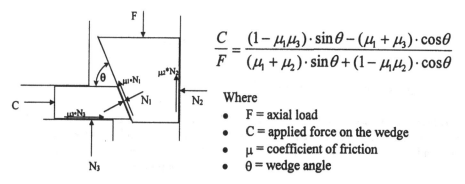

$$\frac{C}{F} = \frac{(1 - \mu_1\mu_3) \cdot \sin\theta - (\mu_1 + \mu_3) \cdot \cos\theta}{(\mu_1 + \mu_2) \cdot \sin\theta + (1 - \mu_1\mu_2) \cdot \cos\theta}$$

Where
- F = axial load
- C = applied force on the wedge
- μ = coefficient of friction
- θ = wedge angle

Figure 3 Self-locking wedge equilibrium equation

Note that in the equation, this means that μ_1 is replaced with μ_S, the friction coefficient for teflon. There are five variables in the problem: the wedge angle and four friction coefficients (three original contacting surfaces and one for teflon). Shown in Figure 4, the permissible range for the wedge angle is from 0 to 90 degrees. The chosen range for the friction coefficients is 0.002 to 0.78, selected as maximum and minimum realistic friction values from handbooks. This range has since been shown to be difficult to realize in practice for machined and/or coated surfaces, particularly the upper limit on friction. The program uses a non-linear optimization technique to maximize the wedge angle by varying the four friction coefficient values. The solution always indicated that the coefficients μ_1 and μ_3 should be as high as possible, while the teflon (slipping) coefficient μ_S should be as low as possible (Figure 5a). Interestingly, μ_2, the vertical friction surface, also went to the lower bound, indicating that the vertical sliding surface should be lubricated. With a reasonably high safety factor, the maximized wedge angle was just over 65 degrees (Figure 5b), so this was chosen as the actuator wedge angle. To ensure a smooth horizontal sliding surface for the wedges, hardened steel rods were integrated into the wedgeworm frame (Figure 6). The device is 12 cm in length, but could be made smaller with a structurally optimized structure.

$$\max(\theta) \qquad\qquad \min(-\theta)$$

$$g_1(x) = \frac{(1 - \mu1 \cdot \mu3)\cdot \sin\theta - (\mu1 + \mu3)\cdot \cos\theta}{(\mu1 + \mu2)\cdot \sin\theta + (1 - \mu1 \cdot \mu2)\cdot \cos\theta} < SF$$

$$g_2(x) = \frac{(1 - \mu S \cdot \mu3)\cdot \sin\theta - (\mu S + \mu3)\cdot \cos\theta}{(\mu S + \mu2)\cdot \sin\theta + (1 - \mu S \cdot \mu2)\cdot \cos\theta} > SF$$

$$0.0 < \theta < 90.0$$
$$0.002 < \mu3 < 0.78$$
$$0.002 < \mu S < 0.78$$
$$0.002 < \mu1 < 0.78$$
$$0.002 < \mu2 < 0.78$$

Figure 4 Optimization problem statement

3. Testing the Prototype Actuator

The prototype wedgeworm actuator was tested to characterize its performance. Since the wedgeworm clamping mechanism is passive, the drive signal is a single frequency sinusoid that could be generated with a simple circuit. This is one of the this is that the only values to vary in experimental characterization experiments are

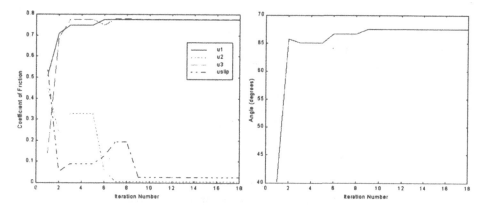

Figure 5 Optimization results showing a) friction coefficients, and b) wedge angle

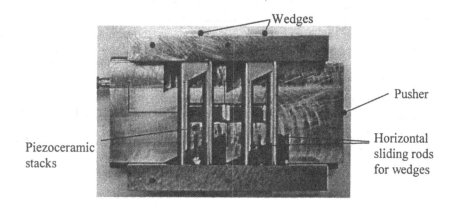

Figure 6 The prototype wedgeworm actuator

the signal voltage and frequency. To characterize the wedgeworm prototype, the free velocity (driving against no load) is first measured as a function of the drive signal frequency. The drive voltage (peak-to-peak, positively biased) is held constant at 144 V. This corresponds to an electric field of 11.5 kV/cm. As seen in Figure 7, the actuator velocity increases linearly with drive frequency. This is expected since the step size should remain constant with frequency, while more steps per second are taken as the drive frequency is increased. At a drive frequency of 200 Hz, the measured free velocity is 8.1 mm/s. It seems that higher speeds would be possible if the frequency were increased further, but the amplifier power limit was reached at 200 Hz. A velocity of 8.1 mm/s at 200 Hz corresponds to a step size of 40.5 microns, which approaches the theoretical free displacement of the stack, 42 microns. Thus, it seems that the backlash under no load is almost zero for the wedgeworm prototype. Next the maximum actuation force, or stall load, was measured as a function of drive voltage. The actuator was driven with a 100 Hz signal with a positively biased peak-to-peak voltage varying from 65 V to 150.

388

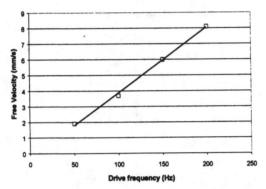

Figure 7 Free velocity of the wedgeworm vs. drive frequency

V (the maximum drive voltage for the stacks). The actuator pushed against a compression spring, and was allowed to push until a stall condition occurred Using the original wedge surfaces (stainless steel on stainless steel with $\mu = 0.3$), the maximum actuation force was 240 N for a 150 V drive signal. In an attempt to increase the stall load, a thin, hard coating of aluminum oxide was applied to the wedge surfaces. The intent was to increase the coefficient of friction at the wedge interface, thereby increasing the actuation force. As seen in Figure 8, the maximum actuation force increased with drive voltage to 246 N at 150 V, so the actuation force was not increased significantly with the aluminum oxide coating. The relationship between stall load and drive voltage is not quite linear, and seems in fact to be approaching a limit around 275 N. This indicates that as the actuation force becomes very high there may be some slip in the wedge clamps, limiting the stall load.

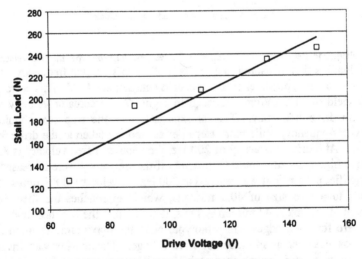

Figure 8 Maximum actuation force vs. drive signal voltage

4. Conclusions

A linear wedgeworm actuator was developed for structural shape control applications. Features of the subject device include very simple drive signals, high actuation force, high speed, and relaxed machining tolerances. An optimization procedure is used to determine the optimal friction surfaces and wedge angle to use in the design. A prototype actuator was built and shown to have a maximum actuation force of 250 N and a free velocity of 8 mm/s at a drive frequency of 200 Hz.

5. Acknowledgements

This work was supported by DARPA under the SAMPSON project, and performed in collaboration with engineers at Boeing St. Louis, General Dynamics, Penn State's Applied Research Laboratory, and PCB Piezotronics.

6. References

1. Stibitz, R. (1964) Incremental Feed Mechanisms, *U.S. Patent:* 3,138,749.
2. McNancy, J.T. (1964) Piezoelectric transducer force to motion converter, *U.S. Patent:* 2,154,700.
3. Hsu, K. and Biatter, A. (1966) Transducer, *U.S. Patent:* 3,292,019.
4. Locher, G.L. (1967) Micrometric linear actuator, *U.S. Patent:* 3,296,467.
5. Brisbane, .A.D (1968) Position control device, *U.S. Patent:* 3,377,489.
6. Galutva, G.V. (1972) Device for precision displacement of a solid body, *U.S. Patent:* 3,684,904.
7. Bizzigotti, R.A. (1975) Electromechanical translational apparatus, *U.S. Patent:* 3,902,085.
8. Sakitani, Y. (1976) Stepwise fine adjustment, *U.S. Patent:* 3,952,215.
9. Ishikawa, and Sakitani, Y. (1979) Two-directional piezoelectric driven fine adjustment device", *U.S. Patent:* 4,163,168.
10. O'Neill, G. (1980) Electromotive actuator, *U.S. Patent:* 4,219,755.
11. Taniguchi, T. (1984) Piezoelectric driving apparatus, *U.S. Patent:* 4,454,441.
12. Hara, A. Takao, H. Kunio, Y. Sadayuki, T. and Keiji, N. (1986), Electromechanical translation device comprising an electrostrictive drive of a stacked ceramic capacitor type, *U.S. Patent:* 4,570,096.
13. Staufenberg, C.W. Jr., and Hubbell, R.J. (1986) Piezoelectric electromechanical translation apparatus, *U.S. Patent:* 4,622,483.
14. Fujimoto, T. (1987), Piezo-electric actuator and stepping device using same, *U.S. Patent:* 4,714,855.
15. Murata, T. (1990), Drive apparatus and motor unit using the same, *U.S. Patent:* 4,974,077.
16. Rennex, G. (1994), Inchworm actuator, *U.S. Patent:* 5,3323,942.
17. Meisner J.E. and Teter, J.P. (1994) Piezoelectric/magnetostrictive resonant inchworm motor. *SPIE,* Vol. 2190, pp. 520-527.
18. Lee S.K. and Esachi, M. (1995) Design of the electrostatic linear microactuator based on the inchworm motion, *Mechatronics.* Vol. 5, No. 8, pp. 9653-972.
19. Funakubo, T. Tsubata, T. Tanigughi, Y. and Kumei, K. (1995), "Ultrasonic linear motor using multilayer piezoelectric actuators", *J. J. of Appl. Phys.,* Vol. 34, Part 1, No. 5B, pp. 2756-2759.
20. Pandell T. and Garcia, E. (1996), Design of a piezoelectric caterpillar motor, *Proceedings of the ASME aerospace division.* AD-Vol. 52, pp. 627-648.

21. Galante, T. (1997), Design and fabrication of a high-force linear piezoceramic actuator, M.S. thesis, Penn State University..

22. Newton, D., Garcia, E., Horner, G.C. (1998) A linear piezoelectric motor *Smart Materials and Structures* 7, 3, 295-304.

23. J. Frank, G.H. Koopmann, G.A. Lesieutre, and W. Chen, (1999), "Design and performance of a high force piezoelectric inchworm-type actuator", Proceedings of SPIE's 6th Annual International Symposium on Smart Structures and Materials, Vol. 3668.

PIEZOELECTRIC COMPONENTS FOR TECHNICAL APPLICATIONS

C. SCHUH, K. LUBITZ, TH. STEINKOPFF, A. WOLFF
Siemens AG, Corporate Technology, Munich, Germany

1. Introduction

Piezoceramic components are well established in many fields of large scale applications due to their fast electromechanical response and compact size, e.g. as buzzers, telephone diaphragms or ultrasonic transducers. In the last few years, complex multilayer actuators made by cofiring technique also gained increasing interest due to their specific advantages in performance like fast switching time, high stiffness and blocking force and low driving voltage (Fig. 1). These features will open a wide field of industrial applications for example in the automotive area for fast injection valves (Fig. 2) and in textile industry for automated weaver's looms (electronic Jacquard technology, Fig. 3).

However, besides the performance, the production costs and the long-term reliability of the components play a decisive role for practical application. The multilayer technology is well kown from ceramic capacitors (MLC's), but has to be carefully adapted for the manufacturing of actuators. Using the capacitor design with inactive regions, the durability of the electrical contact connecting the inner electrodes requires special solutions for dynamic driving conditions. Moreover, piezoelectric ceramics represent an extra challenge for the treatment and description of reliability because of their nonlinear and coupled electrical and mechanical material behavior.

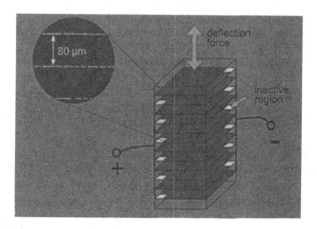

Figure 1. Design of a cofired multilayer stack actuator (grey regions represent electrodes, transparent regions represent PZT layers).

C. Galassi et al. (eds.), Piezoelectric Materials: Advances in Science, Technology and Applications, 391–399.
© 2000 *Kluwer Academic Publishers. Printed in the Netherlands.*

392

Figure 2. Multilayer piezoceramic stack actuators for fast injection valves in automotive applications.

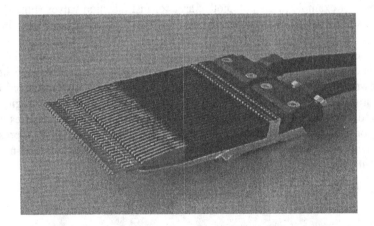

Figure 3. Weaver's loom control module for textile machines build up with piezoceramic trimorph bending elements.

2. Some relevant challenges in the processing of monolithic multilayer actuators

In order to build up cofired multilayer stack actuators with dimensions of 7x7x30 mm³ and PZT single layer thicknesses of 80 μm a manufacturing technology similar to the fabrication of multilayer ceramic capacitors (MLC's) has to be applied. A typical process flow chart is given in Fig. 4.

In practice lamination, debinding and sintering of the stacks are the most critical manufacturing steps. Each actuator consists of about 350 alternating PZT single layers and inner electrodes which have to be joined homogeneously without dislocations of the electrode patterns or initiation of weak interfaces.

In the debinding step, a large amount of organic material has to be removed from the green body. In the case of thermal decomposition, the chemical cracking of the binder phases and the evaporation of the crack products comprise both thermodynamically and kinetically controlled processes. With increasing stack volume, the operation time necessary for complete binder removal raises dramatically because of long diffusion pathes to the surface. Therefore, a sophisticated optimization of the thermal and chemical process conditions must be carried out, in order to avoid mechanical damage or overheating of the ceramic body (Fig. 5).

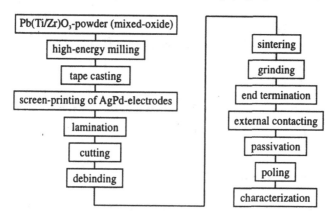

Figure 4. Manufacturing process of cofired multilayer actuators.

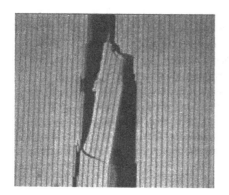
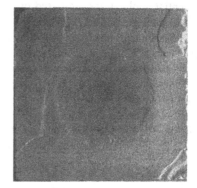

Figure 5. Crack initiation in a multilayer body during the debinding step, caused by unadapted heating rates. Left picture: side view; right picture: top view onto a delamination plane after separation.

In order to obtain PZT ceramics suitable for multilayer technology, investigations of lowering the sintering temperature, controlling the stoichiometry and designing the microstructural evolution were carried out sucessfully [1]. However, during cofiring of multilayer stacks additionally strong reactions between the PZT ceramics and the AgPd inner electrode paste take place (Fig. 6). It has been proved that Ag atoms are incorporated substitutionally in the PZT lattice at the A-sites leading to a shift of the morphotropic phase boundary [2,3]. Hence, the temperature regime and the PbO-partial pressure have to be controlled carefully during sintering in order to achieve a dense monolithic multilayer ceramics without cracks, pores and other non-homogeneities.

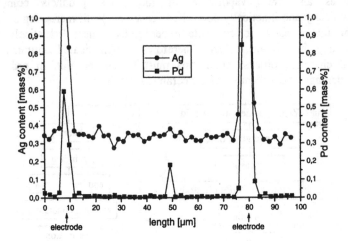

Figure 6. Ag and Pd distribution measured by electron probe microanalysis (EPMA), taken from a cross section of a cofired multilayer actuator.

Although several new interesting actuator concepts have recently been proposed [4-6], multilayer stack actuators exhibit some specific advantages such as high blocking force and stiffness, low driving voltage and high resonance frequencies. Hence, some potential applications are based on the availability of longer stack actuators to achieve higher deflections, but with small cross areas to obtain low enough capacitances, which will be suitable for driving electronics even under large-signal conditions. These stack geometries require a special handling concept during the production process, which should not raise the production costs in a substantial manner.

However, the Pd content of the inner electrode material is one of the highest cost factors of cofired multilayer actuators. The amount of electrode material necessary for a stack depends on the stack size, the PZT single layer thickness and the inner electrode thickness. Therefore, a balance must be found between the costs for the actuator and the costs for the driving unit in order to keep the final system price as low as possible.

3. Performance of monolithic multilayer actuators

Component optimization. Finite-Element methods are proven and powerful tools for loading analyses of complex structural and functional components. Especially in the sense of parametrization and optimization with regard to sizes, shapes, and loading conditions the Finite-Element method is its own recommendation. For cofired multilayer actuators, the dependence of deflection on shape and size of inactive regions is of special interest (Fig. 7).

In most cases the linear piezoelectric FE analysis of ferroelectric components is used for both the small and the large signal simulation. There are first attempts to insert nonlinear properties of the piezoelectric material, for example in the field of high precision applications. Some aspects of the coupled nonlinear behavior of ferroelectric ceramics will be discussed later.

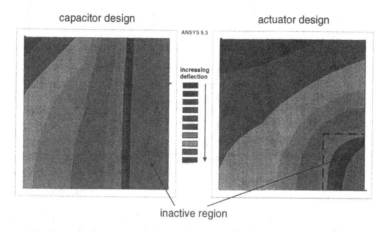

Figure 7. Finite-Element modelling of deflection profiles of piezoceramic multilayer actuators with different designs of the inactive region.

Component characterization. Ferroelectric components within devices are both mechanically and electrically interacting with other components. In the case of multilayer actuators the clamping and driving conditions must be carefully adjusted in order to set up the device performance. In particular, the static and dynamic actuator´s behavior is strongly influenced by clamping force and stiffness. Fig. 8 shows typical force vs. deflection curves of an actuator-spring unit. Note, that the limits of linear description will be reached for sufficiently high force values. The considerable dependance of energy conversion efficiency on stiffness of the load is given in Fig. 9.

On the other hand, the dynamic actuator´s behavior is essentially determined by pulse shape and duration.

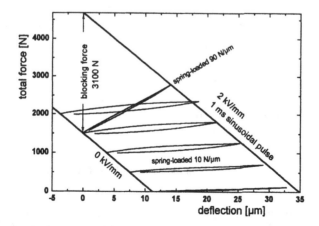

Figure 8. Measured deflection and force generation of a cofired multilayer stack actuator as function of the applied spring load [7].

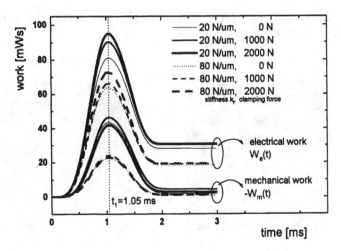

Figure 9. Behavior of electrical work $W_e(t)$ and mechanical work $-W_m(t)$ of an actuator with dimensions of $7 \times 7 \times 30$ mm³ (actuator stiffness $k_a = 60 N/\mu m$) for different stiffness' k_l of load and clamping forces driven with amplitude of single sinusoidal pulse of 2 kV/mm.

Nonlinear effects. In piezoelectric ceramics, domain switching causes additional contributions to both strain and polarization. These domain processes can be induced by mechanical or electrical loading. In analogy to the definition of the coercive field strength in the case of ferroelectricity, a coercive stress is defined for the ferroelastic behavior. This coercive stress depends on the electrical field, and vice versa, the coercive field strength is a function of the applied mechanical stress.

As a consequence of ferroelasticity, at total strain levels of $2 \cdot 10^{-3}$ the material stiffness may decrease to half the linear-elastic value (Fig. 10).

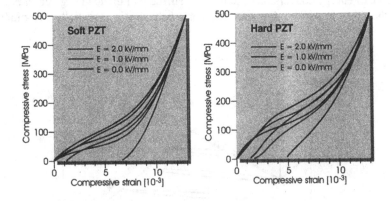

Figure 10. The electric field and composition dependence of the ferroelastic behavior of piezoelectric ceramics [8].

To examine the influence of coupled mechanical and electrical loading on actuator characteristics, a nonlinear piezo-element has been implemented into ANSYS using a microscopic domain switching model [9]. As a result, the calculated loadings generated

in the inactive regions of the multilayer actuator are considerably lower than estimated with the linear analysis (Fig. 11).

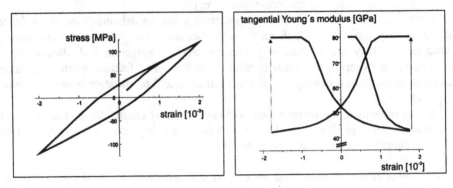

Figure 11. Nonlinear FE simulation (ANSYS) showing the influence of ferroelasticity on material stiffness.

4. Reliability considerations

In order to meet the customers' needs in automotive and industrial applications, piezoceramic multilayer actuators have to fulfill several strong demands. Under extreme dynamic large signal driving conditions and hostile environment, life times of more than 10^9 cycles and long-term failure rates lower than 10^{-5} must be guaranteed. Besides this, in automotive applications operable temperature ranges between -40 and +150°C are often required.

During operation piezoceramic components show new types of damage phenomena as a consequence of coupled electromechanical loading. Careful and systematic analyses of damage mechanisms (Fig. 12) must be carried out in order to receive failure road maps.

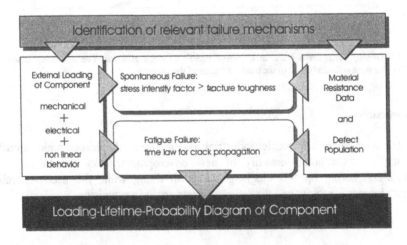

Figure 12. Postulated comprehensive concept for lifetime prediction.

First of all, piezoceramic materials sometimes show typical degradation of piezoelectric properties during lifetime which is caused by domain wall pinning, domain reorientation, microcracking and stress relaxation [10,11].

During poling, in large-scale multilayer stack actuators delaminations of the inner electrodes may take place which result in the occurrence of so-called poling cracks. These defects may act as starting points for crack propagation and other damage mechanisms as observed at further cycling. With the aid of fracture mechanics, some efforts in mathematical description of crack initiation and propagation have been made (Fig. 13).

On the other hand, cyclic operation of the actuator may cause fatigue of the external electrical connections. The prevention of such deterioration processes is a very decisive point in achieving the required device life time.

To meet all reliability requirements, much cost-intensive work has to be done. Whereas detailed damage analyses could give valuable hints for manufacturing optimization, future work must focus on the guarantee of extremely low failure rates by means of accelerated life and proof tests.

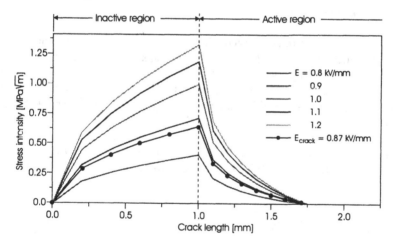

Figure 13. Driving force for a crack in a cofired multilayer stack actuator. The modelling bases on the formation of a penny-shaped crack in the inactive region of the actuator.

5. Conclusions

In order to realize large-scale industrial applications of piezoceramic multilayer actuators, an exceptional quantity of new development tasks has to be solved. Nevertheless, taking up this challenge is worth doing while the need for reliable actuators increases rapidly and the future markets are very promising.

6. References

[1]   Lubitz, K., Hellebrand, H., Cramer, D. and Probst, I. (1994) Low sintering PZT for multilayer actuators, in G. Ziegler, H. Hausner (eds.), *Proceedings Euro-Ceramics II*, DKG e.V., Köln, pp. 1955-1959

[2]   Wen, J., Hellebrand, H., Cramer, D., Lubitz, K., and Tomandl, G. (1994) Grain growth in multilayer PZT, in R. Waser (ed.), *Proceedings Electroceramics IV*, Verlag der Augustinus Buchhandlung, Aachen, pp. 247-252

[3]   Schuh, C., Kulig, M., and Lubitz, K. (1996) Ag doping of rare earth modified PZT, in J.L. Baptista, J.A. Labrincha, P.M. Vilarinho (eds.), *Proceedings Electroceramics V*, Fundacao Joao Jacinto de Magalhaes, Aveiro, pp. 201-204

[4]   Sugawara, Y., Onitsuka, K., Yoshikawa, S., Xu, Q., Newnham, R.E. and Uchino, K. (1992) Metal-ceramic composite actuators, *J. Am. Ceram. Soc.* **75** [4], 996-998

[5]   Haertling, G.H. (1994) Rainbow ceramics - a new type of ultra-high-displacement actuator, *Am. Ceram. Soc. Bull.* **73** [1], 93-96

[6]   Dogan, A., Uchino, K. and Newnham, R.E. (1997) Composite piezoelectric transducer with truncated conical endcaps "cymbal", *IEEE Trans. Ultrason., Ferroelect., Freq. Contr.*, **44** [3], 597-605

[7]   Wolff, A., Cramer, D., Hellebrand, H., Schuh, C. Steinkopff, T. and Lubitz, K. (1996) Dynamic behaviour of PZT multilayer actuators, in H. Borgmann (ed.), *Proceedings Actuator 96*, AXON Technoloie Consult GmbH, Bremen, pp. 193-195

[8]   Schäufele, A.B. and Härdtl, K.H. (1996) Ferroelastic properties of lead zirconate titanate ceramics, *J. Am. Ceram. Soc.* **79** [10], 2637-2640

[9]   Steinkopff, T. (1998) Finite-Element modelling of ferroic domain switching in piezoelectric ceramics, *Proc. Electroceramics VI*, (submitted to the *J. Europ.Ceram. Soc.*, 24 August 1998)

[10]   Sakai, T., Terai, Y. and Ishikiriyama, M. (1995) Improvement in durability of piezoelectric ceramics for actuator, *Jpn. J. Appl. Phys., Part 1* **34** [9B], 5276-5278

[11]   Dausch, D.E. (1997) Ferroelectric polarization fatigue in PZT-based rainbows and bulk ceramics, *J. Am. Ceram. Soc.* **80** [9], 2355-2360

AUTHOR INDEX

SUBJECT INDEX